**Operator Theory**
**Advances and Applications**
**Vol. 64**

**Editor**
**I. Gohberg**

# New Aspects in Interpolation and Completion Theories

Edited by

I. Gohberg

Birkhäuser Verlag
Basel · Boston · Berlin

Volume Editorial Office:

Raymond and Beverly Sackler
Faculty of Exact Sciences
School of Mathematical Sciences
Tel Aviv University
69978 Tel Aviv
Israel

A CIP catalogue record for this book is available from the Library of Congress, Washington D.C., USA

Deutsche Bibliothek Cataloging-in-Publication Data
**New aspects in interpolation and completion theory** / ed. by I.
Gohberg. – Basel ; Boston ; Berlin : Birkhäuser, 1993
(Operator theory ; Vol. 64)
ISBN 3-7643-2948-3
NE: Gochberg, Izrail' [Hrsg.] ; GT


Camera-ready copy prepared by the editor
Printed on acid-free paper produced from chlorine-free pulp
Cover design: Heinz Hiltbrunner, Basel
Printed in Germany
ISBN 3-7643-2948-3
ISBN 0-8176-2948-3

9 8 7 6 5 4 3 2 1

**This book is dedicated to Ciprian Foias on the occasion of his sixtieth birthday.**

The authors, editor and publisher highly appreciate his strong influence in the development of operator theory and its applications, his leadership and friendship. They wish him very many fruitful years to come.

# Contents

## EDITORIAL INTRODUCTION

This volume consists of eight papers containing recent advances in interpolation theory for matrix functions and completion theory for matrices and operators.

In the first paper, D. Alpay and P. Loubaton, "The tangential trigonometric moment problem on an interval and related topics" a trigonometric moment problem on an interval for matrix valued functions is studied. The realization approach plays an important role in solving this problem.

The second paper, M. Bakonyi, V.G. Kaftal, G. Weiss and H.J. Woerdeman, "Maximum entropy and joint norm bounds for operator extensions" is dedicated to a matrix completion problem. In it is considered the problem when only the lower triangular part of the operator entries of a matrix is identified. Completions which have simultaneously a small usual norm and a small Hilbert-Schmidt norm are considered. Bounds for these norms are obtained. The analysis of the maximum entropy extension plays a special role. The paper contains applications to nest algebras and integral operators.

The third paper, J.A. Ball, I. Gohberg and M.A. Kaashoek, "Bitangential interpolation for input-output operators of time varying systems: the discrete time case" contains solutions of time varying interpolation problems. The main attention is focused on the time varying analog of the Nevanlinna-Pick tangential problem in the case where the interpolation conditions appear from two sides. The state space theory of time varying systems play an important role.

The fourth paper, J.A. Ball, I. Gohberg and L. Rodman, "Two-sided tangential interpolation of real rational matrix functions" contains results on interpolation of real rational matrix functions which are similar to those proven earlier by the authors for complex rational matrix functions. In this paper the state space method for dealing with different real interpolation problems is developed systematically.

The fifth paper, H.Du and C. Gu, "On the spectra of operator completion problems" contains an analysis of the spectra of 2 x 2 block matrices in which three blocks except the lower left block are identified. The paper focuses on the study of the union and intersections of the spectra of all completions of such matrices. The finite and infinite dimensional cases are considered separately.

In the sixth paper, C. Foias, A.E. Frazho and W.S. Li, "The exact $H^2$ estimate for the central $H^\infty$ interpolation" is considered the Nehari interpolation problem when the norm of the extension is estimated in $H^2$ and $H^\infty$ simultaneously. The central intertwining lifting for the commutant lifting theorem is used essentially. New bounds are obtained and applications to other interpolation problems are deduced.

In the seventh paper, A.E. Frazho and S.M. Kherat, "On mixed $H^2$ - $H^\infty$ tangential interpolation" tangential interpolation problems with simultaneous estimation of the extension in the $H^2$ and $H^\infty$ norms are considered. Here Nevanlinna-Pick, Carathéodory and Hermite-Fejer problems are analysed. The approach is based on the use of the central solution for commutant lifting theorem. The paper also contains explicit computational formulas in state space realization terms.

In the final paper, I. Gohberg and C. Gu, "On a completion problem for matrices" are considered scalar matrices in which the last row is unidentified except the main diagonal entry. All other entries are identified. One of the problems solved describes all the invariant polynomials of all possible completions of this matrix. It also contains infinite dimensional generalizations.

The topics which are discussed in this volume belong to areas of operator theory in which the influence of Ciprian Foias was very strongly felt, starting with commutant lifting theorems, and going up until present day interpolation problems connected with modern system theory. It is a special pleasure to dedicate this volume to a friend and colleague who did so much and was so successful in the development of this area.

I. Gohberg

Operator Theory:
Advances and Applications, Vol. 64




# The tangential trigonometric moment problem on an interval and related topics

Daniel Alpay and Philippe Loubaton

We study interpolation problems for Carathéodory functions when the support of the underlying measure in the Riesz–Herglotz representation is carried by a symmetric subinterval of $[-\pi, \pi]$. Our approach is based on the realization theory of matrix–valued functions.

## 1 Introduction

In this paper we study the tangential trigonometric moment problem when the underlying measure is carried by a symmetric closed subinterval of $[-\pi, \pi]$. To solve this problem, we first look at a related Carathéodory–Féjèr interpolation problem. An associated tangential Nevanlinna–Pick interpolation problem for Carathéodory functions of the open unit disk will also be considered. Let us first introduce some notations and definitions. The symbol $\mathbb{C}^{p\times\ell}$ denotes the set of $p$ rows and $\ell$ columns matrices with complex entries; the identity in $\mathbb{C}^{p\times p}$ is denoted by $I_p$ and $\mathbb{C}^p$ stands for the set of $p$ columns vectors with complex entries. The adjoint of a matrix $A$ will be denoted by $A^*$.
A Carathéodory function is a $\mathbb{C}^{p\times p}$ valued function analytic in the open unit disk $\mathbb{D}$ with nonnegative real part in $\mathbb{D}$. The family of such functions will be denoted by $\mathcal{C}_p$. By the Riesz–Herglotz representation theorem (see e.g. [22]), an element $\phi \in \mathcal{C}_p$ can be written in a unique way as

$$\phi(z) = i\alpha + \int_{-\pi}^{\pi} \frac{e^{it} + z}{e^{it} - z} d\mu(t), \tag{1.1}$$

where $\alpha$ is a hermitian matrix and $\mu$ is a $\mathbb{C}^{p\times p}$–valued measure on the interval $[-\pi, +\pi]$ which is positive and integrable.

We will denote by $\mathcal{C}_p(\omega)$ the family of Carathéodory functions whose associated measure $\mu$ is carried by the interval $[-\omega, \omega]$.
We will consider the following version of the tangential Carathéodory–Féjèr interpolation problem:

**Problem 1.1** *Given matrices $\phi_0 = \phi_0^*$ and $\phi_1$ in $\mathbb{C}^{p\times p}$ and vectors $\xi_0, \eta_2, \ldots, \eta_N$ in $\mathbb{C}^p$, find all functions $\phi \in \mathcal{C}_p$ such that:*

$$\phi(0) = \phi_0 \tag{1.2}$$

$$\phi^{(1)}(0) = 2\phi_1 \tag{1.3}$$

$$\phi^{(j)}(0)^*\xi_0 = 2j!\eta_j\,,\ \ j = 2,\ldots,N. \tag{1.4}$$

Since $\phi_0$ is selfadjoint, it follows that the matrix $\alpha$ in the representation (1.1) is equal to zero, and that the conditions (1.2)–(1.4) are equivalent to the following moment problem

**Problem 1.2** *Given matrices $\phi_0 = \phi_0^*$ and $\phi_1$ in $\mathbb{C}^{p\times p}$ and vectors $\xi_0, \eta_2, \ldots, \eta_N$ in $\mathbb{C}^p$, find all measures $d\mu$ on $[-\pi, \pi]$ such that*

$$\phi_0 = \int_{-\pi}^{\pi} d\mu(t), \tag{1.5}$$

$$\phi_1 = \int_{-\pi}^{\pi} e^{-it} d\mu(t), \tag{1.6}$$

$$\eta_k = \left(\int_{-\pi}^{\pi} e^{-ikt} d\mu(t)\right)^* \xi_0 \qquad k = 2,\ldots N. \tag{1.7}$$

When $p = 1$, these are classical interpolation problems, which originate with the works of Carathéodory, Féjèr, Toeplitz and Schur (see [1] for references).

The trigonometric moment problem on a symmetric interval consists in finding all positive measures $d\mu$ satisfying conditions (1.5)–(1.7) with support in the interval $[-\omega, \omega]$, where $0 < \omega < \pi$ is preassigned. When $p = 1$, this problem (and more generally, when constraints are set on the support of the measure) seems to have been first considered by Akhiezer and Krein (see [2] and [22]). They did not give a description of the set of all solutions, but proved the following theorem, by using a convex analysis result :

**Theorem 1.3 ([22, Theorem 2.3])** *Let $\phi_0, \ldots, \phi_N$ be complex numbers, and let $0 < \omega < \pi$. Then there exists a positive measure $\mu$ carried by the interval $[-\omega, \omega]$ such that*

$$\phi_k = \int_{-\omega}^{\omega} e^{-ikt} d\mu(t)\,,\ \ k = 0,\ldots N, \tag{1.8}$$

*if and only if the Toeplitz matrices $\Phi_N$ and $\tilde{\Phi}_{N-1}$ defined by $(\Phi_N)_{ij} = \phi_{i-j}$ and*

$$(\tilde{\Phi}_{N-1})_{ij} = \phi_{i-j+1} + \phi_{i-j-1} - 2\phi_{i-j}\cos\omega \tag{1.9}$$

*are nonnegative, where, in the first case, $i, j$ run through $0,\ldots N$ and in the second case through $0,\ldots N-1$. Finally, the solution is unique if and only if $\Phi_N$ or $\tilde{\Phi}_{N-1}$ is singular.*

When the Toeplitz matrices $\Phi_N$ and $\tilde{\Phi}_{N-1}$ are strictly positive, the description of all the solutions of the trigonometric moment problem with measures carried in the interval $[-\omega, \omega]$ was recently given by T. Chovanel and Ph. Loubaton in [12]. The approach in [12] used the concept of canonical solutions defined in [22] and does not seem to be adaptable to the case $p > 1$. Our present approach is based on the realization theory of matrix–valued rational functions, and is close in spirit to the solution of the tangential interpolation problem for Stieltjes functions, as developped in [4] and [3]. Another approach, based on the theory of orthogonal polynomials, will be presented in [8].

A key observation for the present method is the following (see [13]): the measure $d\mu$ has support on $[-\omega,\ \omega]$ if and only if the measure $d\tilde{\mu}$ is also positive, where

$$d\tilde{\mu}(t) = 2(\cos t - \cos\omega)d\mu(t) \tag{1.10}$$

Let $\tilde{\phi}$ be defined by:

$$\tilde{\phi}(z) = (z + z^{-1} - 2\cos\omega)\phi(z) + \phi_0(z - z^{-1}) + \phi_{-1} - \phi_1. \tag{1.11}$$

where $\phi_{-1} = \phi_1^*$. Since $\tilde{\phi}(z) = \int_{-\pi}^{\pi} \frac{e^{it}+z}{e^{it}-z} d\tilde{\mu}(e^{it})$, the function $\tilde{\phi}$ is in $\mathcal{C}_p$ if and only if the measure $d\tilde{\mu}$ is positive. Moreover, it is easily checked that

$$\tilde{\phi}(z) = \tilde{\phi}_0 + 2\sum_{1}^{N-1} \tilde{\phi}_i z^i + z^N O(z) \tag{1.12}$$

where

$$\tilde{\phi}_i = \phi_{i+1} + \phi_{i-1} - 2\phi_i \cos\omega. \tag{1.13}$$

This allows us to associate to Problem 1.1 another interpolation problem in $\mathcal{C}_p$, defined in Section 3. Relating the resolvent matrices of these two problems leads to the solution of Problem 1.1 in the class $\mathcal{C}_p(\omega)$.

The paper consists of four sections. This introduction is the first. In the second section, we develop some lemmas on matrix–valued $J$–unitary rational functions. The Carathéodory–Féjèr problem is discussed in the third section, as well as the tangential trigonometric moment problem. In section four we consider the Nevanlinna Pick interpolation problem.

# 2 Some lemmas on matrix–valued rational functions

We first review some facts on realization theory of matrix valued rational functions. Let $\Theta$ be a $\mathbb{C}^{m\times m}$ valued rational function analytic at the origin. It can be written as

$$\Theta(z) = D + zC(I_n - zA)^{-1}B \tag{2.1}$$

where $C \in \mathbb{C}^{m\times n}$, $A \in \mathbb{C}^{n\times n}$, $B \in \mathbb{C}^{n\times m}$ and $D = \Theta(0) \in \mathbb{C}^{m\times m}$.

An expression of the form (2.1) is called a realization of $\Theta$. The realization is called minimal if $n$ is minimal in (2.1). The realization is minimal if the pair $(C, A)$ is observable and the pair $(A, B)$ is controllable, i.e. if it respectively holds that

$$\cap_{k=0}^{\infty} \ker CA^k = \{0\} \tag{2.2}$$

and

$$\cup_{k=0}^{\infty} \operatorname{ran} A^k B = \mathbb{C}^n. \tag{2.3}$$

We refer to the books [21], [10], [9], [19] and [20] for further information on the theory of matrix–valued rational functions and their realizations. These works consider essentially functions analytic at infinity; a transformation $z \to 1/z$ permits to translate the results for the functions under consideration here. The pair $(C, A)$ characterizes the left pole strucure of $\Theta(1/z)$.

Let

$$J = \begin{pmatrix} 0 & -I_p \\ -I_p & 0 \end{pmatrix}. \tag{2.4}$$

The function $\Theta$ is called $J$-inner if $\Theta(z)J\Theta(z)^* \leq J$ for all points of analyticity of $\Theta$ in $\mathbb{D}$ and if equality holds on the unit circle $[-\pi, \pi]$. The function is called $J$–unitary (on the unit circle) if only this last condition is in force. For more information on $J$–inner and $J$–unitary rational functions, we refer to [6] and [17]. The next proposition considers how to reconstruct $\Theta$ from the left pole structure of $\Theta(1/z)$.

**Proposition 2.1** *Let $(C, A) \in \mathbb{C}^{m\times n} \times \mathbb{C}^{m\times m}$ be an observable pair of matrices. Then, there exists a $J$–unitary (resp. $J$–inner) function $\Theta$ with minimal realization (2.1) if and only if the Stein equation*

$$P - A^*PA = C^*JC \tag{2.5}$$

*has a nonsingular hermitian (resp. strictly positive) solution. For a given choice of $P$, the matrices $B$ and $D$ are then uniquely defined, up to a multiplicative $J$-unitary constant factor on the right. Let $x_0 \in \mathbb{T}$ be such that $(I_n - x_0A)$ is invertible; $B$ and $D$ may be chosen to be equal to*

$$\begin{aligned} B_0 &= (x_0^*I_n - A)P^{-1}(I_n - x_0A)^{-*}C^* & (2.6)\\ D_0 &= J - CP^{-1}(I_n - x_0A)^{-*}C^*. & (2.7)\end{aligned}$$

*For the above choice of $B$ and $D$, we have*

$$\Theta_0(z) = J - (1 - zx_0^*)C(I_n - zA)^{-1}P^{-1}(I_n - x_0A)^{-*}C^*. \tag{2.8}$$

*Any other $\Theta$ differs from $\Theta_0$ by a multiplicative $J$–unitary constant on the left.*

*Proof.* Let us suppose first that the Stein equation (2.5) has a nonsingular hermitian solution $P$ and define $B_0$, $D_0$ and $\Theta_0$ as in the statement. Then, it is readily seen that

$$\frac{J - \Theta_0(z)J\Theta_0(z)^*}{1 - |z|^2} = C(I_n - zA)^{-1}P^{-1}(I_n - zA)^{-*}C^* \tag{2.9}$$

at the points $z$ for which $(I_n - zA)$ is invertible. The various claims drop easily from this formula. To prove the converse, one can, e.g., look at the realization with state space $H(\Theta)$, the reproducing kernel space with reproducing kernel $\frac{J-\Theta(z)J\Theta(\omega)^*}{1-zw^*}$. The realization (2.1) is then given by $D = \Theta(0)$ and

$$\begin{array}{ll} Af(z) = \frac{f(z)-f(0)}{z} & H(\Theta) \rightarrow H(\Theta) \\ Bc(z) = \frac{\Theta(z)-\Theta(0)}{z}c & \mathbb{C}^m \rightarrow H(\Theta) \\ Cf = f(0) & H(\Theta) \rightarrow \mathbb{C}^m. \end{array}$$

The matrix $\begin{pmatrix} A & B \\ C & D \end{pmatrix}$ is then a unitary operator from $H(\Theta) \oplus \mathbb{C}^m$ into itself. This allows to deduce the Stein equation. We refer to [6] or [17] for further details. When $A$ is invertible, we also refer to [7]. ■

The next fact is easily verified and is stated as a lemma for future reference.

**Lemma 2.2** *Let $a$ and $b$ be $\mathbb{C}^{p\times p}$ matrices, respectively selfadjoint and skewselfadjoint, and let $\omega \in [0, \pi]$. Let $M$ and $N$ be defined by*

$$M(z) = \begin{pmatrix} (z + z^{-1} - 2\cos\omega)I_p & a(z - z^{-1}) + b \\ 0 & I_p \end{pmatrix} \tag{2.10}$$

*and*

$$N(z) = \begin{pmatrix} I_p/(z - e^{i\omega}) & 0 \\ 0 & I_p/(1 - ze^{i\omega}) \end{pmatrix}. \tag{2.11}$$

*Then, the rational matrix-valued function $\Theta$ is $J$-unitary on the unit circle if and only if $\tilde{\Theta} = M\Theta N$ is $J$-unitary on the unit circle.*

In general, the function $\tilde{\Theta}$ need not be analytic at the origin and therefore lacks realizations of the form (2.1). Moreover, even when $\Theta$ is $J$-inner, $\tilde{\Theta}$ need not be $J$-inner.
We now relate minimal realizations of $\Theta$ and $\tilde{\Theta}$. In the following proposition, $\Theta$ is not assumed $J$-unitary on $\mathbb{T}$.

**Proposition 2.3** *Let $\omega \in ]0, \pi[$, let $\Theta$ be a $\mathbb{C}^{2p\times 2p}$ valued rational function analytic for $z \in \{0, e^{i\omega}, e^{-i\omega}\}$ and let*

$$\Theta(z) = \begin{pmatrix} D_{11} & D_{12} \\ D_{21} & D_{22} \end{pmatrix} + z \begin{pmatrix} C_1 \\ C_2 \end{pmatrix} (I_n - zA)^{-1}(B_1 \ B_2) \tag{2.12}$$

*be a minimal realization of $\Theta$, with $D_{ij} \in \mathbb{C}^{p\times p}$, $C_i \in \mathbb{C}^{p\times n}$ and $B_i \in \mathbb{C}^{n\times p}$, $i \in \{1,2\}$, and $A \in \mathbb{C}^{n\times n}$. Let $a$ and $b$ be $\mathbb{C}^{p\times p}$ matrices respectively selfadjoint and skewselfadjoint and let $M$ and $N$ be defined by (2.10) and (2.11) respectively. Then, the function $\tilde{\Theta} = M\Theta N$ is analytic for $z \in \{0, e^{i\omega}, e^{-i\omega}\}$ if and only if the following conditions hold:*

$$D_{21} = -e^{i\omega}C_2(I_n - e^{i\omega}A)^{-1}B_1 \tag{2.13}$$

$$D_{22} = -e^{-i\omega}C_2(I_n - e^{-i\omega}A)^{-1}B_2 \tag{2.14}$$

$$D_{12} = aD_{22} \tag{2.15}$$

$$D_{11} = aD_{21}. \tag{2.16}$$

*Then, a (in general nonminimal) realization of $\tilde{\Theta}$ is given by*

$$\tilde{\Theta}(z) = \begin{pmatrix} H_{11} & H_{12} \\ H_{21} & H_{22} \end{pmatrix} + z \begin{pmatrix} G_1 \\ G_2 \end{pmatrix} (I_n - zA)^{-1}(F_1 \ F_2) \tag{2.17}$$

*where*

$$\begin{aligned}
H_{11} &= ((-aC_2e^{i\omega}+bC_2-aC_2A)(I_n-e^{i\omega}A)^{-1}-e^{-i\omega}C_1)B_1\\
H_{12} &= -e^{-i\omega}((-aC_2e^{-i\omega}+bC_2-aC_2A)(I_n-e^{-i\omega}A)^{-1}-e^{-i\omega}C_1)B_2\\
H_{21} &= C_2(I_n-e^{i\omega}A)^{-1}B_1,\\
H_{22} &= -e^{-i\omega}C_2(I_n-e^{-i\omega}A)^{-1}B_2,\\
G_1 &= C_1(I_n-e^{i\omega}A)(I_n-e^{-i\omega}A)+aC_2+bC_2A-aC_2A^2\\
G_2 &= C_2A\\
F_1 &= (I_n-e^{i\omega}A)^{-1}B_1\\
F_2 &= -e^{-i\omega}(I_n-e^{-i\omega}A)^{-1}B_2
\end{aligned}$$

*Proof.* Since the realization (2.1) is minimal, both $(I_n-e^{i\omega}A)$ and $(I_n-e^{-i\omega}A)$ are invertible. Let $(M\Theta N)_{ij}$, $i,j\in\{1,2\}$, be the block decomposition of $M\Theta N$ into four $\mathbb{C}^{p\times p}$ valued functions. Then,

$$(M\Theta N)_{21}(z)=\frac{D_{21}+zC_2(I_n-zA)^{-1}B_1}{z-e^{i\omega}} \tag{2.18}$$

and

$$(M\Theta N)_{22}(z)=\frac{D_{22}+zC_2(I_n-zA)^{-1}B_2}{1-ze^{i\omega}}. \tag{2.19}$$

Thus, $(M\Theta N)_{21}$ is analytic at $z=e^{i\omega}$ if and only if

$$D_{21}+e^{i\omega}C_2(I_n-e^{i\omega}A)^{-1}B_1=0. \tag{2.20}$$

When this condition holds, we can rewrite $(M\Theta N)_{21}$ as

$$\begin{aligned}
(M\Theta N)_{21}(z) &= \frac{-e^{i\omega}C_2(I_n-e^{i\omega}A)^{-1}B_1+zC_2(I_n-zA)^{-1}B_1}{z-e^{i\omega}}\\
&= C_2(I_n-e^{i\omega}A)^{-1}(I_n-zA)^{-1}B_1\\
&= C_2(I_n-e^{i\omega}A)^{-1}B_1+zC_2A(I_n-zA)^{-1}(I_n-e^{i\omega}A)^{-1}B_1, && (2.21)
\end{aligned}$$

which can be written as $H_{21}+zG_2(I_n-zA)^{-1}F_1$.

Similarly, the block $(M\Theta N)_{22}$ is analytic at $z=e^{-i\omega}$ if and only if

$$D_{22}+e^{-i\omega}C_2(I_n-e^{-i\omega}A)^{-1}B_2=0, \tag{2.22}$$

and thus,

$$\begin{aligned}
(M\Theta N)_{22}(z) &= \frac{-e^{-i\omega}C_2(I_n-e^{-i\omega}A)^{-1}B_2+zC_2(I_n-zA)^{-1}B_2}{1-ze^{i\omega}}\\
&= -e^{-i\omega}C_2(I_n-e^{-i\omega}A)^{-1}(I_n-zA)^{-1}B_2\\
&= -C_2(e^{i\omega}I_n-A)^{-1}B_2-zC_2A(I_n-zA)^{-1}(e^{i\omega}I_n-A)^{-1}B_2 && (2.23)\\
&= H_{22}+zG_2(I_n-zA)^{-1}F_1. && (2.24)
\end{aligned}$$

To obtain (2.21) and (2.24) we have used the identity

$$(I_n-zA)^{-1}=I_n+zA(I_n-zA)^{-1}.$$

Next, we turn to $(M\Theta N)_{11}$:

$$\begin{aligned}
(M\Theta N)_{11}(z) &= \frac{(z-e^{i\omega})(z-e^{-i\omega})}{z}\frac{D_{11}+zC_1(I_n-zA)^{-1}B_1}{z-e^{i\omega}}+\\
&\quad +(a(z-z^{-1})+b)\frac{D_{21}+zC_2(I_n-zA)^{-1}B_1}{z-e^{i\omega}}\\
&= \frac{z-e^{-i\omega}}{z}(D_{11}+zC_1(I_n-zA)^{-1}B_1)+\\
&\quad +(a(z-z^{-1})+b)\frac{D_{21}+zC_2(I_n-zA)^{-1}B_1}{z-e^{i\omega}}.
\end{aligned}$$

Thus, $(M\Theta N)_{11}$ is analytic at $z=e^{-i\omega}$. It is also analytic at $z=e^{i\omega}$ thanks to (2.20). Computing the coefficient of the power $z^{-1}$, we see that $(M\Theta N)_{11}$ is analytic at the origin if and only if $D_{11}=aD_{21}$. When this condition holds, we can write

$$\begin{aligned}
(M\Theta N)_{11}(z) &= \left(\frac{z-e^{-i\omega}}{z}\right)\left(D_{11}+zC_1(I_n-zA)^{-1}B_1\right)\\
&\quad +(a(z-z^{-1})+b)\left(\frac{D_{21}+zC_2(I_n-zA)^{-1}B_2}{z-e^{i\omega}}\right)\\
&= D_{11}+zC_1(I_n-zA)^{-1}B_1\\
&\quad -e^{-i\omega}C_1(I_n-zA)^{-1}B_1\\
&\quad +(az+b)C_2(I_n-zA)^{-1}(I_n-e^{i\omega}A)^{-1}B_1\\
&\quad -z^{-1}(e^{-i\omega}D_{11}+aC_2(I_n-e^{i\omega}A)^{-1}B_1\\
&\quad -aC_2A(I_n-zA)^{-1}(I_n-e^{i\omega}A)^{-1}B_1\\
&= aD_{21}+zC_1(I_n-zA)^{-1}B_1-e^{-i\omega}C_1(I_n-zA)^{-1}B_1\\
&\quad +(az+b)C_2(I_n-zA)^{-1}(I_n-e^{i\omega}A)^{-1}B_1\\
&\quad -aC_2A(I_n-zA)^{-1}(I_n-e^{i\omega}A)^{-1}B_1,
\end{aligned}$$

and therefore,

$$\begin{aligned}
(M\Theta N)_{11}(z) &= aD_{21}-e^{-i\omega}C_1B_1+bC_2(I_n-e^{i\omega}A)^{-1}B_1\\
&\quad -aC_2A(I_n-e^{i\omega}A)^{-1}B_1\\
&\quad +zC_1(I_n-zA)^{-1}B_1-ze^{-i\omega}C_1A(I_n-zA)^{-1}B_1\\
&\quad +zaC_2(I_n-zA)^{-1}(I_n-e^{i\omega}A)^{-1}B_1\\
&\quad +zbC_2A(I_n-zA)^{-1}(I_n-e^{i\omega}A)^{-1}B_1\\
&\quad -zaC_2A^2(I_n-zA)^{-1}(I_n-e^{i\omega}A)^{-1}B_1,
\end{aligned}$$

so that $(M\Theta N)_{11}(z)=H_{11}+zG_1(I_n-zA)^{-1}F_1$.
The case of the block $(M\Theta N)_{12}$ is treated in a similar way. ∎

We note that conditions (2.15)–(2.16) may be rewritten as

$$(I_p-a)\Theta(0)=0 \tag{2.25}$$

while (2.13) and (2.14) can be rewritten respectively as

$$(\Theta(e^{i\omega}))_{21} = 0 \tag{2.26}$$

and

$$(\Theta(e^{-i\omega}))_{22} = 0. \tag{2.27}$$

Let us now consider $\Theta$ of the form $\Theta = \Theta_0 U$ where $\Theta_0$ is given by formula (2.8) with $x_0 = e^{i\omega}$ and where $U$ is a $J$–unitary constant. In general, $\Theta$ does not meet conditions (2.25)–(2.27). Clearly, condition (2.25) cannot be forced by a suitable choice of $U$; it either holds simultaneously for all $\Theta_0 U$ or for none. On the other hand, we show now how, under adequate hypothesis, conditions (2.26) and (2.27) can be achieved by chosing appropriately $U$.

Since $\Theta(e^{i\omega}) = JU$ is $J$–unitary, condition (2.26) will be achieved if and only if $U$ is of the form

$$U = \begin{pmatrix} 0 & X \\ X^{-*} & Z \end{pmatrix} \tag{2.28}$$

where $X$ and $Z$ are in $\mathbb{C}^{p\times p}$, and are such that $X$ is invertible and

$$Z^*X + X^*Z = 0. \tag{2.29}$$

Let us write

$$\Theta_0(e^{-i\omega}) = \begin{pmatrix} a_0 & b_0 \\ c_0 & d_0 \end{pmatrix}(\omega).$$

Since $\Theta_0(e^{-i\omega})$ is $J$–unitary, it holds that $c_0(\omega)d_0(\omega)^* + d_0(\omega)c_0(\omega)^* = 0$. On the other hand, condition (2.27) can be rewritten as

$$c_0(\omega)X + d_0(\omega)Z = 0.$$

Thus, when $d_0(\omega)$ is invertible, we may chose

$$X = d_0(\omega)^* \tag{2.30}$$

$$Z = c_0(\omega)^*. \tag{2.31}$$

**Proposition 2.4** *If* $\det C_2P^{-1}C_2^* \neq 0$, *then* $d_0(\omega)$ *is invertible for all* $\omega \in \mathbb{T}$, *at the possible exception of finitely many points.*

*Proof.* It suffices to remark that the function

$$-d_0(\omega)/(1-e^{-2i\omega}) = C_2(I_n - e^{-i\omega}A)^{-1}P^{-1}(I_n - e^{i\omega}A)^{-*}C_2^* \tag{2.32}$$

is analytic and tends to the value $C_2P^{-1}C_2^*$ as $\omega = ix$, $x \in \mathbb{R}$ and $x \to -\infty$. ■

# 3 The main result

We first discuss the Carathéodory–Féjèr problem 1.1. Let $A$ denote the block diagonal matrix with $p-1$ first diagonal blocks equal to

$$\begin{pmatrix} 0 & 1 \\ 0 & 0 \end{pmatrix}$$

and $p$–th block equal to a $(N+1)\times(N+1)$ Jordan block with eigenvalue 0, and let $C$ be defined by

$$C = (C_1, C_2, \dots C_{p-1}, C_p) \tag{3.1}$$

where, for $i \in \{1, \dots, p-1\}$,

$$C_i = \begin{pmatrix} -\phi_0 e_i & -2\phi_1^* e_i \\ e_i & 0 \end{pmatrix} \tag{3.2}$$

where $\{\zeta_0, e_1, \dots, e_{p-1}\}$ is a basis of $\mathbb{C}^p$ and

$$C_p = \begin{pmatrix} -\phi_0\xi_0 & -2\phi_1^*\xi_0 & -2\eta_2 & \dots & -2\eta_N \\ \xi_0 & 0 & 0 & \dots & 0 \end{pmatrix} \tag{3.3}$$

We note that the matrix $(0, I_p)C$ has full rank. Therefore, as is well known, a necessary and sufficient condition for the Carathéodory–Féjèr problem 1.1 to be solvable is that the (unique) solution of the Stein equation (2.5) is nonnegative. For a proof a these facts, as well as for the description (3.4) of all the solutions, we refer to [5]. In that work, a more general interpolation problem is stated for functions analytic off the real line and with nonnegative imaginary part in the open upper half–plane, but the transition to the disk case is easily done using a Cayley transform. For the case where full values of the derivatives are given, we refer to [16] and [14].

When $P > 0$, formula (2.8) defines a $J$ inner function $\Theta_0$ with $x_0 = e^{i\omega} \in \mathbb{T}$ (still with $J$ as in (2.4)). The set of all solutions to the Carathéodory Féjèr problem is then described by a linear fractional transformation as follows: fix any $J$–unitary constant $U$ and set $\Theta = \Theta_0 U$. A function $\phi \in \mathcal{C}_p$ is a solution if and only if it can be written as

$$\phi(z) = (\Theta_{11}(z)\alpha(z) + \Theta_{12}(z)\beta(z))(\Theta_{21}(z)\alpha(z) + \Theta_{22}(z)\beta(z))^{-1} \tag{3.4}$$

where $(\alpha, \beta)$ spans the set of the Carathéodory pairs. A pair $(\alpha, \beta)$ of $\mathbb{C}^{p\times p}$–valued functions meromorphic in $\mathbb{C}\backslash\mathbb{T}$ is called a Carathéodory pair if the function

$$K(z, w) = \frac{\beta(w)^*\alpha(z) + \alpha(w)^*\beta(z)}{(1 - zw^*)} \tag{3.5}$$

is positive for $z, w$ in the domain of analyticity of $(\alpha, \beta)$ and if moreover, the matrix $(\alpha(z), \beta(z))$ has rank $p$ for all $z$ where $(\alpha, \beta)$ is analytic at the possible exception of a zero set, and for $z$ where the expression makes sense,

$$\beta(1/z^*)^*\alpha(z) + \alpha(1/z^*)\beta(z) = 0.$$

For instance, if $\psi$ is a Carathéodory function, the pair $(I_p, \psi)$ is a Carathéodory pair.

The functions $\Theta$ are called resolvent matrices of the interpolation problem. In (3.4), the choice of $U$ is irrelevant. Indeed, if $(\alpha, \beta)$ is a Carathéodory pair so are all the pairs $(\alpha_U, \beta_U)$ defined by

$$\begin{pmatrix} \alpha_U \\ \beta_U \end{pmatrix} = U \begin{pmatrix} \alpha \\ \beta \end{pmatrix}. \tag{3.6}$$

As we will see soon, it will be necessary to fix $U$ in order to solve Problem 1.1 in $\mathcal{C}_p(\omega)$. We note that (3.4) can be rewritten as

$$\begin{pmatrix} \phi \\ I_p \end{pmatrix} = \Theta \begin{pmatrix} \gamma \\ \delta \end{pmatrix}, \tag{3.7}$$

where $(\gamma, \delta)$ is a Carathéodory pair.

The functions $\Theta$ are such that (2.25) holds (see e.g. [17]). On the other hand, $C_2$ has the property that ran $C_2 = \mathbb{C}^p$, so that $\ker C_2^* = \{0\}$. In particular, $C_2 P^{-1} C_2^* > 0$, and Proposition 2.4 insures that, for all $\omega \in ]0, \pi[$ at the possible exception of a finite set, the $J$–unitary constant may be fixed so that (2.26) and (2.27) hold.

Let now $\phi$ be a solution of the Carathéodory–Féjèr problem and let (1.1) denote its Riesz–Herglotz representation. As already noted, $\alpha = 0$ since $\phi_0$ is self–adjoint. The support of the measure $d\mu$ is in the interval $[-\omega, \omega]$ if and only if the function $\tilde{\phi}$ defined in (1.11) is in $\mathcal{C}_p$. From the Taylor expansion (1.12) it follows that $\tilde{\phi}$ satisfies the interpolation problem:

$$\begin{aligned} \tilde{\phi}(0) &= \tilde{\phi}_0 &(3.8)\\ \tilde{\phi}^{(j)}(0)^* \xi_0 &= 2j! \tilde{\eta}_j\,, \quad j = 1, \dots, N-1. &(3.9) \end{aligned}$$

where $\tilde{\phi}_i$ are defined in (1.9) and where

$$\begin{aligned} \tilde{\eta}_1 &= \eta_2 + \phi_0 \xi_0 - 2\phi_1^* \xi_0 \cos \omega &(3.10)\\ \tilde{\eta}_i &= \eta_{i+1} + \eta_{i-1} - 2\eta_i \cos \omega\,, \ i = 2, \dots, N-1. &(3.11) \end{aligned}$$

We are thus lead to another interpolation problem, namely:

**Problem 3.1** *Given matrices $\phi_0 = \phi_0^*$ and $\phi_1$ in $\mathbb{C}^{p \times p}$ and vectors $\xi_0, \eta_2, \dots, \eta_N$ in $\mathbb{C}^p$, find all functions $\psi \in \mathcal{C}_p$ such that:*

$$\begin{aligned} \psi(0) &= \tilde{\phi}_0 &(3.12)\\ \psi^{(j)}(0)^* \xi_0 &= 2j! \tilde{\eta}_j\,, \quad j = 1, \dots, N-1. &(3.13) \end{aligned}$$

*where $\tilde{\phi}_0$ and the $\tilde{\eta}_i$ are defined as above in terms of the original interpolation data.*

Let $\hat{A}$ denote the block diagonal matrix with first block equal to the zero element of $\mathbb{C}^{p-1 \times p-1}$ and second block equal to a $N \times N$ Jordan block with eigenvalue 0, and let $\hat{C}$ be defined by

$$\hat{C} = (\hat{C}_1, \hat{C}_2, \dots, \hat{C}_p)$$

with

$$\hat{C}_i = \begin{pmatrix} -\tilde{\phi}_0 e_i \\ e_i \end{pmatrix}, \; i = 1 \ldots p-1,$$

and

$$\hat{C}_p = \begin{pmatrix} -\tilde{\phi}_0 \xi_0 & -2\tilde{\eta}_1 & \ldots & -2\tilde{\eta}_{N-1} \\ \xi_0 & 0 & \ldots & 0 \end{pmatrix}.$$

The Carathéodory–Féjèr problem 3.1 has a solution if and only if the Stein equation

$$\hat{P} - \hat{A}^* \hat{P} \hat{A} = \hat{C}^* J \hat{C} \tag{3.14}$$

has a nonnegative solution $\hat{P}$. When $\hat{P} > 0$, all solutions are given via a linear fractional transformation with a resolvent matrix $\hat{\Theta}_0$ defined as in (2.8), but with $\hat{C}$ and $\hat{A}$ in place of $C$ and $A$.

We note that, in the scalar case, the matrices $P$ and $\hat{P}$ coincide respectively with the matrices $\Phi_N$ and $\tilde{\Phi}_{N-1}$ defined in Theorem 1.3.

We show that the resolvent matrices $\Theta_0$ and $\hat{\Theta}_0$ may be related by the formula $\hat{\Theta}_0 V = M\Theta_0 UN$, where $M$ and $N$ are defined by (2.10) and (2.11), with $a = \phi_0$ and $b = \phi_1^* - \phi_1$ and where $U$ and $V$ are $J$–unitary constants, $U$ being chosen in such a way that (2.26) and (2.27) hold. We first need a preliminary computation.

**Lemma 3.2** *Let*

$$C = \begin{pmatrix} C^{(1)} \\ C^{(2)} \end{pmatrix}$$

*be the decomposition of the matrix $C$ defined in (3.1)–(3.3) into two $\mathbb{C}^{p\times(2(p-1)+N-1)}$ matrices and let $\tilde{C}^{(1)}$ and $\tilde{C}^{(2)}$ be defined by*

$$\tilde{C}^{(1)} = -(0, \tilde{\phi}_0 e_1, 0, \tilde{\phi}_0 e_2, \ldots, 0, \tilde{\phi}_0 \xi_0, 2\tilde{\eta}_1, 2\tilde{\eta}_2, \ldots, 2\tilde{\eta}_{N-1}) \tag{3.15}$$

*and*

$$\tilde{C}^{(2)} = (0, e_1, 0, e_2, \ldots, 0, e_{p-1}, 0, \xi_0, 0, 0, \ldots). \tag{3.16}$$

*Then,*

$$\begin{aligned} \tilde{C}^{(1)} &= C^{(1)}(I_n - e^{i\omega}A)(I_n - e^{-i\omega}A) \\ &\quad + aC^{(2)} + bC^{(2)}A - aC^{(2)}A^2 \end{aligned} \tag{3.17}$$

$$\tilde{C}^{(2)} = C^{(2)}A. \tag{3.18}$$

*Proof.* We have

$$C^{(1)} = -(\phi_0 e_1, 2\phi_1^* e_1, \phi_0 e_2, 2\phi_1^* e_2, \ldots, \phi_o \xi_0, 2\phi_1^* \xi_0, 2\eta_2, \ldots, 2\eta_N) \tag{3.19}$$

and

$$C^{(2)} = (e_1, 0, \ldots, \xi_0, 0, \ldots). \tag{3.20}$$

Therefore,

$$C^{(1)}A = -(0, \phi_0 e_1, 0, \phi_0 e_2, \ldots, 0, \phi_0 \xi_0, 2\phi_1^* \xi_0, 2\eta_2, \ldots, 2\eta_{N-1}) \tag{3.21}$$

and

(3.22) $$C^{(1)}A^2 = -(0,0,\ldots,\ldots,0,0,\phi_0\xi_0, 2\phi_1^*\xi_0, 2\eta_2,\ldots,2\eta_{N-2}).$$

Similarly,

(3.23) $$C^{(2)}A = (0,e_1,0,e_2,\ldots,0,\xi_0,0,0,\ldots)$$

and

(3.24) $$C^{(2)}A^2 = (0,0,\ldots,\ldots,0,0,\xi_0,\ldots).$$

Let us write the righthandside of equation (3.17) as

$$(t_1, v_1,\ldots,t_{p-1},v_{p-1},s_0,s_1,\ldots,s_N).$$

From the above equations, we obtain:

$$\begin{aligned} t_1 &= -\phi_0 e_1 + a e_1 \\ &= 0 \\ v_1 &= -2\phi_1^* e_1 + 2\cos\omega\phi_0 e_1 + b e_1 \\ &= -2\phi_1^* e_1 + 2\cos\omega\phi_0 e_1 + (\phi_1^* - \phi_1)e_1 \\ &= = -(\phi_1^* + \phi_1)e_1 + 2\cos\omega\phi_0 e_1 \\ &= -\tilde{\phi}_0^* e_1 \end{aligned}$$

and similarly for $t_2, v_2, \ldots,$ and for $s_0, s_1$. Moreover,

$$\begin{aligned} s_2 &= -2\eta_2 + 2\cos\omega\phi_1^*\xi_0 - \phi_0\xi_0 - a\xi_0 \\ &= -2\tilde{\eta}_2, \end{aligned}$$

and for $j \geq 3$, we have:

$$\begin{aligned} s_j &= -2\eta_j - 2\eta_{j-2} + 4\cos\omega\eta_{j-1} \\ &= -2\tilde{\eta}_{j-1} \end{aligned}$$

so that (3.17) holds. The proof of (3.18) is immediate. ■

Let $\tilde{C}$ be defined by (3.17)–(3.18), let $C, A, \hat{A}$ be as above and let $T$ be the diagonal block matrix defined by

$$T = \text{diag}\,(T_1,\ldots T_{p-1}, T_p)$$

where, for $j \in \{1,\ldots,p-1\}$,

$$T_j = \begin{pmatrix} 0 \\ 1 \end{pmatrix},$$

and $T_p$ is the $N \times N$ matrix defined by:

$$T_p = \begin{pmatrix} 0 & 0 & \ldots & \\ 1 & 0 & \ldots & \\ 0 & 1 & 0\ldots & \\ \ldots & & & \end{pmatrix}.$$

Then,

$$\hat{C}(I_n - z\hat{A})^{-1} \equiv \tilde{C}(I_n - zA)^{-1}T \tag{3.25}$$

so that the functions $M\Theta_0 UN$ and $\hat{\Theta}_0$ have the same left zero pair. By Proposition 2.4, these two functions differ by a right multiplicative $J$-unitary constant. ■

Before stating the main result of this section we present a matrix-valued generalization of a result of [22, Appendix, P.A. 13, p.397]. In the lemma, $[e^{-i\omega}, e^{i\omega}]$ denotes the arc of circle $\{e^{it}, t \in [-\omega, \omega]\}$, and its complement in $\mathbb{T}$ is $]e^{i\omega}, e^{-i\omega}[$.

**Lemma 3.3** *Let $\Psi$ be an element of $\mathcal{C}_p$. Then, the function*

$$z \longrightarrow \frac{z - e^{i\omega}}{1 - ze^{i\omega}}\Psi(z) \tag{3.26}$$

*belongs to $\mathcal{C}_p$ if and only if $\Psi$ is holomorphic on $\mathbb{C} \setminus [e^{-i\omega}, e^{i\omega}]$ and*

$$\frac{1}{2i}(\Psi(e^{it}) - \Psi(e^{it})^*) \geq 0$$

*on $]e^{i\omega}, e^{-i\omega}[$, or equivalently if and only if the Riesz–Herglotz representation of $\Psi$ can be written as*

$$\Psi(z) = i\beta + \int_{-\omega}^{\omega} \frac{e^{it} + z}{e^{it} - z} d\lambda(t), \tag{3.27}$$

*where $\beta$ and $\lambda$ satisfy*

$$\beta - \int_{-\omega}^{\omega} \cot((\omega + t)/2) d\lambda(t) \geq 0. \tag{3.28}$$

*Proof.* Let $\Gamma$ be the function defined by (3.26), and let us suppose that $\Gamma$ belongs to $\mathcal{C}_p$. Then, for $r < 1$, it is easily seen that

$$\begin{aligned} \Gamma(re^{i\theta}) + \Gamma(re^{i\theta})^* &= \frac{2(r\cos\theta - \cos\omega)(\Psi(re^{i\theta}) + \Psi(re^{i\theta})^*)}{|1 - re^{i(\theta+\omega)}|^2} \\ &+ \frac{(1 - r^2)(e^{-i\omega}\Psi(re^{i\theta}) + e^{i\omega}\Psi(re^{i\theta})^*)}{|1 - re^{i(\theta+\omega)}|^2} \end{aligned} \tag{3.29}$$

Let $\lambda$ and $\nu$ be the positive matrix-valued measures associated to $\Psi$ and $\Gamma$ through the Riesz–Herglotz representation. Let us begin by showing that if the arc $[e^{i\alpha_1}, e^{i\alpha_2}]$ is included in $]e^{i\omega}, e^{-i\omega}[$, then $\lambda([\alpha_1, \alpha_2]) = \nu([\alpha_1, \alpha_2]) = 0$. For this purpose, we calculate $\nu([\alpha_1, \alpha_2])$ by using the fact that

$$\lim_{r\to 1} \int_{\alpha_1}^{\alpha_2} (\Gamma(re^{i\theta}) + \Gamma(re^{i\theta})^*) d\theta - \nu([\alpha_1, \alpha_2]) \tag{3.30}$$

Put

$$g(r, \theta) = \frac{2(r\cos\theta - \cos\omega)}{|1 - re^{i(\theta+\omega)}|^2} - \frac{2(\cos\theta - \cos\omega)}{|1 - e^{i(\theta+\omega)}|^2} \tag{3.31}$$

Then, by the uniform continuity of $g(r, \theta)$ on compact sets such as $\alpha_1 \leq \theta \leq \alpha_2$ and $r_0 \leq r \leq 1$, it is clear that for each $\epsilon > 0$, there exists $\eta > 0$ for which $1 - \eta < r < 1$ implies $|g(r, \theta)| < \epsilon$, $\forall \theta \in [\alpha_1, \alpha_2]$. From this, we get immediatly that

$$\lim_{r\to 1} \int_{\alpha_1}^{\alpha_2} g(r, \theta)(\Psi(re^{i\theta}) + \Psi(re^{i\theta})^*) d\theta = 0 \tag{3.32}$$

i.e.,

$$\lim_{r\to 1}\int_{\alpha_1}^{\alpha_2}\frac{2(r\cos\theta-\cos\omega)}{|1-re^{i(\theta+\omega)}|^2}(\Psi(re^{i\theta})+\Psi(re^{i\theta})^*)d\theta =$$

$$\int_{\alpha_1}^{\alpha_2}\frac{2(\cos\theta-\cos\omega)}{|1-e^{i(\theta+\omega)}|^2}d\lambda(\theta) \tag{3.33}$$

Let us now study the contribution of the second term of the righthandside of (3.29). Let us first observe that, as $r\cos\theta-\cos\omega$ is negative $\forall\theta\in[\alpha_1,\alpha_2]$ for $r$ close enough from 1, the first term of the righthandside of (3.29) defines a negative matrix.Therefore, the condition $\Gamma(re^{i\theta})+\Gamma(re^{i\theta})^*\geq 0$ implies that

$$\frac{(1-r^2)(e^{-i\omega}\Psi(re^{i\theta})+e^{i\omega}\Psi(re^{i\theta})^*)}{|1-re^{i(\theta+\omega)}|^2}\geq 0 \tag{3.34}$$

$\forall\theta\in[\alpha_1,\alpha_2]$ for $r$ close enough from 1. If $x^2=\inf_{r<1,\theta\in[\alpha_1,\alpha_2]}|1-re^{i(\theta+\omega)}|^2$, we get that

$$\frac{(1-r^2)(e^{-i\omega}\Psi(re^{i\theta})+e^{i\omega}\Psi(re^{i\theta})^*)}{|1-re^{i(\theta+\omega)}|^2}\leq \tag{3.35}$$

$$\frac{(1-r^2)(e^{-i\omega}\Psi(re^{i\theta})+e^{i\omega}\Psi(re^{i\theta})^*)}{x^2} \tag{3.36}$$

By a trivial modification of [15, Theorem 2-2]

$$\lim_{r\to 1}(1-r)\Psi(re^{i\theta})=\lambda\{\theta\} \tag{3.37}$$

from which we deduce that $(1-r)\|\Psi(re^{i\theta})\|$ converges to zero almost everywhere (w.r.t. the Lebesgue measure) as $r\to 1$. But, it is easily seen that $(1-r)\|\Psi(re^{i\theta})\|$ is bounded, and by the Lebesgue dominated convergence Theorem, we get that

$$\lim_{r\to 1}\int_{\alpha_1}^{\alpha_2}(1-r)\|\Psi(re^{i\theta})\|d\theta=0 \tag{3.38}$$

In view of (3.35), this implies that

$$\lim_{r\to 1}\int_{\alpha_1}^{\alpha_2}\frac{(1-r^2)(e^{-i\omega}\Psi(re^{i\theta})+e^{i\omega}\Psi(re^{i\theta})^*)}{|1-re^{i(\theta+\omega)}|^2}d\theta=0 \tag{3.39}$$

Finally, it appears that

$$\nu([\alpha_1,\alpha_2])=\int_{\alpha_1}^{\alpha_2}\frac{2(\cos\theta-\cos\omega)}{|1-e^{i(\theta+\omega)}|^2}d\lambda(\theta) \tag{3.40}$$

As $\cos\theta-\cos\omega<0$ on $[\alpha_1,\alpha_2]$, we deduce that $\nu([\alpha_1,\alpha_2])=\lambda([\alpha_1,\alpha_2])=0$. Therefore, $\Psi$ can be written as in (3.27) from which it follows that $\Psi$ is holomorphic on $\mathbb{C}\setminus[e^{-i\omega},e^{i\omega}]$ and that $\Psi(e^{i\theta})+\Psi(e^{i\theta})^*=0$ on $]e^{i\omega},e^{-i\omega}[$. As $e^{-i\omega}\Psi(re^{i\theta})+e^{i\omega}\Psi(re^{i\theta})^*=\cos\omega(\Psi(re^{i\theta})+\Psi(re^{i\theta})^*)+\sin\omega(\Psi(re^{i\theta})-\Psi(re^{i\theta})^*)/i$ is positive for $r$ close enough from 1 on $]e^{i\omega},e^{-i\omega}[$, it follows that $1/i(\Psi(e^{i\theta})-\Psi(e^{i\theta})^*)\geq 0$.

Conversely, suppose that $\Psi$ is holomorphic on $\mathbb{C}\setminus[e^{-i\omega},e^{i\omega}]$ and that $1/i(\Psi(e^{i\theta})-\Psi(e^{i\theta})^*)$ is positive on $]e^{i\omega},e^{-i\omega}[$. By analytic continuation arguments, $\Psi$ satisfies $\Psi(z)+(\Psi(1/z^*))^*=0$ on $\mathbb{C}\setminus[e^{-i\omega},e^{i\omega}]$, which in particular implies that $\Psi(e^{i\theta})+\Psi(e^{i\theta})^*=0$ on $]e^{i\omega},e^{-i\omega}[$. Therefore, the Riesz–Herglotz representation of $\Psi$ can be written as in (3.27). Moreover, (3.28) follows immediately from the properties of the function $\theta\to\cot\theta$. Let us denote by $\beta_0$ the positive matrix defined by the left hand side of (3.28). Then, (3.27) can be written as

$$\begin{aligned}\Psi(z) &= i\beta_0+\int_{-\omega}^{\omega}\left(\frac{e^{it}+z}{e^{it}-z}-\frac{e^{it}+e^{-i\omega}}{e^{it}-e^{-i\omega}}\right)d\lambda(t)\\ &= i\beta_0+2\int_{-\omega}^{\omega}\frac{e^{it}(z-e^{-i\omega})}{(e^{it}-z)(e^{it}-e^{-i\omega})}d\lambda(t)\end{aligned}\tag{3.41}$$

Thus, $\Gamma(z)$ is given by

$$\Gamma(z)=i\beta_0\frac{z-e^{i\omega}}{1-ze^{i\omega}}+2\int_{-\omega}^{\omega}\frac{z-e^{i\omega}}{(e^{it}-z)(e^{-it}-e^{i\omega})}d\lambda(t)\tag{3.42}$$

from which we get that

$$\begin{aligned}\Gamma(z)+\Gamma(z)^* &= 2\beta_0\sin\omega\frac{1-|z|^2}{|1-ze^{i\omega}|^2}\\ &+ 2\int_{-\omega}^{\omega}\frac{(\cos t-\cos\omega)(1-|z|^2)}{|e^{it}-z|^2|e^{-it}-e^{i\omega}|^2}d\lambda(t).\end{aligned}\tag{3.43}$$

Therefore, $\Gamma\in\mathcal{C}_p(\omega)$.

■

This result implies that there always exist Carathéodory pairs $(\alpha,\beta)$ for which $(\alpha,\beta)N^{-1}$ still defines a Carathéodory pair. We are now in position to give the main result of this section.

**Theorem 3.4** *Let $C, A, \hat{C}, \hat{A}$ be as above and suppose that both $P$ and $\hat{P}$, the solutions of the Stein equations (2.5) and (3.14) are strictly positive. Let $\omega$ such that $\det d_0(\omega)$ is not equal to zero, where $d_0$ is defined by (2.32). Finally, let $U$ and $V$ be such that $M\Theta_0UN=\hat{\Theta}_0V$. Then, Problem 1.1 is solvable in $\mathcal{C}_p(\omega)$. An element $\phi\in\mathcal{C}_p(\omega)$ is a solution to Problem 1.1 if and only if it can be written as in (3.4) with $\Theta=\Theta_0U$ and parameter $(\alpha,\beta)$ such that $(\alpha,\beta)N^{-1}$ still defines a Carathéodory pair.*

*Proof.* Let $(\gamma,\delta)$ be a Carathéodory pair such that $(\tilde{\gamma},\tilde{\delta})=(\gamma,\delta)N^{-1}$ still defines a Carathéodory pair. Let $\phi$ be defined by (3.7). Then, it is a solution of Problem 1.1. Let us show that $\tilde{\phi}$ defined by (1.11) is a Carathéodory function solution to Problem 3.1. Let us put $\tilde{\Theta}=M\Theta N$. Then, as

$$\begin{pmatrix}\tilde{\phi}\\ I_p\end{pmatrix}=M\begin{pmatrix}\phi\\ I_p\end{pmatrix},\tag{3.44}$$

we get immediately that

$$\begin{pmatrix} \tilde{\phi} \\ I_p \end{pmatrix} = M\Theta N N^{-1} \begin{pmatrix} \gamma \\ \delta \end{pmatrix} = \tilde{\Theta} \begin{pmatrix} \tilde{\gamma} \\ \tilde{\delta} \end{pmatrix}.$$

Therefore, $\phi$ belongs to $\mathcal{C}_p(\omega)$, so that the Problem 1.1 is solvable in $\mathcal{C}_p(\omega)$.

Conversely, let $\phi$ be a element of $\mathcal{C}_p(\omega)$ solution of the Carathéodory–Féjèr problem 1.1. Then, there exists a Carathéodory pair $(\gamma, \delta)$ such that (3.7) holds. The function $\tilde{\phi}$ is a solution to the Carathéodory–Féjèr problem 3.1, and therefore there exists a Carathéodory pair $(\tilde{\gamma}, \tilde{\delta})$ such that

$$\begin{pmatrix} \tilde{\phi} \\ I_p \end{pmatrix} = \tilde{\Theta} \begin{pmatrix} \tilde{\gamma} \\ \tilde{\delta} \end{pmatrix}. \tag{3.45}$$

On the other hand, by using (3.44), we get immediately that

$$\begin{pmatrix} \tilde{\phi} \\ I_p \end{pmatrix} = M\Theta \begin{pmatrix} \gamma \\ \delta \end{pmatrix} = M\Theta N N^{-1} \begin{pmatrix} \gamma \\ \delta \end{pmatrix}.$$

Comparing this last equation with (3.45) it follows that

$$\begin{pmatrix} \tilde{\gamma} \\ \tilde{\delta} \end{pmatrix} = N^{-1} \begin{pmatrix} \gamma \\ \delta \end{pmatrix}. \tag{3.46}$$

■

A similar approach was used in the papers [4], [3] in the study of interpolation in the Stieltjes class. A $\mathbb{C}^{p\times p}$ valued function $S$ is in the Stieltjes class $\mathcal{S}_p$ if it is analytic in the open upper plane $\mathbb{C}_+$ and if both $S$ and $\tilde{S}$ have positive imaginary parts in $\mathbb{C}_+$, with $\tilde{S}(z) = -z^{-1}S(z)$, i.e. if both $S$ and $\tilde{S}$ are Nevanlinna functions. Therefore, to any interpolation problem for $S$ is associated another natural interpolation problem for $\tilde{S}$. The core of the method in [3] is to compare, in the nondegenerate case, the corresponding resolvent matrices. We note two differences with [4] and [3]. There, the relation between the two resolvent matrices is of the form $\tilde{\Theta} = M\Theta M^{-1}$ (with $M(z) = \begin{pmatrix} -z^{-1}I_p & 0 \\ 0 & I_p \end{pmatrix}$), i.e. $\Theta$ is linked to $\tilde{\Theta}$ by an automorphism of the group of nonsingular $\mathbb{C}^{p\times p}$–valued rational functions. Furthermore, in the case of the resolvent matrices intervening in the solution of the interpolation problem in $\mathcal{S}_p$, the function $\Theta$ and $\tilde{\Theta}$ have same McMillan degree. Here, we have a drop of degree. For further references on the Stieltjes interpolation problem, we refer to [22], [18] and [11].

In the previous theorem, we supposed that $d_0$ was nonsingular at $\omega$. If $d_0$ is singular at a given $\omega_0$ but $P$ and $\hat{P}(\omega_0)$ are still supposed strictly positive, we remark that the problem 1.1 still has solutions. Indeed, $\hat{P}$ is a continuous function of $\omega$. It is still positive in a small neighbourhood of $\omega_0$. This neighbourhood can be chosen in such a way that $d_0(\omega)$ does not vanish in it for $\omega$ different from $\omega_0$. For such $\omega$, the above theorem is applicable, and allows to give solutions in $\mathcal{C}_p(\omega_0-\varepsilon)$ for $\varepsilon$ small enough. This implies the existence of solutions in $\mathcal{C}_p(\omega_0)\supset\mathcal{C}_p(\omega_0-\varepsilon)$. The precise description of all the solutions eludes us in this case. We conclude this section with:

**Proposition 3.5** *Suppose that the matrices $P$ and $\hat{P}$ are nonnegative. Then, Problem 1.1 is solvable in $\mathcal{C}_p(\omega)$.*

*Proof.* We proceed by approximation. Let us set $h_k=\int_{-\omega}^{\omega}e^{ikt}dt=\frac{2sin(k\omega)}{k}$ for $k=0,1,\dots,N$ and $\tilde{h}_k=h_{k-1}+h_{k+1}-2h_k\cos\omega$ (with $h_{-1}=h_1$). Let $\varepsilon>0$. We consider Problem 1.1 associated to the perturbed data

$$\begin{aligned}\phi_0(\varepsilon)&=\phi_0+\varepsilon h_0 I_p\\ \phi_1(\varepsilon)&=\phi_1+\varepsilon h_1 I_p\\ \xi_0(\varepsilon)&=\xi_0\\ \eta_i(\varepsilon)&=\eta_i+\varepsilon h_i\xi_0,\quad i=2,\dots,N.\end{aligned}$$

Without loss of generality we will suppose that $\{\xi_0,e_1,\dots,e_{p-1}\}$ is an orthonormal basis of $\mathbb{C}^p$. Let $C(\varepsilon)=C+\varepsilon H$ where

$$H=\begin{pmatrix}-h_0e_1 & -2h_1e_1 & \dots & -h_0\xi_0 & -2h_1\xi_0 & -2h_2\xi_0\dots & -2h_N\xi_0 & \\ 0 & 0 & \dots & 0 & 0 & 0 & \dots & 0\end{pmatrix}.$$

Since $H^*JH=0$, it is readily checked that the solution of the perturbed Stein equation (2.5) (with $C(\varepsilon)$ in place of $C$) is given by

$$P(\varepsilon)=P+2\varepsilon\text{diag}(D_1,\dots D_{p-1},D_p)\tag{3.47}$$

where, for $i\in\{1,\dots,p-1\}$,

$$D_i=\begin{pmatrix}h_0 & h_1\\ h_1 & h_0\end{pmatrix},$$

and

$$D_p=\begin{pmatrix}h_0 & \cdots & h_N\\ \vdots & & \vdots\\ h_N & \cdots & h_0\end{pmatrix}.$$

Note that $P(\varepsilon)$ is strictly positive for $\varepsilon>0$. Similarly, let

$$\begin{aligned}\tilde{\phi}_0(\varepsilon)&=\tilde{\phi}_0+\varepsilon\tilde{h}_0I_p\\ \tilde{\eta}_i(\varepsilon)&=\tilde{\eta}_i+\varepsilon\tilde{h}_i,\xi_0\ i=1,\dots,N.\end{aligned}$$

The solution of the perturbed Stein equation (3.14) is given by

$$\hat{P}(\varepsilon)=\hat{P}+2\varepsilon\text{diag}(\tilde{h}_0,\dots,\tilde{h}_0,\tilde{D})$$

where

$$\tilde{D} = \begin{pmatrix} \tilde{h}_0 & \cdots & \tilde{h}_{N-1} \\ \vdots & & \vdots \\ \tilde{h}_{N-1} & \cdots & \tilde{h}_0 \end{pmatrix}.$$

Since $\hat{P}(\varepsilon) > 0$ for $\varepsilon > 0$ there exits a function $\phi_\varepsilon \in \mathcal{C}_p(\omega)$ solution to the perturbed interpolation problem. Since the set of matrices $\phi_\varepsilon(0)$ is bounded for $\varepsilon \leq 1$, the measures appearing in the Riesz–Herglotz representation (1.1) are bounded and by Helly's theorem it is easily shown that the family $\phi_\varepsilon$ has a convergent subsequence, which is still in $\mathcal{C}_p(\omega)$. The limit is then a solution of Problem 1.1. ■

# 4 The Nevanlinna–Pick problem

In this section we solve a tangential Nevanlinna–Pick problem in the class $\mathcal{C}_p(\omega)$. The method is the same as in Section 3 and will be only outlined.

**Problem 4.1** *Given matrices $\phi_0 = \phi_0^*$ and $\phi_1$ in $\mathbb{C}^{p\times p}$ and given $\xi_i, \eta_i, i = 1, \dots, N$ vectors in $\mathbb{C}^p$ and given $z_1, \dots, z_n$ $N$ points in $\mathbb{D}$ distinct from 0, find all functions in $\phi \in \mathcal{C}_p(\omega)$ such that*

$$\phi(0) = \phi_0 \tag{4.1}$$

$$\phi^{(1)}(0) = \phi_1 \tag{4.2}$$

$$\phi(z_i)^*\xi_i = \eta_i, \quad i = 1, \dots N. \tag{4.3}$$

Let us define by $A$ the block diagonal matrix with first block diagonal equal to

$$\begin{pmatrix} 0 & I_p \\ 0 & 0 \end{pmatrix}$$

and second block diagonal equal to

$$\mathrm{diag}\,(z_1, \dots, z_N),$$

and by $C$ the matrix

$$C = \begin{pmatrix} -\phi_0 & -2\phi_1^* & -\eta_1 & \dots & -\eta_N \\ I_p & 0 & \xi_1 & \dots & \xi_N \end{pmatrix}. \tag{4.4}$$

The problem 4.1 has a solution in $\mathcal{C}_p$ if and only if the Stein equation (2.5) (with this choice of $A$ and $C$) has a nonnegative solution $P$. We will suppose that $P > 0$. Then, the set of all solutions to Problem 4.1 is parametrized as in (3.4) (or (3.7)), with a resolvent matrix $\Theta = \Theta_0 U$. Here, $U$ is an arbitrary $J$–unitary matrix and $\Theta_0$ is obtained form formula (2.8), with $x_0 = e^{i\omega}$ and $C, A$ as fixed in the present section. Since

$$C^{(2)} = (I_p, 0, \xi_1, \dots, \xi_N),$$

we have that ran $C^{(2)} = \mathbb{C}^p$. By Proposition 2.4, for every $\omega \in [0\,,\,\pi]$ at the possible exception of a finite set, $U$ may be chosen so that the conditions (2.25)–(2.26) hold.

Let $\tilde{\eta}_i$ be defined by

$$\tilde{\eta}_i = (1 - 2z_i^* \cos\omega + z_i^{2*})\eta_i + (z_i^{2*}\phi_0 + z_i^*(\phi_1 - \phi_1^*) - \phi_0)\xi_i. \tag{4.5}$$

The function $\tilde{\phi}$ satisfies the interpolation problem

**Problem 4.2** *Given matrices $\phi_0 = \phi_0^*$ and $\phi_1$ in $\mathbb{C}^{p\times p}$ and given $\xi_i, \eta_i, i = 1, \ldots, N$ vectors in $\mathbb{C}^p$ and given $N$ points $z_1, \ldots, z_N$ in $\mathbb{D}$ distinct from 0, find all functions in $\psi \in \mathcal{C}_p$ such that*

$$\psi(0) = \tilde{\phi}_0 \tag{4.6}$$

$$z_i^*\psi(z_i)^*\xi_i = \tilde{\eta}_i, \; i = 1, \ldots N. \tag{4.7}$$

Let $\hat{A}$ and $\hat{C}$ be defined as follows: $\hat{A}$ is the block diagonal matrix with first block the $p \times p$ zero matrix and second block

$$\text{diag}\,(z_1, \ldots, z_N),$$

and

$$\hat{C} = \begin{pmatrix} -\phi_0 & -\tilde{\eta}_1 & \ldots & -\tilde{\eta}_N \\ I_p & z_1^*\xi_1 & \ldots & z_N^*\xi_N \end{pmatrix} \tag{4.8}$$

and suppose that the solution of the equation (2.5) (with these matrices in place of $A$ and $C$) is strictly positive. We denote by $\hat{\Theta}_0$ the corresponding resolvent matrix given by (2.8), with $x_0 = e^{i\omega}$ and $\hat{C}, \hat{A}$ instead of $C$ and $A$.

**Lemma 4.3** *Let*

$$C = \begin{pmatrix} C^{(1)} \\ C^{(2)} \end{pmatrix}$$

*be the decomposition of the matrix $C$ defined in (4.4) into two $\mathbb{C}^{p\times(2p+N)}$ matrices and let $\tilde{C}^{(1)}$ and $\tilde{C}^{(2)}$ be defined by*

$$\tilde{C}^{(1)} = -(0, \tilde{\phi}_0, \tilde{\eta}_1, \tilde{\eta}_2, \ldots) \tag{4.9}$$

*and*

$$\tilde{C}^{(2)} = (0, I_p, z_1\xi_1, z_2\xi_2, \ldots). \tag{4.10}$$

*Then, equations (3.17) and (3.18) hold.*

The proof is a direct computation which will be omitted. This Lemma allows, as in the previous section, to relate the resolvent matrices $\Theta_0$ and $\hat{\Theta}_0$, and we obtain:

**Theorem 4.4** *Let $C, A, \hat{C}, \hat{A}$ be as above and suppose that both $P$ and $\hat{P}$, the solutions of the Stein equations (2.5) and (3.14) are strictly positive. Let $\omega$ such that $\det d_0(\omega)$ is not equal to zero, where $d_0$ is defined by (2.32). Finally, let $U$ and $V$ be such that $M\Theta_0 UN = \hat{\Theta}_0 V$. Then, the problem 4.1 is solvable in $\mathcal{C}_p(\omega)$. An element $\phi \in \mathcal{C}_p(\omega)$ is a solution to Problem 4.1 if and only if it can be written as in (3.4) with a parameter $(\alpha, \beta)$ such that $(\alpha\ \beta)N^{-1}$ still defines a Carathéodory pair.*

More generally, the same method allows to solve the more general interpolation problems considered in [9], [5] and [3]. The existence in the singular cases is then settled as in the previous section by approximation arguments.

## References

[1] N.I. Akhiezer. *The classical moment problem.* Hafner, New-York, 1965.

[2] N.I. Akhiezer and M.G. Kreĭn. *Some questions in the theory of moments*, volume 2 of *Translations of mathematical monographs.* American mathematical society, Rhode Island, 1962.

[3] D. Alpay, J. Ball, I. Gohberg, and L. Rodman. Interpolation in the Stieltjes class. In preparation, 1992.

[4] D. Alpay, J. Ball, I. Gohberg, and L. Rodman. State space theory of automorphisms of rational matrix functions. *Integral equations and operator theory*, 15:349–377, 1992.

[5] D. Alpay, P. Bruinsma, A. Dijksma, and H. de Snoo. *Interpolation problems, extensions of symmetric operators and reproducing kernel spaces I*, pages 35–82. Operator theory: advances and applications OT50. Birkhäuser Verlag, Basel, 1991.

[6] D. Alpay and H. Dym. *On applications of reproducing kernel spaces to the Schur algorithm and rational J-unitary factorization*, pages 89–159. Operator theory: advances and applications OT18. Birkhäuser Verlag, Basel, 1986.

[7] D. Alpay and I. Gohberg. *Unitary rational matrix functions*, pages 175–222. Operator theory: advances and applications OT33. Birkhäuser Verlag, Basel, 1988.

[8] D. Alpay and P. Loubaton. The trigonometric moment problem on an interval: the matrix case. In preparation.

[9] J. Ball, I. Gohberg, and L. Rodman. *Interpolation of rational matrix functions.* Birkhäuser Verlag, Basel, 1990.

[10] H. Bart, I. Gohberg, and M. Kaashoek. *Minimal factorization of matrix and operator functions.* Birkhäuser Verlag, Basel, 1979.

[11] V. Bolotnikov. Two sided interpolation in the Stieltjes class. To appear in Integral equation and operator theory.

[12] T. Chovanel and P. Loubaton. Le problème des moments trigonométriques sur un intervalle. *Comptes-rendus de l'académie des sciences de Paris, série A*, 314:41–44, 1992.

[13] T. Chovanel and P. Loubaton. On the realization of band-limited power spectra from partial covariance sequences. Submitted to IEEE–IT, 1992.

[14] P. Delsarte, Y. Genin, and Y. Kamp. Schur parametrization of positive definite block-Toeplitz systems. *SIAM journal in applied mathematics*, pages 34–46, 1979.

[15] P. Dewilde and H. Dym. Lossless inverse scattering, digital filters and estimation theory. *IEEE Transactions on Information theory*, 30:644–662, 1984.

[16] V. Dubovoy, B. Fritzsche, and B. Kirstein. *Matricial version of the classical Schur problem*, volume 129 of *Teubner–Texte zur Mathematik*. B.G. Teubner Verlagsgesellschaft, Stuttgart–Leipzig, 1992.

[17] H. Dym. *J-contractive matrix functions, reproducing kernel spaces and inteprolation*, volume 71 of *CBMS lecture notes*. American mathematical society, Rhodes island, 1989.

[18] Yu. Dyukarev and V.E. Katznelson. Multiplicative and additive classes of Stieltjes analytic matrix–valued functions and interpolation problems associated with them. I. *American Mathematical Society Translations*, 131:55–70, 1986.

[19] P.A. Fuhrmann. *Linear systems and operators in Hilbert space*. McGraw-Hill international book company, 1981.

[20] I. Gohberg, P. Lancaster, and L. Rodman. *Invariant subspaces of matrices and applications*. Wiley, New–York, 1986.

[21] R.E. Kalman, P.L. Falb, and M.A.K. Arbib. *Topics in mathematical system theory*. Mc Graw–Hill, New-York, 1969.

[22] M.G. Kreĭn and A.A. Nudelman. *The Markov moment problem and extremal problems*, volume 50 of *Translations of mathematical monographs*. American mathematical society, Providence, Rhode Island, 1977.

Daniel Alpay
Department of Mathematics
Ben–Gurion University of the Negev
POB 653. 84105 Beer-Sheva
Israel

Philippe Loubaton
Département Signal
Ecole Nationale Supérieure des Télécommunications
46 rue Barrault
75634 Paris Cedex 13
France

MSC: 30E05, 42A70, 44A60,47A56, 47A57.

Operator Theory:
Advances and Applications, Vol. 64




# MAXIMUM ENTROPY AND JOINT NORM BOUNDS FOR OPERATOR EXTENSIONS

Mihály Bakonyi[1], Victor G. Kaftal[2], Gary Weiss[2] and Hugo J. Woerdeman[3]

The maximum entropy principle is applied to give a new proof of the Kaftal, Larson and Weiss joint operator norm/Hilbert-Schmidt norm theorem for operator extensions. Using this technique, new local bounds for the joint norm extension problem are obtained.

**1. Introduction.** Consider the lower triangular partial operator matrices

$$F = \begin{pmatrix} F_{11} & X_{12} & \dots & X_{1n} \\ F_{21} & F_{22} & \dots & X_{2n} \\ \dots & \dots & \dots & \dots \\ F_{n1} & F_{n2} & \dots & F_{nn} \end{pmatrix}, \tag{1.1}$$

in which $F_{ij} \in \mathcal{B}(\mathcal{H}_j, \mathcal{H}_i)$ (the bounded linear operators from the Hilbert space $\mathcal{H}_j$ to $\mathcal{H}_i$) are specified for $1 \leq j \leq i \leq n$ and $X_{ij} \in \mathcal{B}(\mathcal{H}_j, \mathcal{H}_i)$ are unspecified for $1 \leq i < j \leq n$. For any choice of all the $X_{ij}$, the resulting operator matrix is called a *completion* of $F$. In this paper we shall concentrate on the central completion described below. The case $n = 2$ serves both as a tool and to illustrate the general case.

It is well known that ([11]) the clearly necessary conditions $\left\| \begin{pmatrix} B \\ A \end{pmatrix} \right\| \leq \delta$ and $\|[A\ C]\| \leq \delta$ are also sufficient for the existence of a completion of $F = \begin{pmatrix} B & X \\ A & C \end{pmatrix}$ with norm $\leq \delta$. The condition $\left\| \begin{pmatrix} B \\ A \end{pmatrix} \right\| \leq \delta$ implies the existence of a unique contraction

[1] Partially supported by Georgia State University.
[2] Partially supported by NSF grant DMS-9123249 and by a grant from the Charles Phelps Taft Memorial Foundation.
[3] Partially supported by NASA Contract NAS1-18347.

$G_1 : \mathcal{D}_{A,\delta} \rightarrow \mathcal{H}_1$ such that $B = G_1 D_{A,\delta}$. Here $D_{A,\delta} = (\delta^2 I - A^*A)^{\frac{1}{2}}$ and $\mathcal{D}_{A,\delta}$ is the closure of the range of $D_{A,\delta}$. We suppress the reference to $\delta$ in case $\delta = 1$. Similarly, $\|[A\ C]\| \leq \delta$ implies the existence of a unique contraction $G_2 : \mathcal{H}_2 \rightarrow \mathcal{D}_{A^*,\delta}$ such that $C = D_{A^*,\delta} G_2$. Then, as proved in [1], there exists a one-to-one correspondence between the set of all completions with norm $\leq \delta$ of the partial matrix $F = \begin{pmatrix} B & X \\ A & C \end{pmatrix}$ and the set of all $G : \mathcal{D}_{G_2} \rightarrow \mathcal{D}_{G_1^*}$ with norm $\leq \delta$. This one-to-one correspondence is given by $X = -G_1 A^* G_2 + D_{G_1^*} G D_{G_2}$. The completion $F_0$ with $X_0 = -G_1 A^* G_2$ is called *the central completion of $F$ with respect to $\delta$*. Note that if $\|A\| < \delta$, then $X_0 = -BA^*(\delta^2 I - AA^*)^{-1} C$.

Given an $n \times n$ partial matrix (1.1), denote $F_k = \begin{pmatrix} F_{k1} & \dots & F_{kk} \\ .. & \dots & \dots \\ F_{n1} & \dots & F_{nk} \end{pmatrix}$, $k = 1, \dots, n$ and $d_\infty(F) = \sup_k \|F_k\|$, the so-called Arveson distance of $F$ (see [2]). It is known ([2]) that $F$ admits a completion of norm $d_\infty(F)$. For every $\delta \geq d_\infty(F)$, the *central completion of $F$ with respect to $\delta$* is constructed as follows: choose first $X_{12}$ to obtain the central completion of $(F_{ij})_{i=1,j=1}^{n,\ 2}$ with respect to $\delta$. Then, continue the process for $k = 3, \dots, n$ by choosing $\begin{pmatrix} X_{1k} \\ \vdots \\ X_{k-1,k} \end{pmatrix}$ to obtain the central completion of $(F_{ij})_{i=1,j=1}^{n,\ k}$ with respect to $\delta$.

In this paper we consider partial operator matrices (1.1) with entries in the Hilbert-Schmidt class $\mathcal{C}_2(\mathcal{H}_j, \mathcal{H}_i)$. The distance in the Hilbert-Schmidt norm of $F$ to the block upper triangular matrices is denoted by $d_2(F)$. One easily finds that $d_2(F) = (\sum_{1 \leq j \leq i \leq n} \|F_{ij}\|_2^2)^{\frac{1}{2}}$. In [10] the following was proved.

THEOREM 1.1. *Let a partial matrix (1.1) be given with $d_2(F) < \infty$. Then, for any $\delta > 1$, the central completion $F_0$ of $F$ with respect to $\delta d_\infty(F)$ satisfies:*

$$\|F_0\| \leq \delta d_\infty(F),$$

*and*

$$\|F_0\|_2 \leq \frac{\delta}{\sqrt{\delta^2 - 1}} d_2(F).$$

As a consequence of Theorem 1.1, any partial matrix (1.1) with $d_2(F) < \infty$ has a completion $G$ such that

$$\|G\|_\infty \leq \sqrt{2} d_\infty(F),$$

and

$$||G||_2 \leq \sqrt{2} d_2(F).$$

The constant $\sqrt{2}$ is referred to as a joint norm interpolation constant. In addition, in [10] the authors extended their result to general nest algebras.

Our main result (Theorem 3.1) is a proof of Theorem 1.1 based on the maximum entropy principle, which generalizes the technique of [6] to the non-commutative case. This proof, together with a sharp estimate for the Hilbert-Schmidt norm of a $2 \times 2$ central completion (Proposition 2.1 below) yields better bounds in Theorem 1.1. The improved bounds are not uniform, in that they depend on the particular operator matrix $F$. These results are then extended to the case of general nest algebras (Theorem 3.5). An application to integral operators is given in Theorem 4.1.

This work was initiated while the first and the last author enjoyed the hospitality of their hosts, the middle two authors, and the University of Cincinnati.

We thank Professor C. Foias for his suggestions and support.

**2. A Sharp Bound in the $2 \times 2$ Case.**

PROPOSITION 2.1. *Let* $F = \begin{pmatrix} B & X \\ A & C \end{pmatrix}$ *be a partial operator matrix in* $\mathcal{B}(\mathcal{H}_1 \oplus \mathcal{H}_2)$ *such that*

$$||\begin{pmatrix} B \\ A \end{pmatrix}|| \leq \nu, \tag{2.1}$$

*for some* $\nu \geq 1$, *and*

$$||\begin{pmatrix} A & C \end{pmatrix}|| \leq 1. \tag{2.2}$$

*Let* $F_0 = \begin{pmatrix} B & X_0 \\ A & C \end{pmatrix}$ *be the central completion of* $F$ *with respect to* $\delta > \max\{1, \nu\}$. *If the operator* $C$ *is Hilbert-Schmidt, then*

$$||X_0||_2 \leq \begin{cases} \frac{\nu^2}{2\delta\sqrt{\delta^2-\nu^2}}||C||_2, \nu^2 \leq \frac{2\delta^2}{\delta^2+1} \\ \frac{(\nu^2-1)^{\frac{1}{2}}}{\delta^2-1}||C||_2, \nu^2 \geq \frac{2\delta^2}{\delta^2+1}. \end{cases} \tag{2.3}$$

*Proof.* It is known (e.g., see [1]) that (2.1) and (2.2) imply the existence of the uniquely determined contractions $G_1 : \mathcal{D}_{A,\nu} \rightarrow \mathcal{H}_1$ and $G_2 : \mathcal{H}_2 \rightarrow \mathcal{D}_{A^*}$ such that $B = G_1 D_{A,\nu}$ and $C = D_{A^*} G$. Then ([11]),

$$X_0 = BA^*(\delta^2 I - AA^*)^{-1}C = G_1(\nu^2 I - A^*A)^{\frac{1}{2}} A^*(\delta^2 I - AA^*)^{-1}C. \tag{2.4}$$

By (2.4) we have that

$||X_0||_2 \leq ||(\nu^2 I - A^*A)^{\frac{1}{2}} A^*(\delta^2 I - AA^*)^{-1}|| \, ||C||_2 = ||A^*(\nu^2 I - AA^*)^{\frac{1}{2}}(\delta^2 I - AA^*)^{-1}|| \, ||C||_2.$
(2.5)

Let $f(x) = \frac{x(\nu^2 - x^2)^{\frac{1}{2}}}{\delta^2 - x^2}$ and $|A^*| = (AA^*)^{\frac{1}{2}}$. Then (2.5) implies

$$(2.6) \qquad ||X_0||_2 \leq ||f(|A^*|)|| \, ||C||_2.$$

Solving $f'(x) = 0$ with $x > 0$ yields $x_0 = \frac{\nu\delta}{\sqrt{2\delta^2 - \nu^2}}$. Furthermore, $x_0 \leq 1$ if and only if $\nu^2 \leq \frac{2\delta^2}{\delta^2 + 1}$. It follows that

$$\max_{0 \leq x \leq 1} f(x) = \begin{cases} \frac{\nu^2}{2\delta\sqrt{\delta^2 - \nu^2}}, \nu^2 \leq \frac{2\delta^2}{\delta^2+1} \\ \frac{(\nu^2-1)^{\frac{1}{2}}}{\delta^2 - 1}, \nu^2 \geq \frac{2\delta^2}{\delta^2+1}. \end{cases}$$

Now (2.6) implies (2.3). □

REMARK 2.2. Consider $\nu^2 \leq \frac{2\delta^2}{\delta^2+1}$, $A = \frac{\delta\nu}{\sqrt{2\delta^2 - \nu^2}}$, $B = \sqrt{\nu^2 - A^2}$ and $C = \sqrt{1 - A^2}$. For this choice we have $\frac{X_0}{C} = -\frac{\nu^2}{2\delta\sqrt{\delta^2 - \nu^2}}$. On the other hand, consider $\nu^2 \geq \frac{2\delta^2}{\delta^2+1}$, $A = x$ (where $|x| \leq 1$), $B = \sqrt{\nu^2 - x^2}$, and $C = \sqrt{1 - x^2}$. Then $X_0 = -\frac{x\sqrt{\nu^2 - x^2}\sqrt{1-x^2}}{\delta^2 - x^2}$, and thus $\lim_{x \uparrow 1} \frac{X_0}{C} = -\frac{\sqrt{\nu^2 - 1}}{\delta^2 - 1}$. This proves that the estimates in (2.3) are sharp.

**3. The Maximum Entropy Method.** Inspired by [8] and [9], we next present a proof of Theorem 1.1 which makes use of the maximum entropy principle. This principle, in one of its forms, says that among all Hilbert-Schmidt contractive completions $F$ of (1.1), the central completion $F_0$ is the unique contractive completion that maximizes the quantity $\det(I - F^*F)$. In [6] (see also the appendix of [7]) a similar reasoning may be found that deals with the commutative case.

THEOREM 3.1. *Let $F$ be an $n \times n$ lower triangular operator matrix with entries in $\mathcal{C}_2(\mathcal{H}_j, \mathcal{H}_i)$ with $d_\infty(F) \leq d_\infty < 1$ and $d_2(F) = d_2$. Then the central completion $F_0$ of $F$ with respect to 1 satisfies*

$$||F_0||_2 \leq \frac{d_2}{\sqrt{1 - d_\infty^2}}.$$

*Proof.* From the inequality $\lambda \leq -\log(1 - \lambda)$ we see that

$$(3.1) \qquad ||F_0||_2^2 = \operatorname{tr}(F_0^* F_0) \leq \operatorname{logdet}(I - F_0^* F_0)^{-1}.$$

Denote $F_k = [A_k\ C_k]$, where $A_1 = 0$, $A_k = \begin{pmatrix} F_{k1} & \dots & F_{k,k-1} \\ \vdots & \vdots & \vdots \\ F_{n1} & \dots & F_{n,k-1} \end{pmatrix}$ for $k = 2, \dots, n$ and $C_k = \begin{pmatrix} F_{kk} \\ \vdots \\ F_{nk} \end{pmatrix}$ for $k = 1, \dots, n$. It is known (Corollary II.2.2 in [9]; see also [3]) that

$$\det(I - F_0^* F_0) = \frac{\prod_{k=1}^n \det(I - A_k A_k^* - C_k C_k^*)}{\prod_{k=2}^n \det(I - A_k A_k^*)}. \tag{3.2}$$

Since $||A_k|| \leq ||F_k|| \leq d_\infty < 1$, $D_{A_k^*}$ is invertible, $G_k = D_{A_k^*}^{-1} C_k$ is a contraction, and $C_k = D_{A_k^*} G_k$. Consequently,

$$\det(I - A_k A_k^* - C_k C_k^*) = \det D_{A_k^*}^2 \det D_{G_k^*}^2,$$

and

$$\det(I - F_0^* F_0)^{-1} = \prod_{k=1}^n \det(I - G_k G_k^*)^{-1}. \tag{3.3}$$

Furthermore,

$$(I - G_k^* G_k)^{-1} = I + G_k^*(I - G_k G_k^*)^{-1} G_k = \tag{3.4}$$

$$I + C_k^* D_{A_k^*}^{-1}(I - D_{A_k^*}^{-1} C_k C_k^* D_{A_k^*}^{-1})^{-1} D_{A_k^*}^{-1} C_k = I + C_k^*(I - A_k A_k^* - C_k C_k^*)^{-1} C_k.$$

Using the inequality $1 + \lambda \leq \exp(\lambda)$ and $||A_k A_k^* + C_k C_k^*|| = ||F_k||^2 \leq d_\infty^2$, we have

$$\det(I - G_k^* G_k)^{-1} \leq \exp\{\mathrm{tr}[C_k^*(I - A_k A_k^* - C_k C_k^*)^{-1} C_k]\} \leq \exp(\frac{\mathrm{tr}(C_k^* C_k)}{1 - d_\infty^2}). \tag{3.5}$$

Finally, (3.1), (3.3) and (3.5) imply that

$$||F_0||_2^2 \leq \sum_{k=1}^n \frac{\mathrm{tr}(C_k^* C_k)}{1 - d_\infty^2} - \frac{d_2^2}{1 - d_\infty^2}.$$

□

From the proof of Theorem 3.1, it follows that

$$\det(I - F_0^* F_0) \geq \exp(-\frac{d_2^2}{1 - d_\infty^2}).$$

Since $0 < I - F_0^* F_0 \leq I$, the smallest eigenvalue of $I - F_0^* F_0$ is greater or equal than $\exp(-\frac{d_2^2}{1 - d_\infty^2})$, which implies that $F_0^* F_0 \leq 1 - \exp(-\frac{d_2^2}{1 - d_\infty^2})$, and thus

$$||F_0|| \leq \sqrt{1 - \exp(-\frac{d_2^2}{1 - d_\infty^2})}. \tag{3.6}$$

Let $\delta_0 = \sqrt{1+\frac{1}{x}}$, where $x > 0$ is the solution of the equation $1 + 2x = \exp(\frac{d_2^2}{d_\infty^2}x)$ if $d_2 < \sqrt{2}d_\infty$ and let $\delta_0 = \infty$ if $d_2 \geq \sqrt{2}d_\infty$.

PROPOSITION 3.2. *Let $F$ be an $n \times n$ lower triangular partial operator matrix with $d_\infty(F) \leq d_\infty$ and $d_2(F) = d_2 < \infty$. Then, for any $\delta > 1$, the central completion $F_0$ of $F$ with respect to $\delta d_\infty$ satisfies*

$$||F_0|| \leq \delta d_\infty \sqrt{1 - \exp[-\frac{d_2^2}{d_\infty^2(\delta^2-1)}]} \tag{3.7}$$

*and*

$$||F_0||_2 \leq \begin{cases} \frac{\delta d_2 \sqrt{\delta^2 - 1 - \exp[-\frac{d_2^2}{d_\infty^2(\delta^2-1)}]}}{\delta^2-1} & \text{if } 1 < \delta < \delta_0 \\ d_2 \cosh \frac{d_2^2}{2d_\infty^2(\delta^2-1)} & \text{if } \delta \geq \delta_0. \end{cases} \tag{3.8}$$

*Proof.* The relation (3.7) is obtained by applying (3.6) to the matrix $\frac{F_0}{\delta d_\infty}$ (note that both $d_2$ and $d_\infty$ also need to be divided by $\delta d_\infty$). Let

$$\nu := \delta \sqrt{1 - \exp[-\frac{d_2^2}{d_\infty^2(\delta^2-1)}]}. \tag{3.9}$$

To improve the Hilbert-Schmidt estimate, we compute the central completion of the matrix $\frac{F_0}{d_\infty}$ with respect to $\delta$ using the iterative method described in the Introduction. At each step, the $2 \times 2$ block matrix completion $\begin{pmatrix} B & X_0 \\ A & C \end{pmatrix}$ satisfies the conditions $||[A\ C]|| \leq 1$ since $d_\infty(\frac{F_0}{d_\infty}) = 1$ and $||\begin{pmatrix} B \\ A \end{pmatrix}|| \leq ||\frac{F_0}{d_\infty}|| \leq \nu$, by (3.7) and (3.9). It is easy to verify that $\nu^2 > \frac{2\delta^2}{\delta^2+1}$ if $1 < \delta < \delta_0$ and $\nu^2 \leq \frac{2\delta^2}{\delta^2+1}$ if $\delta \geq \delta_0$. Thus, by applying at each step Proposition 2.1, we obtain

$$||F_0||_2^2 \leq \begin{cases} d_2^2(1 + \frac{\nu^2-1}{(\delta^2-1)^2}) & \text{if } 1 < \delta < \delta_0 \\ d_2^2(1 + \frac{\nu^4}{4\delta^2(\delta^2-\nu^2)}) & \text{if } \delta \geq \delta_0. \end{cases} \tag{3.10}$$

By substituting (3.9) in (3.10) we obtain (3.8). □

REMARK 3.3. Denote $\alpha = \min_k \operatorname{tr}(C_k^* C_k)$. If $\alpha = 0$, we can eliminate all the columns for which $\operatorname{tr}(C_k^* C_k) = 0$ since the central completion produces only zeros in the columns corresponding to any such $k$. Thus we assume that $\alpha > 0$. For any $\lambda \geq \beta > 0$, $1 + \lambda \leq (1+\beta)^{\frac{\lambda}{\beta}}$. Applying this inequality to (3.4) (instead of $1 + \lambda \leq \exp(\lambda)$ ) with $\beta = \frac{\alpha}{1-d_\infty^2}$ and $\lambda = \frac{\operatorname{tr} C_k^* C_k}{1-d_\infty^2}$, we can refine the inequality in Theorem 3.1 to

$$||F_0||_2 \leq \mu \frac{d_2}{\sqrt{1-d_\infty^2}},$$

where $\mu^2 = \frac{1-d_\infty^2}{\alpha} \log(1 + \frac{\alpha}{1-d_\infty^2})$.

REMARK 3.4. The improvement of (3.8) over the bound in Theorem 1.1 disappears for $\frac{d_2}{d_\infty} \to \infty$. However, for "large" $d_2$ even the trivial bound based on the comparison between Hilbert-Schmidt norm and operator norm can provide an improvement for "small" $\delta$. Indeed, consider the case of an $n \times n$ nonzero partial matrix $F$ with scalar entries. Let $\gamma = \frac{d_2}{\sqrt{n} d_\infty}$, then $\gamma \in (0,1]$. Let $F_0$ be the completion of $F$ with respect to $\delta d_\infty$. Then

$$||F_0||_2 \leq \sqrt{n}||F_0|| = \frac{d_2}{\gamma d_\infty}||F_0|| \leq \frac{\delta}{\gamma} d_2 < \frac{\delta}{\sqrt{\delta^2-1}} d_2,$$

whenever $\delta < \sqrt{1+\gamma^2}$.

We can extend Proposition 3.2 to a more general setting as follows. Let $\mathcal{H}$ be a Hilbert space and $\mathcal{N}$ a nest of orthogonal projections in $\mathcal{B}(\mathcal{H})$, i.e., $\mathcal{N}$ is a strongly closed, linearly ordered collection of projections on $\mathcal{H}$, containing 0 and the identity. Recall that the nest algebra of $\mathcal{N}$ is defined as $Alg\mathcal{N} := \{A \in \mathcal{B}(\mathcal{H}) \mid P^\perp A P = 0 \text{ for all } P \in \mathcal{N}\}$, where $P^\perp = I - P$.

THEOREM 3.5. *Let $\mathcal{N}$ be a nest on a Hilbert space $\mathcal{H}$ and $A \in \mathcal{B}(\mathcal{H})$ be a given Hilbert-Schmidt operator. Denote $d_2 = d_2(A) = \inf\{||A - C||_2 : C \in Alg\mathcal{N} \cap \mathcal{C}_2(\mathcal{H})\}$ and $d_\infty(A) = \inf\{||A-C|| : C \in Alg\mathcal{N}\}$. Then, for every $d_\infty \geq d_\infty(A)$ and $\delta > 1$ there exists a $C \in Alg\mathcal{N} \cap \mathcal{C}_2(\mathcal{H})$ such that*

$$||A - C|| \leq \delta d_\infty \sqrt{1 - \exp[-\frac{d_2^2}{d_\infty^2(\delta^2-1)}]} \tag{3.11}$$

*and*

$$||A - C||_2 \leq \begin{cases} \frac{\delta d_2 \sqrt{\delta^2 - 1 - \exp[-\frac{d_2^2}{d_\infty^2(\delta^2-1)}]}}{\delta^2-1} & \text{if } 1 < \delta < \delta_0 \\ d_2 \cosh \frac{d_2^2}{2d_\infty^2(\delta^2-1)} & \text{if } \delta \geq \delta_0, \end{cases} \tag{3.12}$$

*where $\delta_0$ is given before Proposition 3.2.*

*Proof.* Let $\mathcal{M} \subseteq \mathcal{N}$ be a finite subnest of $\mathcal{N}$, then since $Alg\mathcal{N} \subseteq Alg\mathcal{M}$ it follows that $d_2 \geq \inf\{||A - C||_2 : C \in Alg\mathcal{M} \cap \mathcal{C}_2(\mathcal{H})\}$ and $d_\infty \geq \inf\{||A - C|| : C \in Alg\mathcal{M}\}$. Applying Proposition 3.2 for $Alg\mathcal{M}$ and the constants $d_2$ and $d_\infty$, we obtain that the set $C_{\mathrm{M}} := \{A - B : B \in Alg\mathcal{M}$ satisfies (3.11) and (3.12) $\}$ is nonempty. By the Banach-Alaoglu Theorem the set $C_{\mathrm{M}}$ is compact in the weak* topology of $\mathcal{B}(\mathcal{H})$. Furthermore, since $\emptyset \neq C_{\cup M_i} = \cap C_{M_i}$, for $M_1, ..., M_n$ finite subnest of $\mathcal{N}$, the collection $\{C_{\mathrm{M}}$ :

$M$ is a finite subnest of $\mathcal{N}\}$ has the finite intersection property. Thus there exist an operator $D$ in their intersection. But then $B = A - D$ has the required properties. □

The above theorem allows the extension of Proposition 3.2 to the case of infinite partial matrices with $d_2(F) < \infty$. We should remark that even in this latter case, the operator norm of the central completion with respect to $\delta d_\infty$ is strictly less than $\delta d_\infty$. For the Nehari problem, this question was asked by Ciprian Foias and his collaborators and is still open.

**4. An Application to Integral Operators.** For a set $\Omega$, denote by $L_2^{n\times n}(\Omega)$ the set of all square integrable functions on $\Omega$ with values $n \times n$ matrices. We abbreviate $L_2^{n\times 1}(\Omega)$ by $L_2^n(\Omega)$. Let $\Delta = \{(t,s)|0 \le s \le t \le 1\}$, and $\Delta^c = [0,1]^2 - \Delta$. Given $k_-(t,s) \in L_2^{n\times n}(\Delta)$, let $d_2 = ||k_-||_2$, and let $K_{-,\alpha} : L_2^n[0,\alpha] \to L_2^n[\alpha,1]$,

$$(K_{-,\alpha} f)(t) = \int_0^\alpha k_-(t,s) f(s) ds, \ t \in [\alpha, 1].$$

THEOREM 4.1. *Let $\delta > 1$, $k_-(t,s) \in L_2^{n\times n}(\Delta)$, and $d_\infty \ge \sup_{0\le\alpha\le 1} ||K_{-,\alpha}||$. Then for every $\delta > 1$ there exists $k_+ \in L_2^{n\times n}(\Delta^c)$ such that the operator $K$ on $L_2^n[0,1]$, defined by*

$$(Kf)(t) = \int_0^1 (k_-(t,s) + k_+(t,s)) f(s) ds,$$

*satisfies*

$$||K|| \le \delta d_\infty \sqrt{1 - \exp[-\frac{d_2^2}{d_\infty(\delta^2-1)}]} \tag{4.1}$$

*and*

$$||K||_2 \le \begin{cases} \frac{\delta d_2 \sqrt{\delta^2 - 1 - \exp[-\frac{d_2^2}{d_\infty^2(\delta^2-1)}]}}{\delta^2 - 1} & \text{if } 1 < \delta < \delta_0 \\ d_2 \cosh \frac{d_2^2}{2d_\infty^2(\delta^2-1)} & \text{if } \delta \ge \delta_0, \end{cases} \tag{4.2}$$

*where $\delta_0$ is given before Proposition 3.2.*

*Proof.* Let $P_\alpha : L_2^n[0,1] \to L_2^n[0,1]$ be defined by

$$P_\alpha f = \chi_{[0,\alpha]} f.$$

$\mathcal{N} = \{P_\alpha\}_{\alpha\in[0,1]}$ is a nest on $L_2^n[0,1]$. Let $\tilde{k}$ be the extension of $k_-$ to $[0,1]^2$ with 0 on $\Delta^c$ and the operator $K_-$ on $L_2^n[0,1]$, defined by

$$(K_- f)(t) = \int_0^1 \tilde{k}(t,s) f(s) ds,$$

With $K_-$ given above, we see that $||P_\alpha^\perp K_- P_\alpha|| = ||K_{-,\alpha}||$, hence Arveson's distance formula gives

$$d_\infty(K_-, Alg\mathcal{N}) = \sup ||K_{-,\alpha}|| \leq d_\infty.$$

Clearly,

$$d_2(K_-, Alg\mathcal{N} \cap \mathcal{C}_2(L^2([0,1]))) = ||K_-||_2 = ||k_-||_2 = d_2.$$

By Theorem 3.5 with $d_\infty = 1$, there exists a $K : L_2^n[0,1] \to L_2^n[0,1]$ such that (4.1) and (4.2) are satisfied, and moreover $K - K_- \in Alg\mathcal{N}$. Since $K$ is of the form

$$(Kf)(t) - \int_0^1 k(s,t)f(s)ds,$$

for some $k(s,t) \in L_2^{n\times n}([0,1]^2)$, we obtain from $K - K_- \in Alg\mathcal{N}$ that $k_+ = k - k_- \in L_2^{n\times n}(\Delta^c)$. □

REFERENCES


[1] Gr. Arsene and A. Gheondea, Completing Matrix Contractions, *J. Operator Theory*, 7(1982), 179-189.

[2] W. B. Arveson, Interpolation Problems in Nest Algebras, *J. Functional Analysis*, 20(1975), 208-233.

[3] H. Dym and I. Gohberg, Extensions of Band Matrices with Band Inverses, *Linear Algebra Appl.*, 36(1981), 1-24.

[4] C. Foias, A. E. Frazho, Simultaneous $H^\infty$ and $L^2$ Suboptimization, *SIAM J. Math. Anal.*, 23(1992), 984-994.

[5] C. Foias, A. E. Frazho and W.S. Li, The Exact $H^2$ Estimate for the Central $H^\infty$ Interpolant, preprint.

[6] A. E. Frazho and M. A. Rotea, A Remark on Mixed $L^2/L^\infty$ Bounds, *Integral Equations and Operator Theory*, 15(1992), 343-348.

[7] A. E. Frazho and S. M. Kherat, On Mixed $H^2 - H^\infty$ Interpolation, *Integral Equations and Operator Theory*, to appear.

[8] I. Gohberg, M. A. Kaashoek and H. J. Woerdeman, The Band Method for Positive and Contractive Extension Problems, *J. Operator Theory*, 22(1989), 109-155.

[9] I. Gohberg, M. A. Kaashoek and H. J. Woerdeman, A Maximum Entropy Principle in the General Framework of the Band Method, *J. Functional Anal.*, 95(1991), 231-254.

[10] V. G. Kaftal, D. R. Larson and G. Weiss, Quasitriangular Subalgebras of Semifinite von Neuman Algebras are Closed, *J. Functional Analysis*, 107(1992), 387-401.

[11] S. Parrott, On a Quotient Norm of the Sz. Nagy-Foias Lifting Theorem, *J. Functional Analysis*, 30(1978), 311-328.

Mihály Bakonyi
Department of Mathematics
Georgia State University
Atlanta, Georgia 30303

Victor G. Kaftal
Department of Mathematics
University of Cincinnati
Cincinnati, Ohio 45221

Gary Weiss
Department of Mathematics
University of Cincinnati
Cincinnati, Ohio 45221

Hugo J. Woerdeman
Department of Mathematics
The College of William and Mary
Williamsburg, Virginia 23187

MSC: Primary 47A20, Secondary 15A60

Operator Theory:
Advances and Applications, Vol. 64




# BITANGENTIAL INTERPOLATION FOR INPUT-OUT OPERATORS OF TIME-VARYING SYSTEMS: THE DISCRETE TIME CASE

J.A. Ball, I. Gohberg, M.A. Kaashoek

A discrete-time, time-varying analogue of the function theoretic, matricial Lagrange-Sylvester interpolation problem is introduced and solved. The set of all solutions is described via a linear fractional representation and a formula for a particular solution is given. The time-varying generalization of the bitangential matricial Nevanlinna-Pick interpolation problem is also considered. A time-varying version of the state space method plays an important role.

## INTRODUCTION

This paper may be viewed as a continuation of the paper [BGK1] of the authors where a time-varying generalization of the one-sided tangential Nevanlinna-Pick interpolation problem was considered. In the present paper we focus on the two-sided version of the Lagrange-Sylvester interpolation problem (without norm constraints). In the time-varying version, the interpolants are doubly infinite block lower-triangular matrices acting on $\ell_2$-spaces rather than analytic matrix functions on the unit disk; the analogue of rational matrix functions analytic on the disk are input-output maps of linear, finite-dimensional, time-varying systems. A time-varying analogue of the theory of matricial pole-zero structure and associated interpolation problems is introduced and developed. As in the time-invariant case there is a linear fractional description of the set of all solutions. The coefficient matrix for the linear fractional map parametrizing solutions of the time-varying Lagrange-Sylvester or Nevanlinna-Pick interpolation problems arises as the solution of one such time-varying homogeneous interpolation problem. An important notion in our treatment here of bitangential problems is that of dichotomy; in the time-invariant case, this idea corresponds to splitting a rational function with no poles on the unit circle as the sum of a function with all poles inside the unit disk and another function with all pole outside the unit disk. In the time-varying case, the arguments involving winding numbers in the function-theoretic case are replaced by Fredholm index theorems for operators on $\ell_2$-spaces.

We also present a preliminary analysis of the time-varying analogue of the two-sided Nevanlinna-Pick problem. We obtain a complete description of the set of all solutions via a linear fractional formula in terms of a $J$-inner solution of the associated homogeneous interpolation problem. The actual construction of this $J$-inner operator and the Pick matrix condition for its existence will be derived in a subsequent paper (see [BGK3]). The present paper is the first on the two-sided interpolation problem for the time-varying setting. The one-sided case has been considered in [ADD], [BGK1], [CSK], [DD], and [SCK] (see also the dissertations [S] and [vV] and the references therein). The continuous-time analogues of the problems considered in this paper and the subsequent paper [BGK3] will be dealt with in a future publication; the one-sided continuous-time case was considered in [BGK2].

The present paper consists of nine sections. The first two sections contain preliminary material on point evaluation, residue calculus, and one-sided homogeneous problems. Bitangential interpolation data sets are introduced in the third section and the corresponding interpolation problem in the fourth section. The sections 5, 6, 7 provide the solution of the Lagrange-Sylvester interpolation problem. Section 8 introduces the class of input-output operators of linear, time-varying systems. The last section concerns the two-sided Nevanlinna-Pick interpolation problem.

## 1. RESIDUE CALCULUS AND GENERALIZED POINT EVALUATION

In this section we present some preliminaries on shift expansions, time-varying residue calculus and generalized point evaluations. Throughout this paper $m$ and $r$ are given positive integers. We let $\mathcal{B}^{m\times r}$ denote the set of all doubly infinite block matrices $F = [F_{ij}]_{i,j=-\infty}^{\infty}$ with each block $F_{ij}$ a matrix of size $m \times r$ over $\mathbb{C}$ such that matrix multiplication by $F$ on the left defines a bounded linear operator from $\ell_2^r(\mathbb{Z})$ into $\ell_2^m(\mathbb{Z})$. Here $\ell_2^k(\mathbb{Z})$ is the space of norm square summable sequences $(x_n)_{n=-\infty}^{\infty}$ with entries in $\mathbf{C}^k$ written as column vectors. We will have use of the following subspaces:

$$\mathcal{L}^{m\times r} = \text{lower triangular elements of } \mathcal{B}^{m\times r} = \{F = [F_{ij}] \in \mathcal{B}^{m\times r} \mid F_{ij} = 0 \text{ for } i < j\};$$

$$\mathcal{U}^{m\times r} = \text{upper triangular elements of } \mathcal{B}^{m\times r} = \{F = [F_{ij}] \in \mathcal{B}^{m\times r} \mid F_{ij} = 0 \text{ for } i > j\};$$

$$\mathcal{D}^{m\times r} = \text{diagonal elements of } \mathcal{B}^{m\times r} = \mathcal{L}^{m\times r} \cap \mathcal{U}^{m\times r}.$$

**1a. Shift expansions and residue calculus.** Our interest is to develop further the time-varying analogue of the function theory on the unit disc started in [ADD], [BGK1]. The time-invariant case corresponds to the restriction to block Toeplitz matrices $F = [F_{i-j}]_{i,j=-\infty}^{\infty}$, which are associated to the Laurent series of functions $F(z) = \sum_{j=-\infty}^{\infty} z^j F_j$ on the unit circle. A key tool in the function theory analysis is the coefficient $F_{-1}$ of $z^{-1}$ in the Laurent series. Assuming that the function $F(z) = \sum_{j=-\infty}^{\infty} F_j z^j$ is meromorphic in the open unit disk $\mathbf{D}$ we have by elementary complex analysis that

$$F_{-1} = \frac{1}{2\pi i}\int_{\mathbf{T}} F(z)\,dz = \sum_{z_0\in\mathbf{D}} \mathcal{R}\text{es}_{z=z_0} F(z),$$

where $\mathcal{R}\text{es}_{z=z_0} F(z)$ is the residue of $F(z)$ at the point $z_0$ in $\mathbf{D}$. For the time-varying case, where $F = [F_{ij}]_{i,j=-\infty}^{\infty}$ is a general block matrix representing a bounded operator form $\ell_2^r(\mathbf{Z})$ into $\ell_2^m(\mathbf{Z})$, there are two analogues of the Laurent series of interest, namely

$$F = \sum_{\nu=-\infty}^{\infty} S^\nu F_{[\nu]} = \sum_{\nu=-\infty}^{\infty} F_{\{\nu\}} S^\nu. \tag{1.1}$$

Here $S$ stands for the block lower triangular forward shift on $\ell_2$ of the appropriate dimension, and the coefficients $F_{[\nu]}$ and $F_{\{\nu\}}$ are block diagonal operators from $\mathcal{D}^{m\times r}$ of which the main diagonal entries are given by

$$(F_{[\nu]})_{jj} = F_{j+\nu,j}, \qquad (F_{\{\nu\}})_{jj} = F_{j,j-\nu}. \tag{1.2}$$

The two series in (1.1) converge entrywise. The convergence in (1.1) is not necessarily in the weak operator topology (see [Mer], where also a topology is described in which the convergence of the series in (1.1) has to be understood). The coefficients $F_{[\nu]}$ and $F_{\{\nu\}}$ in (1.1) are uniquely determined by the entrywise convergence of the series and the fact that $F_{[\nu]}$ and $F_{\{\nu\}}$ are block diagonal operators. We call the series expansions in (1.1) the *left* and *right shift expansions* of $F$, respectively.

We define the *left* and *right nonstationary total* D*-residue maps* $\mathcal{R}es_L$ and $\mathcal{R}es_R$ from $\mathcal{B}^{m\times r}$ into $\mathcal{D}^{m\times r}$, by

$$\mathcal{R}es_L(F) = F_{[-1]}, \qquad \mathcal{R}es_R(F) = F_{\{-1\}}, \tag{1.3}$$

where $F_{[-1]}$ and $F_{\{-1\}}$ are the block diagonal operators in $\mathcal{D}^{m\times r}$ arising, respectively, from the left and right shift expansions of $F$ as in (1.1).

PROPOSITION 1.1. *Let $F \in \mathcal{L}^{m\times r}$, $D \in \mathcal{D}^{k\times m}$ and $E \in \mathcal{D}^{r\times n}$. Then*

(i) $\mathcal{R}es_L(DF) = (SDS^{-1})\mathcal{R}es_L(F)$,
(ii) $\mathcal{R}es_L(FE) = \{\mathcal{R}es_L(F)\}E$,
(iii) $\mathcal{R}es_R(DF) = D\mathcal{R}es_R(F)$,
(iv) $\mathcal{R}es_R(FE) = \{\mathcal{R}es_R(F)\}S^{-1}ES$,
(v) $\mathcal{R}es_L(S^{-1}FS) = S^{-1}\{\mathcal{R}es_L(F)\}S = \mathcal{R}es_R(F)$.

PROOF. We begin with a general remark. From (1.1) we see that the left and right shift expansions of $DF$ are given by

$$DF = \sum_{\nu=0}^{\infty} S^\nu(S^{-\nu}DS^\nu F_{[\nu]}), \quad DF = \sum_{\nu=1}^{\infty} DF_{\{\nu\}}S^\nu.$$

Hence, by the uniqueness of the coefficients in the shift expansions,

$$(DF)_{[\nu]} = S^{-\nu}DS^\nu F_{[\nu]}, \quad (DF)_{\{\nu\}} = DF_{\{\nu\}}, \quad \nu \in \mathbb{Z}. \tag{1.4}$$

In a similar way one shows that

$$(FE)_{[\nu]} = F_{[\nu]}E, \quad (FE)_{\{\nu\}} = F_{\{\nu\}}S^\nu ES^{-\nu}, \quad \nu \in \mathbb{Z}. \tag{1.5}$$

By taking $\nu = -1$ in (1.4) and (1.5) one obtains (i) - (iv).

Next, observe that $F = \sum_{\nu=-\infty}^{\infty}(S^\nu F_{[\nu]}S^{-\nu})S^\nu$, and hence

$$S^\nu F_{[\nu]}S^{-\nu} = F_{\{\nu\}}, \qquad \nu \in \mathbb{Z}, \tag{1.6}$$

which yields the second identity in (v) by taking $\nu = -1$. Since

$$SFS^{-1} = \sum_{\nu=0}^{\infty} S^\nu(S^{-1}F_{[\nu]}S),$$

we also have

$$(1.7)\qquad (SFS^{-1})_{[\nu]} = SF_{[\nu]}S^{-1}, \qquad \nu \in \mathbb{Z}.$$

In particular (take $\nu = -1$), the first identity in (v) holds. □

**1b. Right and left generalized point evaluations.** If $F(z) = \sum_{j=0}^{\infty} z^j F_j$ is a scalar function which is analytic on the closure of the open unit disc $\mathbb{D}$ and $z_0 \in \mathbb{D}$, then the value $F(z_0)$ at the point $z_0$ is also given by

$$F(z_0) = \frac{1}{2\pi i}\int_{\mathbb{T}} (z - z_0)^{-1} F(z)\,dz = \sum_{w\in\mathbb{D}} \mathrm{Res}_{z=w}(z - z_0)^{-1} F(z).$$

In terms of the Toeplitz matrix $F = [F_{i-j}]_{i,j=-\infty}^{\infty}$ associated with the function $F(\cdot)$ we have therefore

$$F(z_0)I = \mathcal{R}es_L((S - z_0 I)^{-1}F) - \mathcal{R}es_R(F(S - z_0 I)^{-1}).$$

In this form the point evaluation carries over to the time-varying case.

Let $F = [F_{ij}]_{i,j=-\infty}^{\infty}$ be a general element of $\mathcal{L}^{m\times r}$, and let $A \in \mathcal{D}^{r\times r}$ be a diagonal matrix such that the spectral radius $\rho(S^{-1}A)$ of $S^{-1}A$ is strictly less than 1. We may then define $F^{\wedge R}(A)$ (the *right point evaluation* of $F$ at $A$) by

$$(1.8)\qquad F^{\wedge R}(A) = \mathcal{R}es_R(F(S - A)^{-1}).$$

Note that the condition $\rho(S^{-1}A) < 1$ guarantees that $S - A = S(I - S^{-1}A)$ is invertible as an operator on $\ell_2^r$. Similarly, if $Z \in \mathcal{D}^{m\times m}$ is such that $\rho(ZS^{-1}) < 1$, then we may define the *left point evaluation* of $F$ at $Z$ by

$$(1.9)\qquad F^{\wedge L}(Z) = \mathcal{R}es_L((S - Z)^{-1}F).$$

More explicitly,

$$(1.10)\qquad F^{\wedge R}(A) = \sum_{j=0}^{\infty} F_{\{j\}} S^j (S^{-1}A)^j \quad \text{if} \quad F = \sum_{j=0}^{\infty} F_{\{j\}} S^j,$$

$$(1.11)\qquad F^{\wedge L}(Z) = \sum_{j=0}^{\infty} (ZS^{-1})^j S^j F_{[j]} \quad \text{if} \quad F = \sum_{j=0}^{\infty} S^j F_{[j]}.$$

Since the spectral radii of $S^{-1}A$ and $ZS^{-1}$ are strictly less than 1, the series in the left hand sides of (1.10) and (1.11) converge in the operator norm. For point evaluation maps and associated one-sided time-varying interpolation problems we refer to [AD], [ADD], [BGK1], [CSK], [D], [DD] and [SCK].

The following two propositions (see [ADD], and [BGK1], Propositions 1.1 and 1.2) are most useful. The first characterizes right and left point evaluations, and the second provides a rule for point evaluations of products .

PROPOSITION 1.2. *Let $F \in \mathcal{L}^{m\times r}$, $A \in \mathcal{D}^{r\times r}$ and $Z \in \mathcal{D}^{m\times m}$, and assume that the spectral radii of $S^{-1}A$ and $ZS^{-1}$ are strictly less than 1. Then $F^{\wedge_R}(A)$ and $F^{\wedge_L}(Z)$ are the unique elements of $\mathcal{D}^{m\times r}$ such that*

$$(F - F^{\wedge_R}(A))(S-A)^{-1} \in \mathcal{L}^{m\times r}, \quad (S-Z)^{-1}(F - F^{\wedge_L}(Z)) \in \mathcal{L}^{m\times r}.$$

PROPOSITION 1.3. *Let $F \in \mathcal{L}^{m\times r}$, $G \in \mathcal{L}^{k\times m}$ and $H \in \mathcal{L}^{r\times s}$. Let $A \in \mathcal{D}^{r\times r}$ and $Z \in \mathcal{D}^{m\times m}$, and assume that the spectral radii of $S^{-1}A$ and $ZS^{-1}$ are strictly less than 1. Then*

$$(GF)^{\wedge_R}(A) = (GF^{\wedge_R}(A))^{\wedge_R}(A), \tag{1.12}$$

$$(FH)^{\wedge_L}(Z) = (F^{\wedge_L}(Z)H)^{\wedge_L}(Z). \tag{1.13}$$

**1c. Hilbert-Schmidt operators and duality.** We write $\mathcal{B}_{HS}^{m\times r}$ for the set of all $F = [F_{ij}]_{i,j=-\infty}^{\infty} \in \mathcal{B}^{m\times r}$ that are Hilbert-Schmidt operators, i.e.,

$$\text{tr } F^*F = \sum_{i,j=-\infty}^{\infty} \text{tr } F_{ij}^*F_{ij} < \infty.$$

The sets $\mathcal{L}_{HS}^{m\times r}$, $\mathcal{U}_{HS}^{m\times r}$ and $\mathcal{D}_{HS}^{m\times r}$ are defined analogously. The space $\mathcal{B}_{HS}^{m\times r}$ is a Hilbert space with respect to the inner product $\langle F, G\rangle = \text{tr } G^*F$, and the spaces $\mathcal{L}_{HS}^{m\times r}$, $\mathcal{U}_{HS}^{m\times r}$ and $\mathcal{D}_{HS}^{m\times r}$ are closed subspaces. Furthermore, $\mathcal{B}_{HS}^{m\times r}$ admits the orthogonal direct sum decomposition

$$\mathcal{B}_{HS}^{m\times r} = \mathcal{U}_{0HS}^{m\times r} \oplus \mathcal{L}_{HS}^{m\times r}, \tag{1.14}$$

where $\mathcal{U}_{0HS}^{m\times r}$ denotes the subspace of all strictly upper triangular Hilbert-Schmidt operators.

If $F \in \mathcal{B}_{HS}^{m\times r}$, then $F_{[\nu]}$ and $F_{\{\nu\}}$ are in $\mathcal{D}_{HS}^{m\times r}$, and the shift expansions in (1.1) converge in the Hilbert-Schmidt norm. Since the trace involves the diagonal elements only, it follows that for $F$ and $G$ in $\mathcal{B}_{HS}^{m\times r}$

$$
\begin{aligned}
\langle F, G\rangle &= \mathrm{tr}\ G^*F = \mathrm{tr}\ \sum_{\nu=-\infty}^{\infty} (G_{[\nu]})^* F_{[\nu]} \\
&= \mathrm{tr}\ FG^* = \mathrm{tr}\ \sum_{\nu=-\infty}^{\infty} (F_{\{\nu\}})(G_{\{\nu\}})^*,
\end{aligned} \tag{1.15}
$$

with the two series converging in the trace class norm.

PROPOSITION 1.4. *Let $F \in \mathcal{L}_{HS}^{m\times r}$, $A \in \mathcal{D}^{r\times r}$ and $Z \in \mathcal{D}^{m\times m}$, and assume that the spectral radii of $S^{-1}A$ and $ZS^{-1}$ are strictly less than 1. Then $F^{\wedge_R}(A)$ and $F^{\wedge_L}(Z)$ are in $\mathcal{D}_{HS}^{m\times r}$, and the maps*

$$
\begin{aligned}
\bigwedge\nolimits_R &: \mathcal{L}_{HS}^{m\times r} \to \mathcal{D}_{HS}^{m\times r}, \quad \bigwedge\nolimits_R(F) := F^{\wedge_R}(A), \\
\bigwedge\nolimits_L &: \mathcal{L}_{HS}^{m\times r} \to \mathcal{D}_{HS}^{m\times r}, \quad \bigwedge\nolimits_L(F) := F^{\wedge_L}(Z),
\end{aligned}
$$

*are bounded linear operators, whose adjoints are given by*

$$\bigwedge\nolimits_R^*(D) = D(I - A^*S)^{-1}, \quad D \in \mathcal{D}_{HS}^{m\times r}, \tag{1.16}$$

$$\bigwedge\nolimits_L^*(D) = (I - SZ^*)^{-1}D, \quad D \in \mathcal{D}_{HS}^{m\times r}. \tag{1.17}$$

PROOF. We prove (1.16); the proof of (1.17) is similar. Take $F \in \mathcal{L}_{HS}^{m\times r}$ and $D \in \mathcal{D}_{HS}^{m\times r}$. Then

$$\langle \bigwedge\nolimits_R(F), D\rangle = \mathrm{tr}\ F^{\wedge_R}(A)D^* = \mathrm{tr}\ \sum_{\nu=0}^{\infty} F_{\{\nu\}} S^\nu (S^{-1}A)^\nu D^* = \langle F, K\rangle,$$

where (use (1.15))

$$K = \sum_{\nu=0}^{\infty} \{D(A^*S)^\nu S^{-\nu}\} S^\nu = D(I - A^*S)^{-1}. \quad \square$$

The following proposition plays a useful role later (in Section 7).

PROPOSITION 1.5. *If $F \in \mathcal{B}^{r\times r}$ is such that $F\mathcal{L}_{HS}^{r\times 1} = \mathcal{L}_{HS}^{r\times 1}$, then $F$ is invertible and both $F$ and $F^{-1}$ are in $\mathcal{L}^{r\times r}$*

PROOF. Clearly, $F\mathcal{L}_{HS}^{r\times 1} \subset \mathcal{L}_{HS}^{r\times 1}$ implies that $F \in \mathcal{L}_{HS}^{r\times r}$. It remains to show that $F$ is invertible with inverse also in $\mathcal{L}^{r\times r}$.

We claim first that the diagonal entries $F_{kk}$ of $F$ are all invertible. Indeed, note that for any $H \in \mathcal{L}_{HS}^{r\times 1}$,

$$(FH)_{kk} = F_{kk}H_{kk}.$$

Hence $F\mathcal{L}_{HS}^{r\times 1} = \mathcal{L}_{HS}^{r\times 1}$ forces $F_{kk}$ to be onto. Since $F_{kk}$ is a square finite matrix, we conclude that $F_{kk}$ is invertible for each $k \in \mathbf{Z}$.

Next we claim that the operator $L_F$ of left multiplication by $F$ on $\mathcal{L}_{HS}^{r\times 1}$ is one-to-one. Indeed, suppose $0 \neq H \in \operatorname{Ker} L_F$. Suppose column $j$ of $H$ is nonzero, and let $k \geq j$ be the first index for which $H_{kj} \neq 0$. Note that then

$$(FH)_{kj} = F_{kk}H_{kj}.$$

The assumption that $FH = 0$ implies in particular that $(FH)_{kj} = 0$. Thus $H_{kj} \in \operatorname{Ker} F_{kk}$, a contradiction to the claim proved in the previous paragraph. We conclude that $L_F$ in one-to-one, and hence invertible, as an operator on $\mathcal{L}_{HS}^{r\times 1}$. From the open mapping theorem, one also knows that $\|(L_F)^{-1}\| < \infty$.

We next show that $(L_F)^{-1}$ is also a multiplication operator, i.e., $(L_F)^{-1} = L_M$ for some $M \in \mathcal{B}^{r\times r}$. Given $\nu \in \mathbf{Z}$ we let $\ell_2^r([\nu,\infty))$ denote the subspace of $\ell_2^r(\mathbf{Z})$ consisting of all sequences $x = (x_i)_{i=-\infty}^{\infty}$ in $\ell_2^r(\mathbf{Z})$ such that $x_i = 0$ for $i < \nu$. With each $x \in \ell_2^r([\nu,\infty))$ we associate a Hilbert-Schmidt operator $H_x^{(\nu)} \in \mathcal{L}_{HS}^{r\times 1}$ by requiring the $\nu$-th column of $H_x^{(\nu)}$ to be the column vector $x$ and all other columns of $H_x^{(\nu)}$ to be zero. By identifying $x \in \ell_2^r([\nu,\infty))$ with $H_x^{(\nu)} \in \mathcal{L}_{HS}^{r\times 1}$ for each $\nu \in \mathbf{Z}$ we see that

$$\mathcal{L}_{HS}^{r\times 1} = \oplus_{\nu\in\mathbf{Z}}\ell_2^r([\nu,\infty)). \tag{1.18}$$

From the equality (1.18) it follows that $L_F$ is equal to the diagonal operator $\operatorname{diag}(F^{(\nu)})_{\nu=-\infty}^{\infty}$ acting on $\oplus_{\nu\in\mathbf{Z}}\ell_2^r([\nu,\infty))$, where

$$F^{(\nu)} = F|_{\ell_2^r([\nu,\infty))} : \ell_2^r([\nu,\infty)) \to \ell_2^r([\nu,\infty)).$$

Since $L_F$ is boundedly invertible, it follows that each $F^{(\nu)}$ is boundedly invertible and

$$\sup_{\nu\in\mathbf{Z}} \|(F^{(\nu)})^{-1}\| \leq \|L_F^{-1}\|. \tag{1.19}$$

Define $M$ on $\cup_{\nu\in\mathbf{Z}}\ell_2^r([\nu,\infty))$ by setting $Mx=(F^{(\nu)})^{-1}x$ whenever $x\in\ell_2^r([\nu,\infty))$. Then $M$ is well-defined and extends to a bounded linear operator on $\ell_2^r(\mathbf{Z})$ by (1.19) and $L_F^{-1}=L_M$. From $M\mathcal{L}_{HS}^{r\times 1}\subset\mathcal{L}_{HS}^{r\times 1}$ it also follows that $M\in\mathcal{L}^{r\times r}$. □

## 2. PAIRS OF DIAGONAL OPERATORS AND HOMOGENEOUS ONE-SIDED INTERPOLATION

Let $B\in\mathcal{D}^{k\times m}$ and $Z\in\mathcal{D}^{k\times k}$. The pair $(Z,B)$ is called a *left admissible pair of diagonal operators*. Such a pair is said to be *stable* if the spectral radius of $ZS^{-1}$ is strictly less than 1, and it is *uniformly controllable* if there exists $\varepsilon>0$ such that

(2.1) $$G_L:=\sum_{\nu=0}^{\infty}(ZS^{-1})^{\nu}BB^*(SZ^*)^{\nu}\geq\varepsilon I_{\ell_2^k(\mathbf{Z})}.$$

Since $\rho(ZS^{-1})<1$, the series in (2.1) converges in the operator norm. Note that the Gramian $G_L\in\mathcal{D}^{k\times k}$, and hence (2.1) requires that the diagonal elements of $G_L$ are uniformly positive definite.

With a stable left admissible pair $(Z,B)$ we also associate a *controllability operator*, namely

(2.2) $$\mathcal{C}:\mathcal{L}_{HS}^{m\times 1}\to\mathcal{D}_{HS}^{k\times 1},\quad \mathcal{C}h:=(Bh)^{\wedge L}(Z).$$

By using the duality result in Proposition 1.4 one sees that

(2.3) $$\mathcal{C}\mathcal{C}^*d=G_Ld,\qquad d\in\mathcal{D}_{HS}^{k\times 1}.$$

The right hand side of (2.3) is a product of two diagonal operators; its left hand side gives the action of the operator $\mathcal{C}\mathcal{C}^*$ on the vector $d$. From (2.3) we may conclude that

$$\langle\mathcal{C}\mathcal{C}^*d,d\rangle=\operatorname{tr}d^*G_Ld,\quad d\in\mathcal{D}_{HS}^{k\times 1},$$

and from this identity it follows that the pair $(Z,B)$ is uniformly controllable if and only if the operator $\mathcal{C}\mathcal{C}^*$ is positive definite.

Now let us consider the following homogeneous left tangential interpolation problem associated with the stable left admissible pair $(Z,B)$. Find all $F\in\mathcal{L}^{m\times r}$ such that

(2.4) $$(BF)^{\wedge L}(Z)=0.$$

Formula (1.13) shows that the set of all solutions $F$ of (2.4) is a right ideal over $\mathcal{L}^{r\times r}$. The next result implies that this ideal is "principal" provided the pair $(Z, B)$ is uniformly controllable.

PROPOSITION 2.1. *Let $(Z, B)$ be a stable, uniformly controllable, left admissible pair of diagonal operators, and let $G_L$ be the corresponding controllability Gramian in (2.1). Then there exist $C_L \in \mathcal{D}^{m\times k}$ and $D_L \in \mathcal{D}^{m\times m}$ such that*

$$\begin{pmatrix} ZS^{-1} & B \\ C_L S^{-1} & D_L \end{pmatrix}\begin{pmatrix} G_L & 0 \\ 0 & I \end{pmatrix}\begin{pmatrix} SZ^* & SC_L^* \\ B^* & D_L^* \end{pmatrix} = \begin{pmatrix} G_L & 0 \\ 0 & I \end{pmatrix},$$

$$\begin{pmatrix} ZS^* & SC_L^* \\ B^* & D_L^* \end{pmatrix}\begin{pmatrix} G_L^{-1} & 0 \\ 0 & I \end{pmatrix}\begin{pmatrix} ZS^{-1} & B \\ C_L S^{-1} & D_L \end{pmatrix} = \begin{pmatrix} G_L^{-1} & 0 \\ 0 & I \end{pmatrix}.$$

*In particular,*

$$\Psi := D_L^* + B^*(I - SZ^*)^{-1}SC_L^* \in \mathcal{L}^{m\times m} \tag{2.5}$$

*is unitary, and $\Psi\mathcal{L}^{m\times 1} = \{K \in \mathcal{L}^{m\times 1} \mid (BK)^{\wedge_L}(Z) = 0\}$.*

PROOF. It suffices to apply Theorem 4.2 in [BGK1] with $\mathcal{H} = G_L$, $J = I$ and $Y = 0$. □

We also need the transposed version of Proposition 2.1. Let $C \in \mathcal{D}^{r\times n}$ and $A \in \mathcal{D}^{n\times n}$. We call $(C, A)$ a *right admissible pair of diagonal operators.* Such a pair is said to be *stable* if $\rho(S^{-1}A) < 1$, and *uniformly observable* if there exists $\varepsilon > 0$ such that

$$G_R := \sum_{\nu=0}^{\infty}(A^*S)^\nu C^*C(S^{-1}A)^\nu \geq \varepsilon I_{\ell_2^n(\mathbb{Z})}. \tag{2.6}$$

With the pair $(C, A)$ we also associate an *observability operator*, namely

$$\mathcal{O} : \mathcal{D}_{HS}^{n\times 1} \to \mathcal{U}_{0HS}^{r\times 1}, \quad \mathcal{O}d = C(S - A)^{-1}d. \tag{2.7}$$

A computation as in the proof of Proposition 1.4 shows that

$$\mathcal{O}^*G = \sum_{\nu=0}^{\infty} S(A^*S)^\nu C^*CS^{-\nu-1}G_{[-\nu-1]}, \quad G \in \mathcal{U}_{0HS}^{r\times 1}.$$

It follows that

$$\mathcal{O}^*\mathcal{O}d = SG_RS^{-1}d, \quad d \in \mathcal{D}_{HS}^{n\times 1}, \tag{2.8}$$

and hence the pair $(C, A)$ is uniformly observable if and only if the operator $\mathcal{O}^*\mathcal{O}$ is positive definite.

The next proposition may be viewed as the transposed version of Proposition 2.1.

PROPOSITION 2.2. *Let $(C, A)$ be a stable, uniformly observable, right admissible pair of diagonal operators, and let $G_R$ be the corresponding observability Gramian in (2.6). Then there exist $B_R \in \mathcal{D}^{k\times r}$ and $D_R \in \mathcal{D}^{r\times r}$ such that*

$$\begin{pmatrix} A^*S & C^* \\ B_R^*S & D_R^* \end{pmatrix}\begin{pmatrix} G_R & 0 \\ 0 & I \end{pmatrix}\begin{pmatrix} S^{-1}A & S^{-1}B_R \\ C & D_R \end{pmatrix} = \begin{pmatrix} G_R & 0 \\ 0 & I \end{pmatrix},$$

$$\begin{pmatrix} S^{-1}A & S^{-1}B_R \\ C & D_R \end{pmatrix}\begin{pmatrix} G_R^{-1} & 0 \\ 0 & I \end{pmatrix}\begin{pmatrix} A^*S & C^* \\ B_R^*S & D_R^* \end{pmatrix} = \begin{pmatrix} G_R^{-1} & 0 \\ 0 & I \end{pmatrix}.$$

*In particular,*

$$\Phi := D_R^* + B_R^* S(I - A^*S)^{-1}C^* \in \mathcal{L}^{r\times r} \tag{2.9}$$

*is unitary, and $\mathcal{L}^{1\times r}\Phi = \{K \in \mathcal{L}^{1\times r} \mid (KC)^\wedge(A) = 0\}$.*

From Proposition 2.2 it follows that the set of all solutions of the homogeneous right tangential interpolation problem $(FC)^{\wedge_R}(A) = 0$ is a principal left ideal over $\mathcal{L}^{m\times m}$ whenever the pair $(C, A)$ is uniformly observable.

## 3. BITANGENTIAL INTERPOLATION DATA SET

Throughout this section we have given the following diagonal operators:

$$Z \in \mathcal{D}^{k\times k}, \quad B \in \mathcal{D}^{k\times m}, \quad \widetilde{B} \in \mathcal{D}^{k\times r}, \tag{3.1a}$$

$$A \in \mathcal{D}^{n\times n}, \quad C \in \mathcal{D}^{r\times n}, \quad \widetilde{C} \in \mathcal{D}^{m\times n}, \tag{3.1b}$$

The pairs $(Z, B)$ and $(C, A)$ are assumed to be stable, $(Z, B)$ is uniformly controllable, and $(C, A)$ is uniformly observable. We are interested in the following interpolation problem. Find all $F \in \mathcal{L}^{m\times r}$ such that

(IC1) $\quad (BF)^{\wedge_L}(Z) = \widetilde{B}$,

(IC2) $\quad (FC)^{\wedge_R}(A) = \widetilde{C}$.

Later (see section 7) we shall also require the solutions to be contractive; in the present section the problem is considered without norm constraints on the solutions.

Assume we have a particular solution $F_0$ of (IC1), (IC2), and let $F$ be a general solution. Then $K = F - F_0$ satisfies the homogeneous equations

$$(BK)^{\wedge_L}(Z) = 0, \quad (KC)^{\wedge_R}(A) = 0.$$

Since the pairs $(Z, B)$ and $(C, A)$ are uniformly controllable and uniformly observable, respectively, we can apply Propositions 2.1 and 2.2 to show that the general solution $F$ of (IC1), (IC2) is of the form

$$\text{(DR1)} \qquad F = F_0 + \Psi G_1,$$

$$\text{(DR2)} \qquad F = F_0 + G_2 \Phi,$$

for some $G_1$ and $G_2$ in $\mathcal{L}^{m \times r}$. Here $\Psi$ and $\Phi$ are the unitary operators given by (2.5) and (2.9). One way for $F$ to satisfy (DR1) and (DR2) simultaneously, involving a single affine formula, is that $F$ have the form

$$\text{(DR)} \qquad F = F_0 + \Psi G \Phi$$

for some $G \in \mathcal{L}^{m \times r}$. However, conversely, the simultaneous validity of (DR1) and (DR2) in general does not imply (DR). To achieve this, an additional condition on the interpolant $F$ is required.

Indeed, let $F$ be as in (DR). Then

$$(3.2) \qquad (S - Z)^{-1} B(F - F_0) C(S - A)^{-1} = \{(S - Z)^{-1} B \Psi\} G \{\Phi C(S - A)^{-1}\}.$$

Since $(B\Psi)^{\wedge_L}(Z) = 0$ and $(\Phi C)^{\wedge_R}(A) = 0$, Proposition 1.2 tells us that $(S - Z)^{-1} B\Psi$ and $\Phi C(S - A)^{-1}$ are both lower triangular. The same holds true for $G$. Thus the right hand side of (3.2) is in $\mathcal{L}^{k \times n}$. Hence, $F$ must satisfy the additional interpolation condition

$$\text{(IC3)} \qquad \mathcal{R}\text{es}_L\{(S - Z)^{-1} BFC(S - A)^{-1}\} = \Gamma,$$

where

$$(3.3) \qquad \Gamma = \mathcal{R}\text{es}_L\{(S - Z)^{-1} BF_0 C(S - A)^{-1}\}.$$

The next result shows that indeed the set of interpolation conditions (IC1) - (IC3) on $F$ is equivalent to $F$ having the single two-sided divisor-remainder form (DR).

PROPOSITION 3.1. *Let $F_0 \in \mathcal{L}^{m\times r}$ be a particular solution of* (IC1), (IC2), *and define* $\Gamma$ *by* (3.3). *Then $F$ is in $\mathcal{L}^{m\times r}$ and satisfies* (IC1) - (IC3) *if and only if $F \in \mathcal{B}^{m\times r}$ and has the form* (DR) *for some $G$ in $\mathcal{L}^{m\times r}$.*

PROOF. We already noted that $F$ is in $\mathcal{L}^{m\times r}$ and satisfies (IC1) - (IC3) if $F$ has the form (DR) with $G \in \mathcal{L}^{m\times r}$. For the converse, suppose that $F \in \mathcal{L}^{m\times r}$ and satisfies (IC1) - (IC3). Then $K := F - F_0$ satisfies the homogeneous interpolation conditions:

(HIC1) $\quad \mathcal{R}es_L\{(S-Z)^{-1}BK\} = 0$

(HIC2) $\quad \mathcal{R}es_R\{KC(S-A)^{-1}\} = 0$

(HIC3) $\quad \mathcal{R}es_L\{(S-Z)^{-1}BKC(S-A)^{-1}\} = 0.$

It follows form Proposition 1.2 that

$$(S-Z)^{-1}BK \in \mathcal{L}^{k\times r}, \quad KC(S-A)^{-1} \in \mathcal{L}^{m\times n}, \quad (S-Z)^{-1}BKC(S-A) \in \mathcal{L}^{k\times n}.$$

We have to verify that $G := \Psi^{-1}K\Phi^{-1} \in \mathcal{L}^{m\times r}$. For this purpose we use the special representations of $\Phi$ and $\Psi$ in (2.5) and (2.9). Since $\Phi$ and $\Psi$ are unitary operators, we have

$$\Psi^{-1} = D_L + C_L S^{-1}(I - ZS^{-1})^{-1}B, \quad \Phi^{-1} = D_R + C(I - S^{-1}A)^{-1}S^{-1}B_R.$$

Thus

$$\Psi^{-1}K\Phi^{-1} = D_L K D_R + C_L(S-Z)^{-1}BKD_R + D_L KC(S-A)^{-1}B + \\ + C_L(S-Z)^{-1}BKC(S-A)^{-1}B_R,$$

and hence $G = \Psi^{-1}K\Phi^{-1}$ is lower triangular □

The previous proposition assumed the existence of a solution to (IC1), (IC2). We now derive a compatibility condition which is necessary for the existence of solutions to (IC1) - (IC3).

PROPOSITION 3.2. *Let $\Gamma \in \mathcal{D}^{k\times n}$, and assume that with this $\Gamma$ the problem* (IC1) - (IC3) *has a solution $F \in \mathcal{L}^{m\times r}$. Then necessarily $\Gamma$ satisfies the equation*

(3.4) $$\Gamma A - ZS^{-1}\Gamma S = B\widetilde{C} - \widetilde{B}C.$$

PROOF. By (IC3), we have $\Gamma = \mathcal{R}es_L\{(S-Z)^{-1}BFC(S-A)^{-1}\}$. Furthermore, by (IC1) and Proposition 1.1(ii),

$$\mathcal{R}es_L\{(S-Z)^{-1}BFC\} = \widetilde{B}C.$$

It follows (again use Proposition 1.1(ii)) that

$$\begin{aligned}\Gamma A &= \mathcal{R}es_L\{(S-Z)^{-1}BFC(S-A)^{-1}A\}\\ &= -\mathcal{R}es_L\{(S-Z)^{-1}BFC\} + \mathcal{R}es_L\{(S-Z)^{-1}BFC(S-A)^{-1}S\}\\ &= -\widetilde{B}C + \mathcal{R}es_L\{S^{-1}S(S-Z)^{-1}BFC(S-A)^{-1}S\}.\end{aligned}$$

By applying Proposition 1.1(v) twice we see that

$$\begin{aligned}&\mathcal{R}es_L\{S^{-1}S(S-Z)^{-1}BFC(S-A)^{-1}S\} =\\ &= S^{-1}(\mathcal{R}es_L\{S(S-Z)^{-1}BFC(S-A)^{-1}\})S\\ &= S^{-1}(\mathcal{R}es_L\{BFC(S-A)^{-1}\})S + S^{-1}(\mathcal{R}es_L\{Z(S-Z)^{-1}BFC(S-A)^{-1}\})S\\ &= \mathcal{R}es_R\{BFC(S-A)^{-1}\} + S^{-1}(\mathcal{R}es_L\{Z(S-Z)^{-1}BFC(S-A)^{-1}\})S.\end{aligned}$$

From Proposition 1.1(iii) and (IC2) it follows that $\mathcal{R}es_R\{BFC(S-A)^{-1}\} = B\widetilde{C}$, and Proposition 1.1(i) implies that

$$\mathcal{R}es_L\{Z(S-Z)^{-1}BFC(S-A)^{-1}\} = SZS^{-1}\Gamma.$$

We conclude that $\Gamma A = -\widetilde{B}C + B\widetilde{C} + ZS^{-1}\Gamma S$, which proves (3.4). □

We shall see later (in Section 6) that the problem (IC1) - (IC3) always has a solution $F \in \mathcal{L}^{m\times r}$ if $\Gamma$ satisfies the time-varying Sylvester equation (3.4). For this reason we call the septet

$$\tau = (\widetilde{C}, C, A; Z, B, \widetilde{B}; \Gamma), \tag{3.5}$$

which consists of diagonal operators,

$$\begin{aligned}&\widetilde{C} \in \mathcal{D}^{m\times n}, \quad C \in \mathcal{D}^{r\times n}, \quad A \in \mathcal{D}^{n\times m},\\ &\widetilde{B} \in \mathcal{D}^{k\times r}, \quad B \in \mathcal{D}^{k\times m}, \quad Z \in \mathcal{D}^{k\times k},\\ &\Gamma \in \mathcal{D}^{k\times n},\end{aligned} \tag{3.6}$$

an *admissible bitangential interpolation data set* whenever

(i) $(C, A)$ is stable and uniformly observable,

(ii) $(Z, B)$ is stable and uniformly controllable,

(iii) $\Gamma A - ZS^{-1}\Gamma S = B\widetilde{C} - \widetilde{B}C$.

Let $\tau$ in (3.5) be an admissible bitangential interpolation data set. Since the set of all solutions (without a norm constraint) of (IC1) - (IC3) has a simple affine parametrization, $F = F_0 + \Psi G \Phi$ with $G \in \mathcal{L}^{m\times r}$, we expect the set of all solutions $F$ with $\|F\| \leq 1$ to have a single linear fractional parametrization as in the classical Nevanlinna-Pick interpolation theory; for this reason we consider (IC1) - (IC3) as the natural two-sided time-varying interpolation problem.

## 4. BITANGENTIAL INTERPOLATION IN GEOMETRIC TERMS

Throughout this section

(4.1) $$\tau = (\widetilde{C}, C, A; Z, B, \widetilde{B}; \Gamma)$$

is an admissible bitangential interpolation data set (see the previous section for the definition and the sizes of entries of the diagonal operators in (4.1)). We are interested in finding solutions $F \in \mathcal{L}^{m\times r}$ of the interpolation problem

(IC1) $$\mathcal{R}es_L\{(S-Z)^{-1}BF\} = \widetilde{B},$$

(IC2) $$\mathcal{R}es_R\{FC(S-A)^{-1}\} = \widetilde{C},$$

(IC3) $$\mathcal{R}es_L\{(S-Z)^{-1}BFC(S-A)^{-1}\} = \Gamma.$$

In this section we express the interpolation conditions in a geometric language using a projective form, i.e., in terms of $\begin{pmatrix} F \\ I \end{pmatrix}$ rather than simply in terms of $F$.

THEOREM 4.1. *Let $\tau$ be the admissible bitangential interpolation data set (4.1), and consider the following two subspaces associated with $\tau$:*

(4.2) $$\mathcal{S}_{\tau_-} = \{C(S-A)^{-1}x \mid x \in \mathcal{D}_{HS}^{n\times 1}\} + \mathcal{L}_{HS}^{r\times 1},$$

*and*

(4.3) $$\mathcal{S}_{\tau_e} = \Big\{ \begin{pmatrix} \widetilde{C} \\ C \end{pmatrix} (S-A)^{-1}x + \begin{pmatrix} h_1 \\ h_2 \end{pmatrix} \mid x \in \mathcal{D}_{HS}^{n\times 1}, \quad h_1 \in \mathcal{L}_{HS}^{m\times 1}, \quad h_2 \in \mathcal{L}_{HS}^{r\times 1}$$
$$\text{such that } \mathcal{R}es_L\{(S-Z)^{-1} \begin{pmatrix} B & \widetilde{B} \end{pmatrix} \begin{pmatrix} h_1 \\ h_2 \end{pmatrix}\} = \Gamma x \Big\}.$$

*Let $F \in \mathcal{B}^{m\times r}$. Then $F \in \mathcal{L}^{m\times r}$ and $F$ satisfies* (IC1) - (IC3) *if and only if*

(4.4) $$\begin{pmatrix} F \\ I \end{pmatrix} \mathcal{S}_{\tau_-} \subset \mathcal{S}_{\tau_e}.$$

PROOF. We split the proof into two parts.

Part(a). In this part we show that $F \in \mathcal{L}^{m\times r}$ and $F$ satisfies (IC1) if and only if

(4.5) $$\begin{pmatrix} F \\ I \end{pmatrix} \mathcal{L}_{HS}^{r\times 1} \subset \mathcal{S}_{\tau_e}.$$

Assume $F \in \mathcal{L}^{m\times r}$ and $F$ satifies (IC1). Take $h \in \mathcal{L}_{HS}^{r\times 1}$. Then $g := Fh \in \mathcal{L}_{HS}^{r\times 1}$. Furthermore

$$\begin{aligned}\mathcal{R}\mathrm{es}_L\{(S-Z)^{-1}\begin{pmatrix} B & -\widetilde{B}\end{pmatrix}\begin{pmatrix} g \\ h \end{pmatrix}\} &= \mathcal{R}\mathrm{es}_L\{(S-Z)^{-1}BFh\} - \mathcal{R}\mathrm{es}_L\{(S-Z)^{-1}\widetilde{B}h\} \\ &= (BFh)^{\wedge_L}(Z) - (\widetilde{B}h)^{\wedge_L}(Z).\end{aligned}$$

Now use the product rule from Proposition 1.3 and the interpolation condition (IC1). It follows that

$$(BFh)^{\wedge_L}(Z) = \{(BF)^{\wedge_L}(Z)h\}^{\wedge_L}(Z) = (\widetilde{B}h)^{\wedge_L}(Z).$$

We conclude that

$$\mathcal{R}\mathrm{es}_L\{(S-Z)^{-1}\begin{pmatrix} B & -\widetilde{B}\end{pmatrix}\begin{pmatrix} g \\ h \end{pmatrix}\} = 0,$$

and hence $\begin{pmatrix} Fh \\ h \end{pmatrix} \in \mathcal{S}_{\tau_e}$.

Conversely, assume (4.5) holds. Take $h \in \mathcal{L}_{HS}^{r\times 1}$. Then, because of (4.5),

$$\begin{pmatrix} Fh \\ h \end{pmatrix} = \begin{pmatrix} \widetilde{C} \\ C \end{pmatrix}(S-A)^{-1}x + \begin{pmatrix} h_1 \\ h_2 \end{pmatrix},$$

where $x, h_1$ and $h_2$ are as is (4.3). It follows that

$$h - h_2 = C(S-A)^{-1}x = \sum_{\nu=0}^{\infty} C(S^{-1}A)^{\nu}S^{-1}x \in \mathcal{U}_{0HS}^{r\times 1}.$$

On the other hand, $h - h_2 \in \mathcal{L}_{HS}^{r\times 1}$. So $h = h_2$, and $\mathcal{O}x = 0$, where $\mathcal{O}$ is the observability operator (see (2.7)) corresponding to the pair $(C, A)$. Since $(C, A)$ is uniformly observable, $\mathcal{O}x = 0$ implies $x = 0$. Hence $Fh = h_1 \in \mathcal{L}_{HS}^{m\times 1}$ and

(4.6) $$\mathcal{R}\mathrm{es}_L\{(S-Z)^{-1}\begin{pmatrix} B & -\widetilde{B}\end{pmatrix}\begin{pmatrix} Fh \\ h \end{pmatrix}\} = 0.$$

In particular, $Fh$ is lower triangular for each $h \in \mathcal{L}_{HS}^{r\times 1}$. This can only happen when $F$ is lower triangular. Thus $F \in \mathcal{L}^{m\times r}$. In this case, we may rewrite (4.6) as

$$\{BFh - \widetilde{B}h\}^{\wedge_L}(Z),$$

and thus, by Proposition 1.2, the operator $(S-Z)^{-1}\{BFh - \widetilde{B}h\}$ is lower triangular for each $h$, which implies that $(S-Z)^{-1}\{BF - \widetilde{B}\}$ is lower triangular. Hence, again apply Proposition 1.2, condition (IC1) is fulfilled.

Part(b). In this part we show that $F \in \mathcal{B}^{m\times r}$ satisfies (IC2) and (IC3) is equivalent to

$$\text{(4.7)} \qquad \begin{pmatrix} F \\ I \end{pmatrix} C(S-A)^{-1}x \in \mathcal{S}_{\tau_e} \quad (x \in \mathcal{D}_{HS}^{n\times 1}).$$

Assume (IC2) and (IC3) are fulfilled. Take $x \in \mathcal{D}_{HS}^{n\times 1}$. By (IC2) and Proposition 2.1,

$$h_1 := FC(S-A)^{-1}x - \widetilde{C}(S-A)^{-1}x \in \mathcal{L}_{HS}^{r\times 1}.$$

Thus

$$\begin{pmatrix} F \\ I \end{pmatrix} C(S-A)^{-1}x - \begin{pmatrix} \widetilde{C} \\ C \end{pmatrix} (S \quad A)^{-1}x + \begin{pmatrix} h_1 \\ 0 \end{pmatrix}.$$

Furthermore, by (IC3),

$$\begin{aligned}
\mathcal{R}es_L\{(S-Z)^{-1}\begin{pmatrix} B & -\widetilde{B} \end{pmatrix}\begin{pmatrix} h_1 \\ 0 \end{pmatrix}\} &= \\
&= \mathcal{R}es_L\{(S-Z)^{-1}BFC(S-A)^{-1}\}x - \mathcal{R}es_L\{(S-Z)^{-1}B\widetilde{C}(S-A)^{-1}\}x \\
&= \Gamma x - \mathcal{R}es_L\{(S-Z)^{-1}B\widetilde{C}(S-A)^{-1}\}x.
\end{aligned}$$

Note that $(S-Z)^{-1}B$ and $\widetilde{C}(S-A)^{-1}$ are both strictly upper triangular. It follows that the diagonal above the main diagonal in $(S-Z)^{-1}B\widetilde{C}(S-A)^{-1}x$ is zero. Thus

$$\mathcal{R}es_L\{(S-Z)^{-1}\begin{pmatrix} B & -\widetilde{B} \end{pmatrix}\begin{pmatrix} h_1 \\ 0 \end{pmatrix}\} = \Gamma x,$$

and therefore (4.7) holds.

Conversely, assume (4.7) holds. Take $y \in \mathcal{D}_{HS}^{n\times 1}$. Then, by (4.7),

$$\text{(4.8)} \qquad \begin{pmatrix} F \\ I \end{pmatrix} C(S-A)^{-1}y = \begin{pmatrix} \widetilde{C} \\ C \end{pmatrix} (S-A)^{-1}x + \begin{pmatrix} h_1 \\ h_2 \end{pmatrix},$$

where $x, h_1$ and $h_2$ are as in (4.3). Note that

$$\mathcal{U}_{0HS}^{r\times 1} \ni C(S-A)^{-1}(y-x) = h_2 \in \mathcal{L}_{HS}^{r\times 1}.$$

Thus $h_2 = 0$ and $C(S-A)^{-1}(y-x) = 0$. Since the pair $(C, A)$ is uniformly observable it follows that $y = x$. We see that (4.8) reduces to

$$FC(S-A)^{-1}y = \widetilde{C}(S-A)^{-1}y + h_1, \tag{4.9}$$

where

$$\mathcal{R}\mathrm{es}_L\{(S-Z)^{-1}Bh_1\} = \Gamma y. \tag{4.10}$$

From (4.9) we conclude that

$$(FC-\widetilde{C})(S-A)^{-1}y \in \mathcal{L}_{HS}^{n\times 1} \quad (y \in \mathcal{D}_{HS}^{n\times 1}).$$

Now, let $h = \sum_{\nu=0}^{\infty} h_{\{\nu\}}S^{\nu}$ be an arbitrary element of $\mathcal{L}_{HS}^{n\times 1}$. Recall that this right shift expansion of $h$ converges in the Hilbert-Schmidt norm. Since the map

$$h \mapsto (FC-\widetilde{C})(S-A)^{-1}h$$

from $\mathcal{L}_{HS}^{n\times 1}$ into $\mathcal{B}_{HS}^{m\times 1}$ is a bounded linear operator with respect to the Hilbert-Schmidt norm, we see that

$$(FC-\widetilde{C})(S-A)^{-1}h = \sum_{\nu=0}^{\infty}(FC-\widetilde{C})(S-A)^{-1}h_{\{\nu\}}S^{\nu} \in \mathcal{L}_{HS}^{m\times 1}.$$

This holds for each $h \in \mathcal{L}_{HS}^{n\times 1}$. Thus $(FC-\widetilde{C})(S-A)^{-1}$ is lower triangular, and therefore

$$\mathcal{R}\mathrm{es}_L FC(S-A)^{-1} = \mathcal{R}\mathrm{es}_L \widetilde{C}(S-A)^{-1} = \widetilde{C},$$

and thus (IC2) is fulfilled.

Since the first diagonal above the main diagonal of $(S-Z)^{-1}B\widetilde{C}(S-A)^{-1}$ is zero, we can use (4.9) to rewrite (4.10) in the following form:

$$\mathcal{R}\mathrm{es}_L\{(S-Z)^{-1}BFC(S-A)^{-1}\}y = \Gamma y. \tag{4.11}$$

Recall that $\mathcal{R}es_L\{(S-Z)^{-1}BFC(S-A)^{-1}\}$ is a diagonal operator. Thus (4.11) for each $y \in \mathcal{D}_{HS}^{r\times 1}$ implies that (IC3) holds.

By combining the results of Parts (a) and (b) the theorem is proved. □

The next two propositions provide further information about the structure of the spaces $\mathcal{S}_{\tau_-}$ and $\mathcal{S}_{\tau_e}$ in Theorem 4.1.

PROPOSITION 4.2. *Let $\tau$ be the admissible bitangential interpolation data set* (4.1), *and let $\mathcal{S}_{\tau_-}$ and $\mathcal{S}_{\tau_e}$ be the subspaces defined by* (4.2) *and* (4.3). *Then*

$$(0 \quad I)\mathcal{S}_{\tau_e} = \mathcal{S}_{\tau_-}. \tag{4.12}$$

PROOF. The inclusion $(0 \quad I)\mathcal{S}_{\tau_e} \subset \mathcal{S}_{\tau_-}$ is trivial. Let us prove the reverse inclusion. Take $g \in \mathcal{S}_{\tau_-}$. So, $g = C(S-A)^{-1} + h$ with $x \in \mathcal{D}_{HS}^{r\times 1}$ and $h \in \mathcal{L}_{HS}^{r\times 1}$. We have to find $h_1 \in \mathcal{L}_{HS}^{m\times 1}$ so that

$$\mathcal{R}es_L\{(S-Z)^{-1}Bh_1\} = \mathcal{R}es_L\{(S-Z)^{-1}\widetilde{B}h\} + \Gamma x. \tag{4.13}$$

Let $\mathcal{C}$ and $\widetilde{\mathcal{C}}$ be the controllability operators (cf., formula (2.2)) associated with the pairs $(Z, B)$ and $(Z, \widetilde{B})$, respectively. Note that (4.13) can be rewritten in the following equivalent form

$$\mathcal{C}h_1 = \widetilde{\mathcal{C}}h + \Gamma x. \tag{4.14}$$

Recall that $\mathcal{C}\mathcal{C}^*$ is positive definite. Thus

$$h_1 := \mathcal{C}^*(\mathcal{C}\mathcal{C}^*)^{-1}\widetilde{\mathcal{C}}h + \mathcal{C}^*(\mathcal{C}\mathcal{C}^*)^{-1}\Gamma x \tag{4.15}$$

will satisfy (4.14) and therefore also (4.13). From (4.13) we see that

$$\begin{pmatrix} \widetilde{C} \\ C \end{pmatrix}(S-A)^{-1}x + \begin{pmatrix} h_1 \\ h \end{pmatrix} \in \mathcal{S}_{\tau_e},$$

and thus $g = C(S-A)^{-1}x + h \in (0 \quad I)\mathcal{S}_{\tau_e}$ □.

PROPOSITION 4.3. *Let $\tau$ be the admissible bitangential interpolation data set* (4.1), *and let $\mathcal{S}_{\tau_-}$ and $\mathcal{S}_{\tau_e}$ be the subspaces defined by* (4.2) *and* (4.3). *Then*

(i) $\mathcal{S}_{\tau_e} \cap \begin{pmatrix} \mathcal{B}_{HS}^{m\times 1} \\ 0 \end{pmatrix} = \begin{pmatrix} \Psi\mathcal{L}_{HS}^{m\times 1} \\ 0 \end{pmatrix}$,

(ii) $\mathcal{S}_{\tau_-} = \Phi^{-1}\mathcal{L}_{HS}^{r\times 1}$.

*Here* $\Psi$ *and* $\Phi$ *are the unitary operators given by* (2.5) *and* (2.9), *respectively.*

PROOF. (i) Assume that $(\, g \quad 0\,)^T$ belongs to the set defined by the left hand side of (i). Thus

$$\begin{pmatrix} g \\ 0 \end{pmatrix} = \begin{pmatrix} \widetilde{C} \\ C \end{pmatrix} (S-A)^{-1}x + \begin{pmatrix} h_1 \\ h_2 \end{pmatrix},$$

where $x, h_1$ and $h_2$ are as in (4.3). In particular,

$$C(S-A)^{-1}x = -h_2 \in \mathcal{U}_{0HS}^{r\times 1} \cap \mathcal{L}_{HS}^{r\times 1},$$

and hence $C(S-A)^{-1}x = 0$ and $h_2 = 0$. Since the pair $(C, A)$ is uniformly observable, it also follows that $x = 0$. So $g = h_1$, and

$$\mathcal{R}\mathrm{es}_L(S-Z)^{-1}Bh_1 = 0.$$

According to Proposition 2.1 this implies that $g = h_1 \in \Psi\mathcal{L}_{HS}^{r\times 1}$, and hence

$$\mathcal{S}_{\tau_e} \cap \begin{pmatrix} \mathcal{B}_{HS}^{m\times 1} \\ 0 \end{pmatrix} \subset \begin{pmatrix} \Psi\mathcal{L}_{HS}^{m\times 1} \\ 0 \end{pmatrix}.$$

To prove the reverse inclusion, take $g$ in $\Psi\mathcal{L}_{HS}^{m\times 1}$. Then, by Proposition 2.1,

$$\mathcal{R}\mathrm{es}_L(S-Z)^{-1}Bg = (Bg)^{\wedge_L}(Z) = 0,$$

which implies that $(\, g \quad 0\,)^T \in \mathcal{S}_{\tau_e}$.

(ii) Since $\Phi$ is unitary, we have

$$\Phi^{-1} = \Phi^* = D_R + C(S-A)^{-1}B_R.$$

Take $g \in \mathcal{L}_{HS}^{r\times 1}$, and put $x_g = \mathcal{R}\mathrm{es}_L(S-A)^{-1}B_R g$. Thus $x_g \in \mathcal{D}_{HS}^{n\times 1}$, and, by Proposition 1.2,

$$h_g := (S-A)^{-1}\{B_R g - x_g\} \in \mathcal{L}_{HS}^{n\times 1}.$$

We conclude that

$$\Phi^{-1}g = C(S-A)^{-1}x_g + \{D_R g + C h_g\},$$

and hence $\Phi^{-1}g \in \mathcal{S}_{\tau_-}$.

To prove the reverse implication, take

$$f = C(S-A)^{-1}x + h \in \mathcal{S}_{\tau_-}.$$

Thus $x \in \mathcal{D}_{HS}^{n\times 1}$ and $h \in \mathcal{L}_{HS}^{r\times 1}$. We have to show that $g := \Phi f \in \mathcal{L}_{HS}^{r\times 1}$. Obviously, $\Phi h \in \mathcal{L}_{HS}^{r\times 1}$, because both $\Phi$ and $h$ are lower triangular. It remains to show that

$$\Phi C(S-A)^{-1}x \in \mathcal{L}_{HS}^{r\times 1}. \tag{4.16}$$

Let us multiply $\Phi C(S-A)^{-1}x$ on the left by an arbitrary element $h \in \mathcal{L}_{HS}^{1\times r}$. By Proposition 2.2, $(h\Phi C)^\wedge(A) = 0$, and thus we can apply Proposition 2.2 to show that $h\Phi C(S-A)^{-1}x$ is lower triangular for each $h \in \mathcal{L}_{HS}^{1\times r}$. This can only happen when (4.16) holds. □

## 5. INTERMEZZO ABOUT ADMISSIBLE SYLVESTER DATA SETS

Let $\omega = (U, A; Z, V; \Gamma)$ be a set of diagonal operators,

$$U \in \mathcal{D}^{p\times n}, \quad A \in \mathcal{D}^{n\times n}, \quad Z \in \mathcal{D}^{k\times k}, \quad V \in \mathcal{D}^{k\times p}, \quad \Gamma \in \mathcal{D}^{k\times n}.$$

The quintet $\omega$ is called an *admissible Sylvester data set* if

(i) $(U, A)$ is stable and uniformly observable,

(ii) $(Z, V)$ is stable and uniformly controllable,

(iii) $\Gamma A - ZS^{-1}\Gamma S = VU$.

In the definition of an admissible Sylvester data set we are allowed to take one or both of the numbers $k$ and $n$ equal to 0; in this case the Sylvester equation in (iii) is satisfied trivially. With an admissible Sylvester data set we associate the subspace

$$\begin{aligned}\mathcal{S}_\omega =&\{U(S-A)^{-1}x + h \mid x \in \mathcal{D}_{HS}^{n\times 1}, \quad h \in \mathcal{L}_{HS}^{p\times 1} \text{ such} \\ &\text{that } \mathcal{R}es_L\{(S-Z)^{-1}Vh\} = \Gamma x\}.\end{aligned} \tag{5.1}$$

Note that $\mathcal{S}_\omega \subset \mathcal{B}_{HS}^{p\times 1}$. We shall refer to $\mathcal{S}_\omega$ as the *singular subspace* associated with $\omega$.

Let $\tau = (\widetilde{C}, C, A; Z, B, \widetilde{B}; \Gamma)$ be the admissible bitangential interpolation data set considered in the previous section. With $\tau$ one may associate two Sylvester data sets, namely

$$\tau_- = (C, A; 0, 0; 0), \quad \tau_e = \left(\begin{pmatrix}\widetilde{C} \\ C\end{pmatrix}, A; Z, \begin{pmatrix} B & -\widetilde{B}\end{pmatrix}; \Gamma\right).$$

The singular subspace corresponding to $\tau_-$ is precisely the set (4.2), and the singular subspace corresponding to $\tau_e$ is the set (4.3).

The definition of the singular subspace does not require that the Sylvester equation (iii) is satisfied. The next proposition explains what this condition (iii) means for $\mathcal{S}_\omega$.

PROPOSITION 5.1. *Let* $\omega = (U, A; Z, V; \Gamma)$ *satisfy the conditions* (i) *and* (ii) *of an admissible Sylvester data set, and let* $\mathcal{S}_\omega$ *be defined by* (5.1). *Then* $\mathcal{S}_\omega S \subset \mathcal{S}_\omega$ *if and only if the Sylvester equation* (iii) *is satisfied.*

PROOF. Take $g = U(S-A)^{-1}x + h \in \mathcal{S}_\omega$. Here $x \in \mathcal{D}_{HS}^{n\times 1}$ and $h \in \mathcal{L}_{HS}^{p\times 1}$ are such that

$$\mathcal{R}\mathrm{es}_L\{(S-Z)^{-1}Vh\} = \Gamma x. \tag{5.2}$$

Note that

$$\begin{aligned} gS &= U(S-A)^{-1}xS + hS = U(S-A)^{-1}S(S^{-1}xS) + hS \\ &= U(S-A)^{-1}(AS^{-1}xS) + U(S^{-1}xS) + hS. \end{aligned}$$

Next we compute $\mathcal{R}\mathrm{es}_L\{(S-Z)^{-1}V[U(S^{-1}xS) + hS]$. We have

$$\mathcal{R}\mathrm{es}_L\{(S-Z)^{-1}VU(S^{-1}xS)\} = VU(S^{-1}xS),$$

because $VU(S^{-1}xS)$ is a diagonal operator. By Proposition 1.1(v) and Proposition 1.1(i),

$$\begin{aligned} \mathcal{R}\mathrm{es}_L\{(S-Z)^{-1}VhS\} &= \mathcal{R}\mathrm{es}_L\{S^{-1}S(S-Z)^{-1}VhS\} \\ &= S^{-1}(\mathcal{R}\mathrm{es}_L\{S(S-Z)^{-1}Vh\})S^{-1} \\ &= S^{-1}(\mathcal{R}\mathrm{es}_L\{Vh + Z(S-Z)^{-1}Vh\})S \\ &= S^{-1}(\mathcal{R}\mathrm{es}_L Vh)S + S^{-1}(\mathcal{R}\mathrm{es}_L\{Z(S-Z)^{-1}Vh\})S \\ &= S^{-1}(SZS^{-1})(\mathcal{R}\mathrm{es}_L\{(S-Z)^{-1}Vh\})S \\ &= ZS^{-1}\Gamma xS, \end{aligned}$$

because of (5.2). In the above computation, $Vh$ is lower triangular, and therefore $\mathcal{R}\mathrm{es}_L Vh = 0$. By combining the results we see that

$$\mathcal{R}\mathrm{es}_L\{(S-Z)^{-1}V[U(S^{-1}xS) + hS]\} = VUS^{-1}xS + (ZS^{-1})\Gamma xS.$$

It follows that $gS \in \mathcal{S}_\omega$ if and only if

$$VUS^{-1}xS + (ZS^{-1})\Gamma xS = \Gamma(AS^{-1})xS.$$

Since $x$ is an arbitrary element of $\mathcal{D}_{HS}^{1\times 1}$, we conclude that $\mathcal{S}_\omega S \subset \mathcal{S}_\omega$ if and only if (iii) holds. □

THEOREM 5.2. *Let $\mathcal{S}_\omega$ be the singular subspace of the admissible Sylvester data set $\omega$. Then*

$$\mathcal{S}_\omega \mathcal{L}^{1\times 1} \subset \mathcal{S}_\omega. \tag{5.3}$$

PROOF. Let $\omega = (U, A; Z, V; \Gamma)$. First note that because of the Sylvester equation (iii) the following identity holds:

$$\Gamma(I - AS^{-1})^{-1} - (I - ZS^{-1})^{-1}\Gamma = (I - ZS^{-1})^{-1}VU(S - A)^{-1}. \tag{5.4}$$

Take $f \in \mathcal{L}^{1\times 1}$ and $g = C(S - A)^{-1}x + h \in \mathcal{S}_\omega$. Here $x \in \mathcal{D}_{HS}^{n\times 1}$ and $h \in \mathcal{L}_{HS}^{p\times 1}$ are such that (5.2) holds. Put

$$\widetilde{x} := \mathcal{R}es_L(S - A)^{-1}xf \in \mathcal{D}_{HS}^{n\times 1}.$$

Then

$$gf = U(S - A)^{-1}xf + hf = U(S - A)^{-1}\widetilde{x} + \widetilde{h},$$

where

$$\widetilde{h} = U(S - A)^{-1}(xf - \widetilde{x}) + hf \in \mathcal{L}_{HS}^{p\times 1}.$$

Here we used that $(S - A)^{-1}(xf - \widetilde{x})$ is lower triangular because of the definition of $\widetilde{x}$ and Proposition 1.2. In order to show that $gf \in \mathcal{S}_\omega$, it suffices to prove that

$$\mathcal{R}es_L(S - Z)^{-1}V\widetilde{h} = \Gamma\widetilde{x}. \tag{5.5}$$

Now, by the identity (5.4),

$$\begin{aligned}
&\mathcal{R}es_L\{(S - Z)^{-1}VU(S - A)^{-1}(xf - \widetilde{x})\}\\
&\qquad = \mathcal{R}es_L\{S^{-1}(I - ZS^{-1})^{-1}VU(S - A)^{-1}(xf - \widetilde{x})\} =\\
&\qquad = \mathcal{R}es_L\{S^{-1}\Gamma(I - AS^{-1})^{-1}(xf - \widetilde{x})\} - \mathcal{R}es_L\{(S - Z)^{-1}\Gamma(xf - \widetilde{x})\}.
\end{aligned}$$

Since $(S-A)^{-1}(xf-\widetilde{x})$ is lower triangular, we have

$$S^{-1}\Gamma(I-AS^{-1})^{-1}(xf-\widetilde{x})=S^{-1}\Gamma S(S-A)^{-1}(xf-\widetilde{x})\in\mathcal{L}_{HS}^{k\times 1},$$

and therefore $\mathcal{R}es_L\{S^{-1}\Gamma(I-AS^{-1})^{-1}(xf-\widetilde{x})\}=0$. It follows that

$$\begin{aligned}\mathcal{R}es_L\{(S-Z)^{-1}VU(S-A)^{-1}(xf-\widetilde{x})\}&=-\mathcal{R}es_L\{(S-Z)^{-1}(\Gamma xf-\Gamma\widetilde{x})\}\\&=-(\Gamma xf)^{\wedge_L}(Z)+\Gamma\widetilde{x},\end{aligned}$$

because $\Gamma\widetilde{x}$ is a diagonal operator. Next, we use Proposition 1.3 to show that

$$\mathcal{R}es_L\{(S-Z)^{-1}Vhf\}=\{(Vh)^{\wedge_L}(Z)f\}^{\wedge_L}(Z)=\{\Gamma xf\}^{\wedge_L}(Z),$$

by (5.2). By combining the results we obtain (5.5). □

## 6. CONSTRUCTION OF A PARTICULAR SOLUTION

Let $\tau=(\widetilde{C},C,A;Z,B,\widetilde{B};\Gamma)$ be an admissible bitangential interpolation data set. In this section we show that there exists an $F\in\mathcal{L}^{m\times r}$ satisfying

(IC1) $\mathcal{R}es_L\{(S-Z)^{-1}BF\}=\widetilde{B}$,

(IC2) $\mathcal{R}es_R\{FC(S-A)^{-1}\}=\widetilde{C}$,

(IC3) $\mathcal{R}es_L\{(S-Z)^{-1}BFC(S-A)^{-1}\}=\Gamma$,

and we give an explicit formula for one such $F$.

THEOREM 6.1. *Let $\tau=(\widetilde{C},C,A;Z,B,\widetilde{B};\Gamma)$ be an admissible bitangential interpolation data set. Let $G_L$ be the controllability Gramian corresponding to $(Z,B)$, let $G_R$ be the observability Gramian corresponding to $(C,A)$, and let*

$$\Phi=D_R^*+B_R^*S(I-A^*S)^{-1}C^*$$

*be as in (2.9). Then*

$$F:=B^*(I-SZ^*)^{-1}G_L^{-1}(\widetilde{B}D_R+\Gamma B_R)\Phi+\widetilde{C}G_R^{-1}(I-A^*S)^{-1}C^*\in\mathcal{L}^{m\times r},\tag{6.1}$$

*and $F$ satisfies the interpolation conditions* (IC1) - (IC3).

PROOF. Take $x \in \mathcal{D}_{HS}^{r\times 1}$. By Proposition 4.3(ii) the product $\Phi^{-1}x \in \mathcal{S}_{\tau_-}$. Since $\Phi$ is unitary, $\Phi^{-1} = \Phi^*$, and thus

$$\Phi^{-1}x = C(S - A)^{-1}B_R x + D_R x \in \mathcal{S}_{\tau_-}.$$

According to Proposition 4.2, we have $\begin{pmatrix} 0 & I \end{pmatrix}\mathcal{S}_{\tau_e} = \mathcal{S}_{\tau_-}$. Thus we can find $h(x) \in \mathcal{L}_{HS}^{r\times 1}$ such that

$$\begin{pmatrix} \widetilde{C} \\ C \end{pmatrix}(S - A)^{-1}B_R x + \begin{pmatrix} h(x) \\ D_R x \end{pmatrix} \in \mathcal{S}_{\tau_e}. \tag{6.2}$$

The proof of Proposition 4.2 gives a method for constructing $h(x)$. Let $\mathcal{C}$ (resp., $\widetilde{\mathcal{C}}$) be the controllability operator (see formula (2.2)) associated with the pair $(Z, B)$ (resp., $(Z, \widetilde{B})$). Then (6.2) holds with

$$h(x) := \mathcal{C}^*(\mathcal{C}\mathcal{C}^*)^{-1}\widetilde{\mathcal{C}}D_R x + \mathcal{C}^*(\mathcal{C}\mathcal{C}^*)^{-1}\Gamma B_R x.$$

From the definition of the controllability operator in (2.2) it follows that

$$\widetilde{\mathcal{C}}D_R x = \mathcal{R}\mathrm{es}_L\{(S - Z)^{-1}\widetilde{B}D_R x\} = \widetilde{B}D_R x,$$

because $\widetilde{B}D_R x$ is a diagonal operator. Furthermore, formula (2.3) implies that $(\mathcal{C}\mathcal{C}^*)^{-1}d = G_L^{-1}d$ for each $d \in \mathcal{D}_{HS}^{k\times 1}$. We conclude (use the duality result in Proposition 1.4) that

$$h(x) = B^*(I - SZ^*)^{-1}G_L^{-1}(\widetilde{B}D_R + \Gamma B_R)x.$$

Put

$$H := B^*(I - SZ^*)^{-1}G_L^{-1}(\widetilde{B}D_R + \Gamma B_R).$$

Then $H \in \mathcal{L}^{m\times r}$, and

$$\begin{pmatrix} \widetilde{C}(S - A)^{-1}B_R + H \\ \Phi^{-1} \end{pmatrix} x \in \mathcal{S}_{\tau_e}, \quad x \in \mathcal{D}_{HS}^{r\times 1}. \tag{6.3}$$

Recall that $\mathcal{S}_{\tau_e}$ is the singular subspace corresponding to the admissible Sylvester data set

$$\tau_e = (\begin{pmatrix} \widetilde{C} \\ C \end{pmatrix}, A; Z, \begin{pmatrix} B & -\widetilde{B} \end{pmatrix}; \Gamma).$$

Thus we can apply Theorem 5.2 to show that $\mathcal{S}_{\tau_e}\mathcal{L}^{1\times 1} \subset \mathcal{S}_{\tau_e}$. By combining this with (6.3) we see that

$$\begin{pmatrix} \widetilde{C}(S-A)^{-1}B_R + H \\ \Phi^{-1} \end{pmatrix} \mathcal{L}_{HS}^{r\times 1} \subset \mathcal{S}_{\tau_e}. \tag{6.4}$$

Put $F_0 = \{\widetilde{C}(S-A)^{-1}B_R + H\}\Phi$. Then we may rewrite (6.4) as

$$\begin{pmatrix} F_0 \\ I \end{pmatrix} \mathcal{S}_{\tau_-} \subset \mathcal{S}_{\tau_e}. \tag{6.5}$$

because $\Phi^{-1}\mathcal{L}_{HS}^{r\times 1} = \mathcal{S}_{\tau_-}$ by Proposition 4.3(ii). According to Theorem 4.1, the inclusion (6.5) shows that $F_0 \in \mathcal{L}^{m\times r}$ and $F_0$ satisfies (IC1) - (IC3).

To complete the proof we show that $F_0 = F$, where $F$ is given by (6.1). To do this it suffices to prove that

$$\widetilde{C}(S-A)^{-1}B_R = \widetilde{C}G_R^{-1}(I-A^*S)^{-1}C^*\Phi^{-1}. \tag{6.6}$$

Note that

$$(I-A^*S)^{-1}C^*\Phi^{-1} = (I-A^*S)^{-1}C^*D_R + (I-A^*S)^{-1}C^*C(I-S^{-1}A)^{-1}S^{-1}B_R.$$

From the definition $G_R$ in (2.6) we know that

$$C^*C = G_R - (A^*S)G_R(S^{-1}A),$$

and hence

$$\begin{aligned}(I-A^*S)^{-1}C^*C(I-S^{-1}A)^{-1} &= \\ &= (I-A^*S)^{-1}G_R(I-S^{-1}A)^{-1} - (I-A^*S)^{-1}(A^*S)G_R(S^{-1}A)(I-S^{-1}A)^{-1} \\ &= (I-A^*S)^{-1}A^*SG_R + G_R(I-S^{-1}A)^{-1}.\end{aligned}$$

Therefore,

$$\begin{aligned}\widetilde{C}G_R^{-1}(I-A^*S)^{-1}C^*\Phi^{-1} &= \\ &= \widetilde{C}(S-A)^{-1}B_R + \widetilde{C}G_R^{-1}(I-A^*S)^{-1}(C^*D_R + A^*SG_RS^{-1}B_R).\end{aligned}$$

From the first $2\times 2$ operator matrix identity in Proposition 2.2 it follows that

$$A^*SG_RS^{-1}B_R + C^*D_R = 0.$$

Hence (6.6) is proved. □

## 7. PARAMETRIZATION OF ALL SOLUTIONS (WITHOUT NORM CONSTRAINTS)

Let $\tau = (\widetilde{C}, C, A; Z, B, \widetilde{B}; \Gamma)$ be an admissible bitangential interpolation data set. In this section we parametrize all $F \in \mathcal{L}^{m\times r}$ satisfying

(IC1) $\quad \mathcal{R}\mathrm{es}_L\{(S-Z)^{-1}BF\} = \widetilde{B}$,

(IC2) $\quad \mathcal{R}\mathrm{es}_R\{FC(S-A)^{-1}\} = \widetilde{C}$,

(IC3) $\quad \mathcal{R}\mathrm{es}_L\{(S-Z)^{-1}BFC(S-A)^{-1}\} = \Gamma$,

THEOREM 7.1 *Let $\tau = (\widetilde{C}, C, A; Z, B, \widetilde{B}; \Gamma)$ be an admissible bitangential interpolation data set, and consider the singular subspace $\mathcal{S}_{\tau_e}$ defined by (4.3). Assume*

$$\mathcal{S}_{\tau_e} = W\begin{pmatrix} \mathcal{L}_{HS}^{m\times 1} \\ \mathcal{L}_{HS}^{r\times 1} \end{pmatrix}, \tag{7.1}$$

*where*

$$W = \begin{pmatrix} W_{11} & W_{12} \\ W_{21} & W_{22} \end{pmatrix} \in \begin{pmatrix} \mathcal{B}^{m\times m} & \mathcal{B}^{m\times r} \\ \mathcal{B}^{r\times m} & \mathcal{B}^{r\times r} \end{pmatrix},$$

*and $W$ is invertible with $W^{-1} \in \mathcal{B}^{(m+r)\times(m+r)}$. Then $F \in \mathcal{L}^{m\times r}$ and $F$ satisfies* (IC1) - (IC3) *if and only if*

$$F = (W_{11}G_1 + W_{12}G_2)(W_{21}G_1 + W_{22}G_2)^{-1} \tag{7.2}$$

*for some $G_1 \in \mathcal{L}^{m\times r}, G_2 \in \mathcal{L}^{r\times r}$ satisfying*

$$(W_{21}G_1 + W_{22}G_2)\mathcal{L}_{HS}^{r\times 1} = \begin{pmatrix} W_{21} & W_{22} \end{pmatrix}\begin{pmatrix} \mathcal{L}_{HS}^{m\times 1} \\ \mathcal{L}_{HS}^{r\times 1} \end{pmatrix}. \tag{7.3}$$

PROOF. Let $F \in \mathcal{L}^{m\times r}$ satisfy (IC1) - (IC3). So, by Proposition 4.3, Theorem 4.1 and formula (7.1), we have

$$\begin{pmatrix} F \\ I \end{pmatrix}\Phi^{-1}\mathcal{L}_{HS}^{r\times 1} \subset W\begin{pmatrix} \mathcal{L}_{HS}^{m\times 1} \\ \mathcal{L}_{HS}^{r\times 1} \end{pmatrix}. \tag{7.4}$$

where $\Phi$ is given by (2.9). We may rewrite (7.4) as:

$$W^{-1}\begin{pmatrix} F \\ I \end{pmatrix}\Phi^{-1}\mathcal{L}_{HS}^{r\times 1} \subset \begin{pmatrix} \mathcal{L}_{HS}^{m\times 1} \\ \mathcal{L}_{HS}^{r\times 1} \end{pmatrix}.$$

It follows that there exist $G_1 \in \mathcal{L}^{m\times r}$ and $G_2 \in \mathcal{L}^{r\times r}$ such that

$$W^{-1}\begin{pmatrix} F \\ I \end{pmatrix}\Phi^{-1} = \begin{pmatrix} G_1 \\ G_2 \end{pmatrix}.$$

In particular,

$$F\Phi^{-1} = W_{11}G_1 + W_{12}G_2, \quad \Phi^{-1} = W_{21}G_1 + W_{22}G_2,$$

and therefore with this choice of $G_1$ and $G_2$ the interpolant $F$ is given by (7.2). Let us prove that (7.3) also holds.

Since $\Phi^{-1} = W_{21}G_1 + W_{22}G_2$, the left hand side of (7.3) is equal to $\mathcal{S}_{\tau_-}$, by Proposition 4.3(ii). According to formula (7.1), the right hand side of (7.3) is equal to $(\,0 \quad I\,)\mathcal{S}_{\tau_e}$. So (7.3) holds, because of Proposition 4.2.

Conversely, let $G_1 \in \mathcal{L}^{m\times r}$ and $G_2 \in \mathcal{L}^{r\times r}$ be such that (7.2) holds, and let $F$ be given by (7.2). By (7.3) and (7.1),

$$(W_{21}G_1 + W_{22}G_2)\mathcal{L}_{HS}^{r\times 1} = (\,0 \quad I\,)W\begin{pmatrix} \mathcal{L}_{HS}^{m\times 1} \\ \mathcal{L}_{HS}^{r\times 1} \end{pmatrix} = (\,0 \quad I\,)\mathcal{S}_{\tau_e}. \tag{7.5}$$

But then we can use Proposition 4.2 and Proposition 4.3(ii) to show that

$$\Phi(W_{21}G_1 + W_{22}G_2)\mathcal{L}_{HS}^{r\times 1} = \mathcal{L}_{HS}^{r\times 1}.$$

This identity implies (cf., Proposition 1.5) that $\Phi(W_{21}G_1 + W_{22}G_2)$ is lower triangular and invertible. In particular, $W_{21}G_1 + W_{22}G_2$ is invertible, and hence $F$ is well-defined by (7.2). Now, rewrite (7.2) as

$$\begin{pmatrix} F \\ I \end{pmatrix}(W_{21}G_1 + W_{22}G_2) = W\begin{pmatrix} G_1 \\ G_2 \end{pmatrix}.$$

According to (7.5) and Proposition 4.2, we have

$$(W_{21}G_1 + W_{22}G_2)\mathcal{L}_{HS}^{r\times 1} = \mathcal{S}_{\tau_-}.$$

Thus

$$\begin{pmatrix} F \\ I \end{pmatrix}\mathcal{S}_{\tau_-} = \begin{pmatrix} F \\ I \end{pmatrix}(W_{21}G_1 + W_{22}G_2)\mathcal{L}_{HS}^{r\times 1} = W\begin{pmatrix} G_1 \\ G_2 \end{pmatrix}\mathcal{L}_{HS}^{r\times 1} \subset W\begin{pmatrix} \mathcal{L}_{HS}^{m\times 1} \\ \mathcal{L}_{HS}^{r\times 1} \end{pmatrix} = \mathcal{S}_{\tau_e},$$

because of (7.1). We conclude that (4.4) holds, and we can apply Theorem 4.1 to show that $F \in \mathcal{L}^{m\times r}$ and $F$ satisfies (IC1) - (IC3). □

The next theorem shows how one may get a $W$ as in (7.1) given a particular solution of (IC1) - (IC3).

THEOREM 7.2. *Let* $\tau = (\widetilde{C}, C, A; Z, B, \widetilde{B}; \Gamma)$ *be an admissible bitangential interpolation data set, and let* $F_0 \in \mathcal{L}^{m\times r}$ *satisfy* (IC1) - (IC3). *Put*

$$W = \begin{pmatrix} \Psi & F_0\Phi^{-1} \\ 0 & \Phi^{-1} \end{pmatrix} \in \mathcal{B}^{(m+r)\times(m+r)}, \tag{7.6}$$

*where* $\Psi$ *and* $\Phi$ *are given by* (2.5) *and* (2.9), *respectively. Then* $W$ *is invertible and with this choice of* $W$ *the identity* (7.1) *holds.*

PROOF. Obviously, $W$ is invertible. Since $F_0$ satisfies (IC1) - (IC3), formula (4.4) holds with $F_0$ in place of $F$, and hence we can use Proposition 4.3(ii) to show that

$$\begin{pmatrix} F_0\Phi^{-1} \\ \Phi^{-1} \end{pmatrix} \mathcal{L}_{HS}^{r\times 1} = \begin{pmatrix} F_0 \\ I \end{pmatrix} \mathcal{S}_{\tau_-} \subset \mathcal{S}_{\tau_e}.$$

This inclusion, together with Proposition 4.3(i), yields

$$\begin{pmatrix} \Psi & F_0\Phi^{-1} \\ 0 & \Phi^{-1} \end{pmatrix} \begin{pmatrix} \mathcal{L}_{HS}^{m\times 1} \\ \mathcal{L}_{HS}^{r\times 1} \end{pmatrix} = \begin{pmatrix} \Psi\mathcal{L}_{HS}^{m\times 1} \\ 0 \end{pmatrix} + \begin{pmatrix} F_0\Phi^{-1} \\ \Phi^{-1} \end{pmatrix} \mathcal{L}_{HS}^{r\times 1} \subset \mathcal{S}_{\tau_e}.$$

To prove the reverse inclusion, take $(\,g_1 \quad g_2\,)^T$ in $\mathcal{S}_{\tau_e}$. In particular, $g_2 \in (\,0 \quad I\,)\mathcal{S}_{\tau_e} = \mathcal{S}_{\tau_-}$, by Proposition 4.2, and hence, by Proposition 4.3(ii), we have $g_2 = \Phi^{-1}h_2$ for some $h_2 \in \mathcal{L}_{HS}^{r\times 1}$. It follows (use (4.4) with $F_0$ in place of $F$ and Proposition 4.3(ii)) that

$$\begin{pmatrix} F_0\Phi^{-1} \\ \Phi^{-1} \end{pmatrix} h_2 \in \begin{pmatrix} F_0 \\ I \end{pmatrix} \mathcal{S}_{\tau_-} \subset \mathcal{S}_{\tau_e},$$

and therefore

$$\begin{pmatrix} g_1 - F_0 g_2 \\ 0 \end{pmatrix} = \begin{pmatrix} g_1 \\ g_2 \end{pmatrix} - \begin{pmatrix} F_0\Phi^{-1} \\ \Phi^{-1} \end{pmatrix} h_2 \subset \mathcal{S}_{\tau_e}.$$

Thus $g_1 - F_0 g_2 = \Psi h_1$ for some $h_1 \in \mathcal{L}_{HS}^{m\times 1}$, by Proposition 4.3(i). We conclude that

$$\begin{pmatrix} \Psi & F_0\Phi^{-1} \\ 0 & \Phi^{-1} \end{pmatrix} \begin{pmatrix} h_1 \\ h_2 \end{pmatrix} = \begin{pmatrix} g_1 \\ g_2 \end{pmatrix},$$

which proves the theorem. □

Note that for $W$ as in (7.6) the condition (7.3) on a parameter pair $G_1 \in \mathcal{L}^{m\times r}, G_2 \in \mathcal{L}^{r\times r}$ collapses to $G_2\mathcal{L}^{r\times 1}_{HS} = \mathcal{L}^{r\times 1}_{HS}$. The latter identity is equivalent to the requirement that $G_2$ is invertible (by Proposition 1.5) and $G_2^{-1} \in \mathcal{L}^{r\times r}$. In this case, we may write $G = G_1G_2^{-1} \in \mathcal{L}^{m\times r}$, and the parametrization (7.2) of the general solution $F$ of (IC1) - (IC3) has the form

$$F = (\Psi G_1 + F_0\Phi^{-1}G_2)(\Phi^{-1}G_2)^{-1} = F_0 + \Psi G\Phi.$$

Conversely, if $W$ is as in (7.6) and $G \in \mathcal{L}^{m\times r}$, then $G_1 := G$ and $G_2 := I$ satisfy the conditions for a valid parameter pair in Theorem 7.1. Thus $F$ is in $\mathcal{L}^{m\times r}$ and satisfies (IC1) - (IC3) if and only if

$$F = F_0 + \Psi G\Phi \tag{7.7}$$

for some $G \in \mathcal{L}^{m\times r}$. The affine parametrization in (7.7) is the same as the divisor remainder form for the interpolant derived in Proposition 3.1.

## 8. INPUT-OUTPUT OPERATORS OF TIME-VARYING SYSTEMS

We shall consider input-output operators of time-varying singular systems of the form:

$$\Sigma \begin{cases} G_{\nu+1}x_{\nu+1} = A_\nu x_\nu + B_\nu u_\nu, & \nu \in \mathbf{Z}, \\ y_\nu = C_\nu x_\nu + D_\nu u_\nu \end{cases}$$

Here

$$G := \text{diag } (G_\nu)_{\nu=-\infty}^{\infty} \in \mathcal{D}^{N\times N}, \quad A := \text{diag } (A_\nu)_{\nu=-\infty}^{\infty} \in \mathcal{D}^{N\times N},$$

$$B := \text{diag } (B_\nu)_{\nu=-\infty}^{\infty} \in \mathcal{D}^{N\times r}, \quad C := \text{diag } (C_\nu)_{\nu=-\infty}^{\infty} \in \mathcal{D}^{m\times N},$$

$$D := \text{diag } (D_\nu)_{\nu=-\infty}^{\infty} \in \mathcal{D}^{m\times m},$$

and we assume that the operator $S^*G - A$ is invertible. The latter condition is equivalent (see [BeG]) to the existence of a dichotomy for the singular difference equations $G_{\nu+1}x_{\nu+1} = A_\nu x_\nu$, $\nu \in \mathbf{Z}$. From the invertibility of $S^*G - A$ it follows that for each input sequence $\vec{u} = (u_\nu)_{\nu=-\infty}^{\infty}$ from $\ell_2^r(\mathbf{Z})$ there exists a unique sequence $\vec{x} = (x_\nu)_{\nu=-\infty}^{\infty}$ in $\ell_2^N(\mathbf{Z})$ such that

$$G_{\nu+1}x_{\nu+1} = A_\nu x_\nu + B_\nu u_\nu, \quad \nu \in \mathbf{Z}.$$

In fact, $\vec{x} = (S^*G - A)^{-1}B\,\vec{u}$. Now, put $\vec{y} = C\,\vec{x} + D\,\vec{u}$. Then $\vec{y} \in \ell_2^m(\mathbb{Z})$ and

$$y_\nu = C_\nu x_\nu + D_\nu u_\nu, \quad \nu \in \mathbb{Z}.$$

We conclude that the system $\Sigma$ has a well-defined input-output operator $T_\Sigma \in \mathcal{B}^{m\times r}$, namely

$$T_\Sigma = D + C(S^*G - A)^{-1}B. \tag{8.1}$$

The class of all operators $T \in \mathcal{B}^{m\times r}$ appearing in this way is denoted by $\mathcal{R}^{m\times r}$. Thus $T \in \mathcal{R}^{m\times r}$ if and only if one can find a system $\Sigma$ with the properties described above such that $T = T_\Sigma$. In the latter case we call $\Sigma$ a *realization* of $T$.

Operators $T \in \mathcal{R}^{m\times r}$ have a number of interesting properties which we shall mention here below. Let $T = [t_{ij}]_{i,j=-\infty}^{\infty} \in \mathcal{B}^{m\times r}$. For each $\nu \in \mathbb{Z}$ we introduce the following operators:

$$H_\nu^+ = \begin{pmatrix} t_{\nu,\nu-1} & t_{\nu,\nu-2} & t_{\nu,\nu-3} & \cdots \\ t_{\nu+1,\nu-1} & t_{\nu+1,\nu-2} & t_{\nu+1,\nu-3} & \cdots \\ t_{\nu+2,\nu-1} & t_{\nu+2,\nu-2} & t_{\nu+2,\nu-3} & \cdots \\ \vdots & \vdots & \vdots & \end{pmatrix} : \ell_2^r \to \ell_2^m, \tag{8.2}$$

$$H_\nu^- = \begin{pmatrix} t_{\nu-1,\nu} & t_{\nu-1,\nu+1} & t_{\nu-1,\nu+2} & \cdots \\ t_{\nu-2,\nu} & t_{\nu-2,\nu+1} & t_{\nu-2,\nu+2} & \cdots \\ t_{\nu-3,\nu} & t_{\nu-3,\nu+1} & t_{\nu-3,\nu+2} & \cdots \\ \vdots & \vdots & \vdots & \end{pmatrix} : \ell_2^r \to \ell_2^m, \tag{8.3}$$

Here $\ell_2^q$ stands for the space of norm square summable sequences $(x_\nu)_{\nu=0}^{\infty}$ with entries in $\mathbb{C}^q$. Since $T \in \mathcal{B}^{m\times r}$ the operators $H_\nu^+$ and $H_\nu^-$ are well-defined bounded linear operators.

THEOREM 8.1. *Let $T = [t_{ij}]_{i,j=-\infty}^{\infty} \in \mathcal{B}^{m\times r}$, and let the operators $H_\nu^+$ and $H_\nu^-$ be given by (8.2) and (8.3), respectively. Then $T$ admits a realization, i.e., $T \in \mathcal{R}^{m\times r}$, if and only if*

$$\sup_{\nu\in\mathbb{Z}} \operatorname{rank} H_\nu^+ < \infty, \quad \sup_{\nu\in\mathbb{Z}} \operatorname{rank} H_\nu^- < \infty, \tag{8.4}$$

*and there exist constants $M \geq 0$, $0 < a < 1$, such that*

$$\|t_{ij}\| \leq M a^{|i-j|}, \quad i,j \in \mathbb{Z}. \tag{8.5}$$

Theorem 8.1 is the analogue of Theorem 1.4 in [BeGK] for doubly infinite block matrices $[t_{ij}]_{i,j=-\infty}^{\infty}$. To prove Theorem 8.1 one may use the same type of arguments as are used in Section 5 of [BeGK] to prove Theorem 1.4 of [BeGK]. We omit the details.

PROPOSITION 8.2. *Input-output operators have the following properties:*

(i) *if* $T_1 \in \mathcal{R}^{m\times r}$ *and* $T_2 \in \mathcal{R}^{r\times s}$ *, then* $T_1T_2 \in \mathcal{R}^{m\times s}$*;*

(ii) *if* $T \in \mathcal{R}^{r\times r}$ *and* $T$ *is invertible, then* $T^{-1} \in \mathcal{R}^{r\times r}$.

PROOF. (i) Let $T_1 = T_{\Sigma_1}$ and $T_2 = T_{\Sigma_2}$, where for $i = 1,2$

$$\Sigma_i \begin{cases} G_{\nu+1}^{(i)} x_{\nu+1}^{(i)} = A_\nu^{(i)} x_\nu^{(i)} + B_\nu^{(i)} u_\nu^{(i)}, & \nu \in \mathbb{Z}, \\ y_\nu^{(i)} = C_\nu^{(i)} x_\nu^{(i)} + D_\nu^{(i)} u_\nu^{(i)}, \end{cases}$$

is a singular time-varying system of the type considered in the first paragraph of this section. Let $\widetilde{\Sigma}$ be the cascade connection of $\Sigma_2$ and $\Sigma_1$, i.e., $\widetilde{\Sigma}$ is the system which one obtains if at each time instance $\nu$ the input $u_\nu^{(1)}$ of $\Sigma_1$ is taken to be equal to the output $y_\nu^{(2)}$ of $\Sigma_2$. In other words

$$\widetilde{\Sigma} \begin{cases} \widetilde{G}_{\nu+1} x_{\nu+1} = \widetilde{A}_\nu x_\nu + \widetilde{B}_\nu u_\nu, & \nu \in \mathbb{Z}, \\ y_\nu = \widetilde{C}_\nu x_\nu + \widetilde{D}_\nu u_\nu, \end{cases}$$

where

$$\widetilde{G}_\nu = \begin{pmatrix} G_\nu^{(1)} & 0 \\ 0 & G_\nu^{(2)} \end{pmatrix}, \quad \widetilde{A}_\nu = \begin{pmatrix} A_\nu^{(1)} & \widetilde{B}_\nu^{(1)} C_\nu^{(2)} \\ 0 & A_\nu^{(2)} \end{pmatrix},$$

$$\widetilde{B}_\nu = \begin{pmatrix} B_\nu^{(1)} D_\nu^{(2)} \\ B_\nu^{(2)} \end{pmatrix}, \quad \widetilde{C}_\nu = \begin{pmatrix} C_\nu^{(1)} & D_\nu^{(2)} C_\nu^{(2)} \end{pmatrix},$$

$$\widetilde{D}_\nu = D_\nu^{(1)} D_\nu^{(2)}.$$

Put

$$\widetilde{G} = \operatorname{diag}\,(\widetilde{G}_\nu)_{\nu=-\infty}^{\infty}, \quad \widetilde{A} = \operatorname{diag}\,(\widetilde{A}_\nu)_{\nu=-\infty}^{\infty}.$$

Then

$$S^*\widetilde{G} - \widetilde{A} = \begin{pmatrix} S^*G^{(1)} - A^{(1)} & -B^{(1)}C^{(1)} \\ 0 & S^*G^{(2)} - A^{(2)} \end{pmatrix},$$

Here

$$G^{(i)} = \operatorname{diag}\,(G_\nu^{(i)})_{\nu=-\infty}^{\infty}, \quad A^{(i)} = \operatorname{diag}\,(A_\nu^{(i)}),\ i = 1,2,$$

and

$$B^{(i)} = \operatorname{diag}\,(B_\nu^{(i)})_{\nu=-\infty}^{\infty}, \quad C^{(2)} = \operatorname{diag}\,(C_\nu^{(2)})_{\nu=-\infty}^{\infty}.$$

It follows that $S^*\widetilde{G} - \widetilde{A}$ is invertible, and hence $\widetilde{\Sigma}$ is a system of the type considered in the first paragraph of this section. Therefore,

$$T_1 T_2 = T_{\Sigma_1} T_{\Sigma_2} = T_{\widetilde{\Sigma}} \in \mathcal{R}^{m \times s}.$$

(ii) Since $T \in \mathcal{R}^{r \times r}$, we have $T = T_\Sigma$, there $T_\Sigma$ is given by (8.1). First we show that we may assume that $D = I$. Indeed, put

$$\begin{aligned}
\widetilde{A} &= \operatorname{diag}\left(\begin{pmatrix} -I_{\mathbb{C}^r} & 0 \\ 0 & A_\nu \end{pmatrix}\right)_{\nu=-\infty}^{\infty} \in \mathcal{D}^{(r+N)\times(r+N)}, \\
\widetilde{G} &= \operatorname{diag}\left(\begin{pmatrix} 0 & 0 \\ 0 & G_\nu \end{pmatrix}\right)_{\nu=-\infty}^{\infty} \in \mathcal{D}^{(r+N)\times(r+N)}, \\
\widetilde{B} &= \operatorname{diag}\left(\begin{pmatrix} I_{\mathbb{C}^r} \\ B_\nu \end{pmatrix}\right)_{\nu=-\infty}^{\infty} \in \mathcal{D}^{(r+N)\times r}, \\
\widetilde{C} &= \operatorname{diag}\left(\begin{pmatrix} D_\nu - I_{\mathbb{C}^r} & C_\nu \end{pmatrix}\right)_{\nu=-\infty}^{\infty} \in \mathcal{D}^{m\times(r+N)}.
\end{aligned}$$

Then we see from (8.1) that

$$T = I + \widetilde{C}(S^*\widetilde{G} - \widetilde{A})^{-1}\widetilde{B}. \tag{8.6}$$

Next, we use the following general remark. If $B, C$ and $E$ are bounded linear operators with $E$ invertible, then $I + CE^{-1}B$ is invertible if and only if $E + BC$ is invertible, and in this case

$$(I + CE^{-1}B)^{-1} = I - C(E + BC)^{-1}B.$$

Since $T$ in (8.6) is assumed to be invertible, the previous remark implies that $S^*\widetilde{G} - \widetilde{A} + \widetilde{B}\widetilde{C}$ is invertible, and

$$T^{-1} = I - \widetilde{C}(S^*\widetilde{G} - (\widetilde{A} - \widetilde{B}\widetilde{C}))^{-1}\widetilde{B}. \tag{8.7}$$

Note that

$$\widetilde{A} - \widetilde{B}\widetilde{C} = \operatorname{diag}(\widetilde{A}_\nu - \widetilde{B}_\nu\widetilde{C}_\nu)_{\nu=-\infty}^{\infty} \in \mathcal{D}^{(r+N)\times(r+N)}.$$

Hence the right hand side of (8.7) is again an input-output operator. Thus $T^{-1} \in \mathcal{R}^{r\times r}$. □

## 9. PARAMETRIZATION OF ALL CONTRACTIVE INPUT-OUTPUT OPERATORS SATISFYING THE BITANGENTIAL INTERPOLATION CONDITIONS

Let $\tau = (\widetilde{C}, C, A; Z, B, \widetilde{B}; \Gamma)$ be an admissible bitangential interpolation data set. In this section we parametrize all input-output operators $F \in \mathcal{LR}^{m\times r}$ that are contractive (i.e., $\|F\| \le 1$) and satisfy the interpolation conditions

(IC1) $\mathcal{R}\mathrm{es}_L\{(S-Z)^{-1}BF\} = \widetilde{B}$,

(IC2) $\mathcal{R}\mathrm{es}_R\{FC(S-A)^{-1}\} = \widetilde{C}$,

(IC3) $\mathcal{R}\mathrm{es}_L\{(S-Z)^{-1}BFC(S-A)^{-1}\} = \Gamma$.

THEOREM 9.1. *Let $\tau = (\widetilde{C}, C, A; Z, B, \widetilde{B}; \Gamma)$ be an admissible bitangential interpolation data set, and consider the singular subspace $\mathcal{S}_{\tau_e}$ defined by (4.3). Assume*

$$\mathcal{S}_{\tau_e} = \theta \begin{pmatrix} \mathcal{L}_{HS}^{m\times 1} \\ \mathcal{L}_{HS}^{r\times 1} \end{pmatrix}, \tag{9.1}$$

*where*

$$\theta = \begin{pmatrix} \theta_{11} & \theta_{12} \\ \theta_{21} & \theta_{22} \end{pmatrix} \in \begin{pmatrix} \mathcal{R}^{m\times m} & \mathcal{R}^{m\times r} \\ \mathcal{R}^{r\times m} & \mathcal{R}^{r\times r} \end{pmatrix}$$

*satisfies*

$$\theta^* J \theta = J = \theta J \theta^*, \tag{9.2}$$

*with*

$$J = \begin{pmatrix} I & 0 \\ 0 & -I \end{pmatrix} \in \begin{pmatrix} \mathcal{D}^{m\times m} & 0 \\ 0 & \mathcal{D}^{r\times r} \end{pmatrix}$$

*Then there exists $F \in \mathcal{LR}^{m\times r}$ with $\|F\| \le 1$ satisfying* (IC1) - (IC3) *if and only if*

$$\theta_{22}^{-1}\theta_{21} \in \mathcal{L}^{r\times m}. \tag{9.3}$$

*In this case all solutions $F \in \mathcal{LR}^{m\times r}$ of* (IC1) - (IC3) *such that $\|F\| \le 1$ are given by*

$$F = (\theta_{11}G + \theta_{12})(\theta_{21}G + \theta_{22})^{-1}, \tag{9.4}$$

*where the free parameter $G$ is an element of $\mathcal{LR}^{m\times r}$ with $\|G\| \le 1$.*

PROOF. We split the proof into 5 parts.

Part (a). In this part we assume that (9.3) holds. Take $G \in \mathcal{LR}^{m\times r}$ with $\|G\| \leq 1$. We shall show that in this case $F$ in (9.4) is well-defined and $F$ is a solution of our contractive bitangential interpolation problem (i.e., $F \in \mathcal{LR}^{m\times r}$ with $\|F\| \leq 1$, and $F$ satisfies (IC1) - (IC3)).

Formula (9.2) means that $\theta$ is $J$-unitary. So we can apply Theorem 2.1 in [BGK1] to show that $\|\theta_{22}^{-1}\theta_{21}\| < 1$. It follows that $R := \theta_{22}^{-1}\theta_{21}G$ is strictly contractive, and hence $R + I$ is invertible, with $(R+I)^{-1} = \Sigma_{\nu=0}^{\infty}(-R)^{\nu}$. Since both $G$ and $\theta_{22}^{-1}\theta_{21}$ are lower triangular, we conclude that $R$ is lower triangular, and hence $(R+I)^{-1}$ is lower triangular. Therefore, $(\theta_{22}^{-1}\theta_{21}G + I)^{-1} \in \mathcal{L}^{r\times r}$. Thus in this case

$$\begin{aligned}(\theta_{21}G + \theta_{22})\mathcal{L}_{HS}^{r\times 1} &= \theta_{22}[\theta_{22}^{-1}\theta_{21}G + I]\mathcal{L}_{HS}^{r\times 1} = \theta_{22}\mathcal{L}_{HS}^{r\times 1} \\ &= \theta_{22}\begin{pmatrix}\theta_{22}^{-1}\theta_{21} & I\end{pmatrix}\begin{pmatrix}\mathcal{L}_{HS}^{m\times 1} \\ \mathcal{L}_{HS}^{r\times 1}\end{pmatrix} = \begin{pmatrix}\theta_{21} & \theta_{22}\end{pmatrix}\begin{pmatrix}\mathcal{L}_{HS}^{m\times 1} \\ \mathcal{L}_{HS}^{r\times 1}\end{pmatrix}.\end{aligned}$$

Thus with $G_1 = G$ and $G_2 = I$ condition (7.3) in Theorem 7.1 is fulfilled for $\theta$ in place of $W$. Therefore, by Theorem 7.1, the operator $F$ in (9.4) is well-defined, $F \in \mathcal{L}^{m\times r}$ and $F$ satisfies (IC1) - (IC3).

By repeatedly applying Proposition 8.2 one sees that $F \in \mathcal{R}^{m\times r}$, and thus $F \in \mathcal{LR}^{m\times r}$. It remains to check that $\|F\| \leq 1$. For this purpose we rewrite (9.4) as

$$\begin{pmatrix}F \\ I\end{pmatrix} = \theta\begin{pmatrix}G \\ I\end{pmatrix}(\theta_{21}G + \theta_{22})^{-1}. \tag{9.5}$$

Since $\theta^* J\theta = J$, it follows that

$$\begin{aligned}F^*F - I &= \begin{pmatrix}F^* & I\end{pmatrix}J\begin{pmatrix}F \\ I\end{pmatrix} \\ &= (\theta_{21}G + \theta_{22})^{-*}\begin{pmatrix}G^* & I\end{pmatrix}\theta^* J\theta\begin{pmatrix}G \\ I\end{pmatrix}(\theta_{21}G + \theta_{22})^{-1} \\ &= (\theta_{21}G + \theta_{22})^{-*}\begin{pmatrix}G^* & I\end{pmatrix}J\begin{pmatrix}G \\ I\end{pmatrix}(\theta_{21}G + \theta_{22})^{-1} \\ &= (\theta_{21}G + \theta_{22})^{-*}(G^*G - I)(\theta_{21}G + \theta_{22})^{-1}.\end{aligned}$$

Here $A^{-*}$ denote the adjoint of $A^{-1}$. From $\|G\| \leq 1$ it follows that $G^*G - I \leq 0$, and hence the above calculation shows that $F^*F - I \leq 0$, which is equivalent to $\|F\| \leq 1$.

Part (b). In this part we assume that $F \in \mathcal{LR}^{m\times r}$ is contractive and satisfies (IC1) - (IC3). By Proposition 4.3(ii), Theorem 4.1, and our assumption (9.1), the latter implies

$$\begin{pmatrix} F \\ I \end{pmatrix} \Phi^{-1} \mathcal{L}_{HS}^{r\times 1} = \begin{pmatrix} F \\ I \end{pmatrix} \mathcal{S}_{\mathcal{T}_-} \subset \mathcal{S}_{\mathcal{T}_e} = \theta \begin{pmatrix} \mathcal{L}_{HS}^{m\times 1} \\ \mathcal{L}_{HS}^{r\times 1} \end{pmatrix}, \tag{9.6}$$

and therefore

$$\theta^{-1} \begin{pmatrix} F \\ I \end{pmatrix} \Phi^{-1} \mathcal{L}_{HS}^{r\times 1} \subset \begin{pmatrix} \mathcal{L}_{HS}^{m\times 1} \\ \mathcal{L}_{HS}^{r\times 1} \end{pmatrix}. \tag{9.7}$$

Note that $\Phi, F$ and $\theta$ are input-output operators, and thus we can apply Proposition 8.2 to show that

$$\theta^{-1} \begin{pmatrix} F \\ I \end{pmatrix} \Phi^{-1} \in \mathcal{R}^{(m+r)\times r}. \tag{9.8}$$

Together formulas (9.7) and (9.8) show that there exist $G_1 \in \mathcal{LR}^{m\times r}$ and $G_2 \in \mathcal{LR}^{r\times r}$ such that

$$\begin{pmatrix} F \\ I \end{pmatrix} \Phi^{-1} = \theta \begin{pmatrix} G_1 \\ G_2 \end{pmatrix}. \tag{9.9}$$

Part (b1). In this part we show that $G_2$ is invertible in $\mathcal{B}^{r\times r}$ (and hence, by Proposition 8.2(ii), we have $G_2^{-1} \in \mathcal{R}^{r\times r}$). From $\theta^* J\theta = J$ and (9.9) we get

$$J\theta^* J \begin{pmatrix} F \\ I \end{pmatrix} \Phi^{-1} = \begin{pmatrix} G_1 \\ G_2 \end{pmatrix},$$

and hence

$$G_2 = (-\theta_{12}^* F + \theta_{22}^*)\Phi^{-1}. \tag{9.10}$$

Now, by (9.2), the operator $\theta^*$ is also $J$-unitary. Thus, Theorem 2.1 in [BGK1] implies that $\theta_{22}^*$ is invertible and $\|\theta_{22}^{-*}\theta_{12}^*\| < 1$. Here $A^{-*}$ stands for $(A^{-1})^*$. Since $\|F\| \le 1$, we have $\|\theta_{22}^{-*}\theta_{12}^* F\| < 1$, and hence $-\theta_{22}^{-*}\theta_{12}^* F + I$ is invertible. But then the same holds for $-\theta_{12}^* F + \theta_{22}^*$, and thus $G_2$ is invertible.

Part (b2). Set $G = G_1 G_2^{-1}$. In this part we show that $\|G\| \le 1$. From $\|F\| \le 1$, formula (9.9), and the first identity in (9.2) we see that

$$\begin{aligned} 0 \ge F^*F - I = \begin{pmatrix} F^* & I \end{pmatrix} J \begin{pmatrix} F \\ I \end{pmatrix} &= \Phi^* \begin{pmatrix} G_1^* & G_2^* \end{pmatrix} \theta^* J \theta \begin{pmatrix} G_1 \\ G_2 \end{pmatrix} \Phi \\ &= \Phi^* \begin{pmatrix} G_1^* & G_2^* \end{pmatrix} J \begin{pmatrix} G_1 \\ G_2 \end{pmatrix} \Phi \\ &= \Phi^*(G_1^* G_1 - G_2^* G_2)\Phi = \Phi^* G_2^*(G^*G - I)G_2\Phi. \end{aligned}$$

Since $G_2\Phi$ is invertible, we conclude that $G^*G - I \leq 0$, and hence $\|G\| \leq 1$.

Part (b3). In this part we show that $G_2^{-1} \in \mathcal{L}^{r\times r}$ (and hence, by Proposition 8.2, we have $G = G_1G_2^{-1} \in \mathcal{LR}^{m\times r}$). From (9.9) we see that $\Phi^{-1} = \theta_{21}G_1 + \theta_{22}G_2$, and thus

$$I = (\Phi\theta_{22})\{\theta_{22}^{-1}\theta_{21}G + I\}G_2. \tag{9.11}$$

The three terms in the right hand side of (9.11) are all in $\mathcal{R}^{r\times r}$ and are invertible operators on $\ell_2^r(\mathbb{Z})$.

Now, let $T$ be an arbitrary operator from $\mathcal{R}^{r\times r}$. For each $\nu \in \mathbb{Z}$, consider the following partitioning of $T$

$$T = \begin{pmatrix} T_{11}^{(\nu)} & T_{12}^{(\nu)} \\ T_{21}^{(\nu)} & T_{22}^{(\nu)} \end{pmatrix} : \ell_2^r((-\infty,\nu-1]) \oplus \ell_2^r([\nu,\infty)) \to \ell_2^r((-\infty,\nu-1]) \oplus \ell_2^r([\nu,\infty)). \tag{9.12}$$

Here $\ell_2^r([\nu,\infty))$ is the subspace of $\ell_2^r(\mathbb{Z})$ consisting of all sequences $(x_i)_{i=-\infty}^{\infty}$ in $\ell_2^r(\mathbb{Z})$ such that $x_i = 0$ for $i < \nu$, and $\ell_2^r((-\infty,\nu-1])$ is the orthogonal complement of $\ell_2^r([\nu,\infty))$ in $\ell_2^r(\mathbb{Z})$. The fact that $T \in \mathcal{R}^{r\times r}$ implies that $T_{21}^{(\nu)}$ and $T_{12}^{(\nu)}$ are finite rank operators (see Theorem 8.1). Thus, if $T \in \mathcal{R}^{r\times r}$ and $T$ is invertible, then the operators $T_{11}^{(\nu)}$ and $T_{22}^{(\nu)}$ are Fredholm operators. Now, assume that $T \in \mathcal{LR}^{r\times r}$ and $T$ is invertible. Then $T_{22}^{(\nu)}$ has a left inverse and $T_{11}^{(\nu)}$ has a right inverse. In particular, ind $T_{22}^{(\nu)} \leq 0$ and ind $T_{11}^{(\nu)} \geq 0$. Furthermore, if in this case ind $T_{22}^{(\nu)} = 0$, then ind $T_{11}^{(\nu)} = 0$ and both $T_{22}^{(\nu)}$ and $T_{11}^{(\nu)}$ are invertible operators, which implies that $(T^{-1})_{12}^{(\nu)} = 0$. Thus, if $T \in \mathcal{LR}^{r\times r}$, $T$ is invertible and ind $T_{22}^{(\nu)} = 0$ for each $\nu \in \mathbb{Z}$, then $T^{-1} \in \mathcal{LR}^{r\times r}$.

Let us return to (9.11), and let us partition each of the terms in (9.11) according to the partitioning in (9.12). Since each of the three terms in the right hand side of (9.11) belongs to $\mathcal{R}^{r\times r}$ and is invertible, the operators $(\Phi\theta_{22})_{22}^{(\nu)}$, $(\theta_{22}^{-1}\theta_{21}G + I)_{22}^{(\nu)}$ and $(G_2)_{22}^{(\nu)}$ are Fredholm operators. Furthermore, by (9.11), the product

$$(\Phi\theta_{22})_{22}^{(\nu)}\{(\theta_{22}^{-1}\theta_{21}G + I)_{22}^{(\nu)}\}(G_2)_{22}^{(\nu)}$$

is equal to the identity operator on $\ell_2^r([\nu,\infty))$ modulo an operator of finite rank. It follows that

$$\text{ind } (\Phi\theta_{22})_{22}^{(\nu)} + \text{ind } (\theta_{22}^{-1}\theta_{21}G + I)_{22}^{(\nu)} + \text{ind } (G_2)_{22}^{(\nu)} = 0. \tag{9.13}$$

Note that (because of (9.1), Proposition 4.2 and Proposition 4.3(ii))

$$\theta_{22}\mathcal{L}_{HS}^{r\times 1} \subset (\,\theta_{22} \quad \theta_{21}\,)\begin{pmatrix} \mathcal{L}_{HS}^{m\times 1} \\ \mathcal{L}_{HS}^{r\times 1} \end{pmatrix} = \Phi^{-1}\mathcal{L}_{HS}^{r\times 1}.$$

Thus

$$\Phi\theta_{22}\mathcal{L}_{HS}^{r\times 1} \subset \mathcal{L}_{HS}^{r\times 1},$$

which implies that $\Phi\theta_{22} \in \mathcal{LR}^{r\times r}$. Also, $G_2 \in \mathcal{LR}^{r\times r}$. Therefore,

$$\text{(9.14)} \qquad \operatorname{ind}\,(\Phi\theta_{22})_{22}^{(\nu)} \le 0, \quad \operatorname{ind}\,(G_2)_{22}^{(\nu)} \le 0.$$

Since

$$\|(\theta_{22}^{-1}\theta_{22}G)_{22}^{(\nu)}\| \le \|\theta_{22}^{-1}\theta_{21}G\| < 1,$$

the operator $(\theta_{22}^{-1}\theta_{22}G + I)_{22}^{(\nu)}$ is invertible, and thus

$$\text{(9.15)} \qquad \operatorname{ind}\,(\theta_{22}^{-1}\theta_{22}G + I)_{22}^{(\nu)} = 0.$$

By combining (9.13) - (9.15), we see that ind $(G_2)_{22}^{(\nu)} = 0$. This holds for each $\nu$, and hence $G_2^{-1} \in \mathcal{L}^{r\times r}$. We conclude that $G := G_1G_2^{-1} \in \mathcal{LR}^{m\times r}$ and, because of (9.9), the interpolant is given by (9.4).

Part (b4). In this part we show that condition (9.3) is fulfilled. Consider the operator $\Phi\theta_{22}$. We know that $\Phi\theta_{22} \in \mathcal{R}^{r\times r}$ and is invertible. In the previous part we have shown that $\Phi\theta_{22} \in \mathcal{LR}^{r\times r}$ and ind $(\Phi\theta_{22})_{22}^{(\nu)} = 0$ for each $\nu \in \mathbb{Z}$. Thus both $\Phi\theta_{22}$ and its inverse are lower triangular. It follows that $\Phi\theta_{22}\mathcal{L}_{HS}^{r\times 1} = \mathcal{L}_{HS}^{r\times 1}$, and hence, by (9.9),

$$\theta_{22}\mathcal{L}_{HS}^{r\times 1} = \Phi^{-1}\mathcal{L}_{HS}^{r\times 1} = (\,\theta_{21} \quad \theta_{22}\,)\begin{pmatrix} \mathcal{L}_{HS}^{m\times 1} \\ \mathcal{L}_{HS}^{r\times 1} \end{pmatrix} = \theta_{22}\,(\,\theta_{22}^{-1}\theta_{21} \quad I\,)\begin{pmatrix} \mathcal{L}_{HS}^{m\times 1} \\ \mathcal{L}_{HS}^{r\times 1} \end{pmatrix}.$$

We conclude that

$$\mathcal{L}_{HS}^{r\times 1} = (\,\theta_{22}^{-1}\theta_{21} \quad I\,)\begin{pmatrix} \mathcal{L}_{HS}^{m\times 1} \\ \mathcal{L}_{HS}^{r\times 1} \end{pmatrix}.$$

In particular, $\theta_{22}^{-1}\theta_{21}\mathcal{L}_{HS}^{m\times 1} \subset \mathcal{L}_{HS}^{r\times 1}$, and therefore (9.3) holds. □

In a subsequent paper (see [BGK3]) we shall derive a necessary and sufficient condition for the existence of an interpolant $F \in LR^{m\times r}$ with $\|F\| \le 1$ and satisfying (IC1) - (IC3) directly in terms of the interpolation data set $\tau$. Furthermore, in this subsequent paper a $J$-unitary operator $\Theta$ of the type appearing in Theorem 9.1 will be constructed and a state space formula for such a $\Theta$ will be given.

REFERENCES

[AD] D. Alpay and P. Dewilde, Time-varying signal approximation and estimation, in: *Signal processing, scattering and operator theory, and numerical methods*, Proceedings of the international symposium MTNS-89, Vol III (eds. M.A. Kaashoek, J.H. van Schuppen and A.C.M. Ran), Birkhäuser Verlag, Boston, 1990.

[ADD] D. Alpay, P. Dewilde and H. Dym Lossles inverse scattering and reproducing kernels for upper triangular operators, in *Extension and interpolation of linear operators and matrix functions* (ed. I. Gohberg), OT 47, Birkhäuser Verlag, Basel, 1990, pp. 61-135.

[BGK1] J.A. Ball, I. Gohberg and M.A. Kaashoek, Nevanlinna-Pick interpolation for time-varying input-output maps: The discrete case, in: *Time-variant systems and interpolation* (ed. I. Gohberg), OT 56, Birkhäuser Verlag, Basel, 1992, pp. 1-51.

[BGK2] J.A. Ball, I. Gohberg and M.A. Kaashoek, Nevanlinna-Pick interpolation for time-varying input-output maps: The continuous time case, in: *Time-variant systems and interpolation* (ed. I. Gohberg), OT 56, Birkhäuser Verlag, Basel, 1992, pp. 52-89.

[BGK3] J.A. Ball, I. Gohberg and M.A. Kaashoek, Two-sided Nudelman interpolation for time-varying input-output maps: The discrete time case, in preparation.

[BeG] A. Ben-Artzi and I. Gohberg, Band matrices and dichotomy, in: *Topics in matrix and operator Theory* (eds. H. Bart, I. Gohberg, M.A. Kaashoek), OT 50, Birkhäuser Verlag, Basel, 1991; pp. 137-170.

[BeGK] A. Ben-Artzi, I. Gohberg and M.A. Kaashoek, Exponentially dominated infinite block matrices of finite Kronecker rank, Integral Equations and Operator Theory, to appear.

[CSK] T. Constantinescu, A.H. Sayed, and T. Kailath, Displacement structure and some general interpolation problems, submitted for publication.

[D] P. Dewilde, A course on the algebraic Schur and Nevanlinna-Pick interpolation problems, in: *Algorithms and paralled VLSI architectures*, Vol. A: Tutorials (eds. E.F. Deprettere and A.-J. van der Veen), Elsevier, Amsterdam, 1991.

[DD] P. Dewilde and H. Dym, Interpolation for upper triangular operators, in: *Time-variant systems and interpolation* (ed. I. Gohberg), OT 56, Birkhäuser Verlag, Basel, 1992, pp. 153-260.

[Mer] R. Mercer, Convergence of Fourier series in discrete crossed products of von Neumann algebras, Proc. A.M.S. 94 (1985), 254-258.

[S] A.H. Sayed, Displacement structure in signal processing and mathematics, Ph. D. Thesis, Department of Electrical Engineering, Information Systems Laboratory, Stanford University, Stanford CA, August 1992.

[SCK] A.H. Sayed, T. Constantinescu, and T. Kailath, Lattice structures of time-variant interpolation problems, in : *Proc. 31-st IEEE Conf. on Decision and Control*, (Tuscon, AZ), Dec. 1992.

[vV] A.-J. van der Veen, Time-varying system theory and computational modeling, Ph. D. Thesis, Department of Electrical Engineering, Delft University of Technology, The Netherlands, June 1993.

J.A. Ball

Department of Mathematics, Virginia Tech

Blacksburg, VA 24061, U.S.A.

I. Gohberg

Raymond and Beverley Sackler Faculty of Exact Sciences

School of Mathematical Sciences, Tel-Aviv University

Ramat-Aviv, Israel.

M.A. Kaashoek

Faculteit Wiskunde en Informatica, Vrije Universiteit

Amsterdam, The Netherlands.

Operator Theory:
Advances and Applications, Vol. 64



# TWO-SIDED TANGENTIAL INTERPOLATION OF REAL RATIONAL MATRIX FUNCTIONS

Joseph A. Ball,[1] Israel Gohberg, Leiba Rodman[2]

We develop results on interpolation of real rational matrix functions analogous to those proved in earlier work for rational matrix functions with complex coefficients. Other relevant problems are studied as well; one of them: decide (in terms of the interpolation data) if a given interpolation problem admits a real rational matrix functions interpolants. The basic methodology is rooted in the approach exposed in [BGR3].

## 1. INTRODUCTION

Various two-sided tangential interpolation problems for rational matrix functions (over the field of complex numbers) have been a subject of extensive research in recent years. A simple interpolation problem of this kind is of the form

$$x_j F(z_j) = y_j \quad (j = 1, \ldots, m), \tag{1.1}$$

$$F(w_k) u_k = v_k \quad (k = 1, \ldots, n), \tag{1.2}$$

where $z_1, \ldots, z_m$, $w_1, \ldots, w_n$ are distinct points in the complex plane $\mathbf{C}$; $x_1, \ldots, x_m$, $y_1, \ldots, y_m$ are given $1 \times M$ row vectors (over $\mathbf{C}$); $u_1, \ldots, u_n$, $v_1, \ldots, v_n$ are given $n \times 1$ column vectors; and $F(z)$ is analytic in $\{z_1, \ldots, z_m, w_1, \ldots, w_n\}$ and satisfies the interpolation conditions (1.1) and (1.2). A general two-sided tangential interpolation problem (which

[1]Partially supported by NSF Grant DMS-9101400.
[2]Partially supported by NSF Grant DMS-9123841 and by United States-Israel Binational Fund Grant 8800304.

is formally stated as the (TSCII) problem in Section 4) allows some points $z_j$ and $w_k$ to coalesce; it may involve several conditions of the form (1.1) and (1.2) at each point $z_j$ and $w_k$, as well as higher order conditions involving the derivatives of $F(z)$ at $z_j$ and $w_k$, and simultaneous two-sided interpolation conditions (a simple form of which is $x_{j_0}F'(z_{j_0})u_{k_0} = q_{j_0k_0}$ provided $z_{j_0} = w_{k_0}$ for some indices $j_0$ and $k_0$). Such general interpolation problems may also have norm restrictions of the form $\sup_{z\in\Omega} \|F(z)\| < 1$, where $\Omega \subset \mathbf{C}$ is a suitable domain, imposed on the unknown function $F(z)$. A general two-sided tangential interpolation problem with norm restriction represents a far-reaching matrix generalization of the classical Nevanlinna-Pick interpolation problem for scalar functions.

Important applications in modern engineering control (notably, the $H^\infty$ control) provided much of the impetus for the expansion of research in this area in recent years. The theory of interpolation for rational matrix functions, as well as several important applications, is presented in [BGR3] in a self-contained and systematic way; and see [ABKW, BGR5, BKRV, BRa2, ABGR] for some of the subsequent developments that are not reflected in the book [BRG3]. The engineering control applications of rational matrix function interpolation are rather scattered in the literature; we mention here only the papers [BRa1, BRa2, Ki, Gl1, Gl2, VK, BGR7, BGR11] on this topic.

In this paper we study the general two-sided tangential interpolation problems with and without the norm constraints for *real* rational matrix functions $W(z)$, i.e., such that every entry in $W(z)$ is a scalar rational function with real coefficients. This is in contrast to the previous work which almost exclusively focused on rational matrix functions interpolants without regard to the special structure and properties imposed by the function being real. It should be noted that rational matrix functions in many engineering control applications are real.

The basic methodology in this paper is the same as in [BGR1, BGR2, BGR3, ABKW], and we also use the results of [BGR4]. Moreover, in the real case completely new problems appear; for example: decide (in terms of $z_j, x_j, y_j, w_k, u_k, v_k$) if the interpolation problem (1.1), (1.2) admits a real interpolant $F(z)$. We solve this problem (in the framework in general two-sided tangential interplation) as well (Theorems 4.1 and 4.1′).

The solutions of a two-sided tangential interpolation problem for real rational matrix functions are typically given by a linear fractional formula (as in Theorem 5.1, for example). The matrix of coefficients (in Theorem 5.1, this is $\Theta$) of such linear fractional formula is obtained as a solution of an interpolation problem of another type, which we call *homogeneous* interpolation problems. Generally speaking, an homogeneous interpolation problem seeks to construct rational matrix functions (if such exist) when only a part of its null and/or pole structure is prescribed. Many homogeneous interpolation problems and their solutions for rational matrix functions over $\mathbf{C}$ are exposed in [BGR3]. We solve in this paper several homogeneous interpolation problems for *real* rational matrix functions. This material, together with general information on null and pole structure (in the form that is needed later), is presented in Section 3. The main results of the paper are stated and proved in Sections 4-6. These results describe existence of real rational matrix functions interpolants (Theorem 4.1$'$), describe all such interpolants in case the interpolation data are given in terms of real matrices (Theorem 5.1), and identify the possible McMillan degrees of real interpolants (Theorem 6.4). Finally, in the last section we present some results concerning interpolation of real rational matrix functions with norm restrictions. As it turns out, these results follow immediately from the corresponding results on complex interpolation.

We follow the book [BGR3] in terminology and notation. Thus, $\text{Res}_{z=z_0}V(z)$ stands for the residue (= coefficient of $(z-z_0)^{-1}$ in the Laurent series centered at $z_0$) of the vector or matrix rational matrix function $V(z)$. The vector space (over $\mathbf{C}$) of $M\times N$ rational matrix functions without poles in a fixed set $\sigma\subseteq\mathbf{C}\cup\{\infty\}$ is denoted $\mathcal{R}_{M\times N}(\sigma)$. For a (constant) $M\times N$ matrix $X$, we let

$$\text{Ker}X=\{x\in\mathbf{C}^N:\ Xx=0\};\ \text{Im}X=\{Xx\in\mathbf{C}^M:\ x\in\mathbf{C}^N\};$$

here $\mathbf{C}$ is replaced by the field of real numbers $\mathbf{R}$ if $X$ is real. The diagonal matrix with $a_1,\dots,a_n$ on the main diagonal is denoted $\text{diag}(a_1,\dots,a_n)$. For a rational matrix function $V(z)=[v_{ij}(z)]$ we let $\bar{V}(z)=[\bar{v}_{ij}(z)]$, where $\bar{v}_{ij}(z)$ stands for a scalar rational function the coefficients of which are complex conjugates of the corresponding coefficients of $v_{ij}(z)$.

## 2. MINIMAL REALIZATIONS

Let $F(z)$ be $n \times n$ matrix with entries of which are rational functions (with complex coefficients) in the complex variable $z$ (in short, $F(z)$ is a *rational matrix function*). If $F(z)$ is a real matrix for every $z \in \mathbf{R}$ which is not a pole of $F(z)$, then we say that $F(z)$ is a *real* rational matrix function. Equivalently, $F(z)$ is called a real rational matrix function if all its entries are rational functions with real coefficients.

In this section we focus on real rational matrix functions $F(z)$ which are *regular* (i.e., det $F(z) \not\equiv 0$).

Throughout the paper minimal realizations of real rational matrix functions $W(z)$ will play a crucial role. Recall that, under the assumption that $W(z)$ has no pole at infinity, a realization of $W(z)$ is representation in the form

$$W(z) = D + C(zI - A)^{-1}B \tag{2.1}$$

with real matrices $A, B, C, D$; it is called *minimal* if the size of the matrix $A$ is minimal possible among all realizations of $W(z)$. The basic and well-known results on minimal realizations are summarized below.

THEOREM 2.1. (a) *Every real rational $n \times n$ matrix function $W(z)$ which is finite at infinity admits a realization.*

(b) *A realization* (2.1) *is minimal if and only if $(A, B)$ is a full range, or controllable, pair (i.e., the rows of*

$$[B, AB, \dots, A^{p-1}B]$$

*are linearly independent for sufficiently large integer $p$), and $(C, A)$ is a null kernel, or observable, pair (i.e., $\bigcap_{i=0}^{\infty} Ker(CA^i) = \{0\}$).*

(c) *Every realization* (2.1) *of $W(z)$ is a dilation of a minimal realization, i.e., in a suitable choice of bases in the involved vector spaces,*

$$A \begin{bmatrix} * & * & * \\ 0 & A_0 & * \\ 0 & 0 & * \end{bmatrix}, \quad B = \begin{bmatrix} * \\ B_0 \\ 0 \end{bmatrix}, \quad C = [0 \ C_0 \ *], \tag{2.2}$$

*where*

$$W(z) = D + C_0(zI - A_0)^{-1}B_0$$

*is a minimal realization (the stars in* (2.2) *stand for matrix block with no immediate interest).*

The proof of Theorem 2.1 is given exactly as in the complex case (see, e.g., [GLR1]) and therefore is omitted.

## 3. LOCAL DATA

We characterize the local data (left and right null and pole pairs) of real rational matrix functions. We adopt the standard definitions given in [BGR3]; in particular, we allow complex zeros and poles and complex null and pole functions for real rational matrix functions.

PROPOSITION 3.1. *Let $W(z)$ be a real rational matrix function. If $u(z)$ is a left (right) null (pole) function of $W(z)$ corresponding to its zero (pole) function of $W(z)$ corresponding to its zero (pole), $z_0$, of order $r$, then $\overline{u(\bar{z})}$ is a left (right) null (pole) function of $W(z)$ corresponding to its zero (pole) $z_0$ of the same order $r$.*

*Proof.* Say $u(z)$ is a left null function of $W(z)$ at $z_0$. Then $u(z_0) \neq 0$ and $u(z)W(z)$ is analytic at $z_0$ having a zero at $z_0$ of order $r$. Upon taking complex conjugates and using $\overline{W(\bar{z})} = W(z)$ the conclusion follows. □

The next result characterizes the real rational matrix functions in terms of the matrices associated with its local data. We recall the basic definitions: Given a (complex) rational $n \times n$ matrix function $W(z)$, and a nonempty set $\Omega \subseteq \mathbf{C}$, the *left null-pole subspace* $\mathcal{S}_\Omega(W)$ of $W(z)$ with respect to $\Omega$ is defined by

$$\mathcal{S}_\Omega(W) = \{W(z)f(z) : \ f(z) \in \mathcal{R}_n(\Omega)\},$$

where $\mathcal{R}_n(\Omega)$ is the $\mathbf{C}$-vector space of all $n$-dimensional column vectors whose entries are rational functions with poles off $\Omega$. A triple $(C_\pi, A_\pi; A_\zeta, B_\zeta; S)$ is called a *left null pole triple* of $W(z)$ with respect to $\Omega$ if the following properties hold:

(i) the sizes of the matrices $C_\pi$, $A_\pi$, $A_\zeta$, $B_\zeta$, and $S$ are $n \times n_\pi$, $n_\pi \times n_\pi$, $n_\zeta \times n_\zeta$, $n_\zeta \times n$, $n_\zeta \times n_\pi$, respectively, for some $n_\pi$ and $n_\zeta$ (the case when one or both of $n_\pi$ and $n_\zeta$ are zeros is not excluded);

(ii) all the eigenvalues of $A_\pi$ and $A_\zeta$ are in $\Omega$;

(iii) the pair $(C_\pi, A_\pi)$ is a *null kernel* pair:

$$\bigcap_{j=0}^{\infty} \mathrm{Ker}(C_\pi, A_\pi^j) = \{0\};$$

(iv) the pair $(A_\zeta, B_\zeta)$ is a *full range* pair:

$$\sum_{j=0}^{\infty} \mathrm{Im}(A_\zeta^j B_\zeta) = \mathbf{C}^{n_\zeta};$$

(v) the Sylvester equation

$$SA_\pi - A_\zeta S = B_\zeta C_\pi$$

is satisfied;

(vi) $\mathcal{S}_\Omega(W) = \{C_\pi(zI - A_\pi)^{-1}x + h(z) : \ x \in \mathbf{C}^{n_\pi}, \ h(z) \in \mathcal{R}_n(\Omega)$ are such that

$$\sum_{z_0 \in \Omega} \mathrm{Res}_{z=z_0}(zI - A_\zeta)^{-1} B_\zeta h(z) = Sx\}. \tag{3.1}$$

Using the Moebius transformation, the concept of a left null-pole triple of $W(z)$ can be extended to sets $\Omega \subset \mathbf{C} \cup \{\infty\}$. A right null-pole triple $(\tilde{A}_\pi, \tilde{B}_\pi; \tilde{C}_\zeta, \tilde{A}_\zeta; \tilde{S})$ of $W(z)$ with respect to $\Omega$ is defined analogously; note that $(\tilde{A}_\pi, \tilde{B}_\pi; \tilde{C}_\zeta, \tilde{A}_\zeta; \tilde{S})$ is a right null-pole triple of $W(z)$ if and only if $(\tilde{B}_\pi^T, \tilde{A}_\pi^T; \tilde{A}_\zeta^T, \tilde{C}_\zeta^T; \tilde{S}^T)$ is a left null-pole triple of $(W(z))^T$ (with respect to $\Omega$). The four pairs $(C_\pi, A_\pi), (A_\zeta, B_\zeta), (\tilde{A}_\pi, \tilde{B}_\pi)$ and $(\tilde{C}_\zeta, \tilde{A}_\zeta)$ that appear in the left null-pole triple and right null-pole triple of $W(z)$ with respect to $\Omega$ are known as the *right pole pair, left null pair, left pole pair* and *right null pair*, respectively, of $W(z)$ (with respect to $\Omega$). We refer the reader to the book [BGR3] for various equivalent definitions of these concepts and their properties (in particular, (3.1) is found in Chapter 12 in [BGR3]).

A set $\Omega$ in the extended complex plane $\mathbf{C} \cup \{\infty\}$ will be called *symmetric* if $\lambda_0 \in \Omega \Rightarrow \bar{\lambda}_0 \in \Omega$ (by definition, $\bar{\infty} = \infty$).

We state the following fact which will be especially useful.

PROPOSITION 3.2. *A left null-pole triple of* $W(z)$ *with respect to* $\Omega$ *is unique up to similarity: if* $(C_\pi^{(i)}, A_\pi^{(i)}; A_\zeta^{(i)}, B_\zeta^{(i)}; S^{(i)})$, $i = 1, 2$, *are left null-pole triples of* $W(z)$ *with respect* *to $\Omega$, then there exist unique invertible matrices $T_1$ and $T_2$ such that*

$$C_\pi^{(1)} = C_\pi^{(2)} T_1, \quad A_\pi^{(1)} = T_1^{-1} A_\pi^{(2)} T_1,$$
$$B_\zeta^{(1)} = T_2 B_\zeta^{(2)}, \quad A_\zeta^{(1)} = T_2 A_\zeta^{(2)} T_2^{-1}, \quad S^{(1)} = T_2 S^{(2)} T_1.$$

The next proposition will be needed later as well.

PROPOSITION 3.3. *$(C, A_\pi; A_\zeta, B; \Gamma)$ is a left null-pole triple of $W(z)$ with respect to $\Omega$ if and only if $(\bar{C}, \bar{A}_\pi; \bar{A}_\zeta, \bar{B}; \bar{\Gamma})$ is a left null-pole triple of $\bar{W}(z)$ with respect to $\bar{\Omega} = \{\bar{\lambda} : \lambda \in \Omega\}$.*

The proof of Proposition 3.3 is immediate using the definition of left null-pole triples. Results analogous to Propositions 3.2 and 3.3 hold for right null-pole triples.

The rest of this section will be devoted to the specific properties of left null-pole triples for *real* rational matrix functions.

THEOREM 3.4. (a) *Let $\Omega \subseteq \mathbf{C} \cup \{\infty\}$ be a symmetric set, and let $W(z)$ be a real rational matrix function. Then the let and right null and pole pairs and the left and right null-pole triples of $W(z)$ with respect to $\Omega$ can be chosen to consist of real matrices.*

(b) *Conversely, if a rational matrix function $W(z)$ has a left (right) null-pole triple over $\mathbf{C} \cup \{\infty\}$ that consists of real matrices, then $W(z)$ is real, up to multiplication by a constant invertible matrix on the right (left).*

*Proof.* We will assume that $\Omega \subseteq \mathbf{C}$ (the general case is easily reduced to this one by a suitable Moebius map). Let $(C, A)$ be a right pole pair of $W(z)$ over $\Omega$. Since $\Omega$ is symmetric, by Proposition 3.1 $(\bar{C}, \bar{A})$ is also a right pole pair of $W(z)$ over $\Omega$ (at this point we use the definition of a right pole pair in terms of pole functions; see Sections 3.2 in [BGR3]). By Proposition 3.2, the pairs $(C, A)$ and $(\bar{C}, \bar{A})$ are similar with the unique similarity matrix $T$:

$$C = \bar{C}T, \quad A = T^{-1}\bar{A}T.$$

Upon taking complex conjugates, we see that

$$C = \bar{C}\bar{T}^{-1}, \quad A = \bar{T}\bar{A}\bar{T}^{-1}.$$

The uniqueness of $T$ now implies $T = \bar{T}^{-1}$. It is known (see, e.g., Section 1a in [BH] or Lemma 4.6.9 in [HJ]) that such a $T$ can be written in the form $T = S\bar{S}^{-1}$, for some invertible matrix $S$. One verifies that the matrices $\bar{C}S$ and $S^{-1}\bar{A}S$ are real, and the pair $(\bar{C}S, S^{-1}\bar{A}S)$, being similar to $(\bar{C}, \bar{A})$, is also a right pole pair of $W(z)$ over $\Omega$. Analogously the statement (a) is proved for left pole pairs and left and right null pairs.

Let now $(C_\pi, A_\pi; A_\zeta, B_\zeta; S)$ be a (left) null-pole triple of $W(z)$ over $\Omega$. Without loss of generality assume $\Omega \subseteq \mathbf{C}$ (if $\infty \in \Omega$, apply a suitable Moebius transformation $z \to (kz+\ell)(rz+s)^{-1}$, where $k, \ell, r, s$ are real numbers and $ks - \ell r \neq 0$). By the already proved part of Theorem 3.4(a) we may assume that the matrices $C_\pi, A_\pi, A_\zeta$ and $B_\zeta$ are real. Recall the formula (3.1):

$$\{W(z)f(z):\ f(z) \in \mathcal{R}_n(\Omega)\} = \{C_\pi(zI - A_\pi)^{-1}x + h(z):\ x \in \mathbf{C}^{n_\pi},\ h(z) \in \mathcal{R}_n(\Omega)$$

$$\text{such that } \sum_{z_0 \in \Omega} \operatorname{Res}_{z=z_0}(zI - A_\zeta)^{-1}B_\zeta h(z) = Sx\}. \tag{3.2}$$

Since $W(z)$ is real and $\Omega$ symmetric, the left-hand side of (3.2) is invariant under the transformation $g(z) \to \overline{g(\bar{z})}$. Applying this transformation to the right-hand side of (3.2) and using the fact that $C_\pi, A_\pi, A_\zeta, B_\zeta$ are real, we find

$$\{W(z)f(z):\ f(z) \in \mathcal{R}_n(\Omega)\} = \{C_\pi(zI - A_\pi)^{-1}\bar{x} + \overline{h(\bar{z})}:\ \bar{x} \in \mathbf{C}^{n_\pi},\ \overline{h(\bar{z})} \in \mathcal{R}_n(\Omega)$$

$$\text{such that } \sum_{z_0 \in \Omega} \operatorname{Res}_{z=z_0}(zI - A_\zeta)^{-1}B_\zeta h(z) = Sx\}. \tag{3.3}$$

Taking complex conjugates in (3.2) we obtain

$$\sum_{z_0 \in \Omega} \operatorname{Res}_{z=z_0}(zI - A_\zeta)^{-1}B_\zeta\overline{h(\bar{z})} = \bar{S}\bar{x}.$$

Comparing with (3.3) we see that $(C_\pi, A_\pi; A_\zeta, B_\zeta; \bar{S})$ is also a (left) null-pole triple of $W(z)$ on $\Omega$. It follows easily from Proposition 3.2 that the coupling matrix $S$ is uniquely defined by $C_\pi, A_\pi, A_\zeta$ and $B_\zeta$; so we must have $S = \bar{S}$ as required.

For the right null-pole triples of $W(z)$ the proof is analogous.

For the proof of (b), assume again without loss of generality that $W(z)$ is analytic and invertible at infinity. Multiplying, if necessary, $W(z)$ on the right by a constant invertible matrix, we can further assume that $W(\infty) = I$. If $(C_\pi, A_\pi; A_\zeta, B_\zeta; S)$ is a (left) null-pole triple of $W(z)$ over $\mathbf{C}$ consisting of real matrices, then (see [GKLR])

$$W(z) = I + C_\pi(zI - A_\pi)^{-1}S^{-1}B_\zeta,$$

and $W(z)$ is obviously real. □

The characteristic properties of null and pole pairs and null-pole triples of real rational matrix functions are given in the following theorem.

THEOREM 3.5. *Let $\Omega \subseteq \mathbb{C}\cup\{\infty\}$ be a symmetric set, and let $W(z)$ be a real rational matrix function. Then any left null-pole triple $(C, A_\pi; A_\zeta, B; \Gamma)$ of $W(z)$ with respect to $\Omega$ is similar to its complex conjugate, i.e., there exist invertible matrices $S$ and $T$ such that*

$$\bar{C} = CS, \quad \bar{A}_\pi = S^{-1}A_\pi S, \tag{3.4}$$

$$\bar{A}_\zeta = T^{-1}A_\zeta T, \quad \bar{B} = T^{-1}B, \tag{3.5}$$

$$\bar{\Gamma} = T^{-1}\Gamma S. \tag{3.6}$$

*An analogous statement holds for the right null-pole triple of $W(z)$ with respect to $\Omega$. In particular, the left and right null and pole pairs of $W(\lambda)$ with respect to $\Omega$ are similar to their complex conjugates.*

*Proof.* By Theorem 3.4, $(C, A_\pi; A_\zeta, B; \Gamma)$ is similar to a left null-pole triple $w$ of $W(z)$ with respect to $\Omega$ having all its members real matrices. Taking complex conjugates in this similarity relation, we obtain that $(\bar{C}, \bar{A}_\pi; \bar{A}_\zeta, \bar{B}; \bar{\Gamma})$ is similar to $w$ as well. Thus, $(C, A_\pi; A_\zeta, B; \Gamma)$ is similar to its complex conjugate. □

Observe that the matrices $S$ and $T$ of Theorem 3.5 are uniquely defined by equalities (3.4) and (3.5). Indeed, let $q_\pi$ and $q_\zeta$ be positive integers such that $\bigcap_{j=0}^{q_\pi-1} \operatorname{Ker}(CA_\pi^j) = \{0\}$ and $\sum_{j=0}^{q_\zeta-1} \operatorname{Im}(A_\zeta^j B) = \mathbb{C}^{n_\zeta}$ (where $n_\zeta$ is the size of $A_\zeta$); for example, one can take $q_\zeta = n_\zeta$ and $q_\pi = n_\pi$ (the size of $A_\pi$). (If one of $A_\pi$ or $A_\zeta$ is empty, the corresponding matrix $S$ or $T$ is empty as well.) If $S$ and $T$ satisfy (3.4) and (3.5), then

$$S = \begin{bmatrix} C \\ CA_\pi \\ \vdots \\ CA_\pi^{q_\pi-1} \end{bmatrix}^{-L} \begin{bmatrix} \bar{C} \\ \bar{C}\bar{A}_\pi \\ \vdots \\ \bar{C}\bar{A}_\pi^{q_\pi-1} \end{bmatrix}, \tag{3.7}$$

$$T = \left[B, A_\zeta B, \dots, A_\zeta^{q_\zeta - 1} B\right] \left[\bar{B}, \bar{A}_\zeta \bar{B}, \dots, \bar{A}_\zeta^{q_\zeta - 1} \bar{B}\right]^{-R}, \tag{3.8}$$

where the superscripts "$-L$" and "$-R$" indicate the left and right inverse, respectively. Thus, Theorem 3.5 can be restated as follows.

THEOREM 3.5$'$. *Let $\Omega$, $W(z)$ and $(C, A_\pi; A_\zeta, B; \Gamma)$ be as in Theorem 3.4. Define the matrices $S$ and $T$ by (3.7) and (3.8). Then $S$ and $T$ do not depend on the choice of one-sided inverses in (3.7) and (3.8), both matrices $S$ and $T$ are invertible, and equalities (3.4)-(3.6) hold.*

It turns out that the converse statement to Theorem 3.5 is also true. In other words, given a set $(C, A_\pi; A_\zeta, B; \Gamma)$ which satisfies the obvious necessary conditions given by definition of a left null-pole triple, as well as additional necessary conditions which stem from Theorem 3.5, a real rational matrix function $W(z)$ can be found, with $(C, A_\pi; A_\zeta, B; \Gamma)$ as its left null-pole triple. Only left null-pole triples will be considered in the rest of this section; the corresponding statements for the right null-pole triples are left to the reader.

To formulate precisely the converse statement to Theorem 3.5, we recall the concept of a Sylvester data set. A triple $\tau = (C_\pi, A_\pi; A_\zeta, B_\zeta; S)$ consisting of two pairs of matrices $(C_\pi, A_\pi)$, $(A_\zeta, A_\zeta)$ and of a matrix $S$ is called a *left Sylvester data set* with respect to $\Omega$ if the properties (i)-(v) are valid. The positive integer $n$ is called the *base dimension* of $\tau$ and the matrix $S$ is called the *coupling matrix.* (See [BGR3] for more information on Sylvester data sets.) Similarity between two left Sylvester data sets having the same base dimensions is defined as in Proposition 3.2.

THEOREM 3.6. *Let $\tau = (C, A_\pi; A_\zeta, B; \Gamma)$ be a left Sylvester data set with respect to a symmetric set $\Omega \subseteq \mathbf{C} \cup \{\infty\}$, and assume that $\tau$ is similar to $(\bar{C}, \bar{A}_\pi; \bar{A}_\zeta, \bar{B}; \bar{\Gamma})$. Then:*

(i) *If $\Omega$ does not contain $\mathbf{R} \cup \{\infty\}$, then $\tau$ is a left null-pole triple for some real rational matrix function $W(z)$ with respect to $\Omega$; moreover, $W(z)$ can be taken to have all its poles and zeros in $\Omega \cup \{\lambda_0\}$, where $\lambda_0$ is any fixed point in $(\mathbf{R} \cup \{\infty\}) \setminus \Omega$.*

(ii) *If $\Omega = \mathbf{C} \cup \{\infty\}$, then $\tau$ is a left null-pole triple for some real rational matrix functions $W(z)$ with respect to $\Omega$ if and only if $\Gamma$ is invertible.*

It is an open question whether the part (i) of Theorem 3.6 holds in the situation when $\Omega$ contains $\mathbf{R} \cup \{\infty\}$ but $\Omega \neq \mathbf{C} \cup \{\infty\}$ (in this situation $\lambda_0$ should be replaced by a pair of conjugate complex numbers not belonging to $\Omega$). The answer to this question hinges upon a corresponding analogue (which currently is not available) of Lemma 3.7 below. We point out, however, that Theorem 3.6 will suffice for applications needed in this paper.

For the proof of Theorem 3.6(i) we need a real analogue of a construction given in [GKR1] (see also Section 4.6 in [BGR3]):

LEMMA 3.7. *Let $(C, A_\pi; A_\zeta, B; \Gamma)$ be a left Sylvester data set with respect to a symmetric set $\Omega \subseteq \mathbf{C}$ which does not contain $\mathbf{R}$. Let be given $\lambda_0 \in \mathbf{R}$, $\lambda_0 \notin \Omega$. Then there exist real matrices $C_0, A_{\pi 0}, A_{\zeta 0}, B_0$ and $\Gamma_0$ of appropriate sizes such that the eigenvalues of $A_{\pi 0}$ and $A_{\zeta 0}$ are all in $\{\lambda_0\}$, the set*

$$\tau := \left( [C\ C_0],\ \begin{bmatrix} A_\pi & 0 \\ 0 & A_{\pi 0} \end{bmatrix};\ \begin{bmatrix} A_\zeta & 0 \\ 0 & A_{\zeta 0} \end{bmatrix} \begin{bmatrix} B \\ B_0 \end{bmatrix};\ \begin{bmatrix} \Gamma & \Gamma_{12} \\ \Gamma_{21} & \Gamma_0 \end{bmatrix} \right) \tag{3.9}$$

*is a left Sylvester data set, and the matrix $\tilde{\Gamma} := \begin{bmatrix} \Gamma & \Gamma_{12} \\ \Gamma_{21} & \Gamma_0 \end{bmatrix}$ is invertible. Moreover, matrices $C_0, A_{\pi 0}, A_{\zeta 0}, B_0$ and $\Gamma_0$ as above exist with the additional property that the size of $\tilde{\Gamma}$ is $(n_\pi + n_\zeta - \text{rank}\ \Gamma) \times (n_\pi + n_\zeta - \text{rank}\ \Gamma)$.*

The proof is obtained by using the same construction as in [GKR1].

It is easy to see that, given an $n_\zeta \times n_\pi$ matrix $\Gamma$, the minimal size of an invertible matrix that contains $\Gamma$ as a submatrix is equal to $(n_\pi + n_\zeta - \text{rank}\ \Gamma) \times (n_\pi + n_\zeta - \text{rank}\ \Gamma)$. Thus, the size of the matrix $\tilde{\Gamma}$ is minimal possible among all the left Sylvester data sets having invertible coupling and matrix and having $(C, A_\pi; A_\zeta, B; \Gamma)$ as their corestriction (cf. [GKR1], Theorem 3.3 in [BKRV]). We observe also that the matrices $\Gamma_{12}$ and $\Gamma_{21}$ are uniquely determed as solutions of the Sylvester equations

$$\Gamma_{21} A_\pi - A_{\zeta 0} \Gamma_{21} = B_0 C; \quad \Gamma_{12} A_{\pi 0} - A_\zeta \Gamma_{12} = B C_0.$$

A result analogous to Lemma 3.7 with $\lambda_0$ replaced by a pair $\lambda_0 \pm i\mu_0 \notin \Omega$, where $\lambda_0$ and $\mu_0$ are real and $\mu_0 \neq 0$, is currently not known. We only observe that in this case the sizes of $A_{\pi 0}$ and $A_{\zeta 0}$ are necessarily even, and therefore a left Sylvester data set $\tau$ as in (3.9), with $n_\pi + n_\zeta - \text{rank}\ \Gamma$ the size of $\tilde{\Gamma}$, may not exist (if, for example, one of the numbers $n_\pi - \text{rank}\ \Gamma$ or $n_\zeta - \text{rank}\ \Gamma$ is odd).

The left Sylvester data set $\tau$ given by (3.9) will be called a *completion* of $(C, A_\pi;$ $A_\zeta, B; \Gamma)$ (with respect to $\{\lambda_0\}$).

*Proof of Theorem 3.6.* For the part (ii) observe that one such $W(z)$ is given by the formula ([GKLR], Theorem 4.43 in [BGR3])

$$W(z) = I + C(zI - A_\pi)^{-1}\Gamma^{-1}B.$$

Taking complex conjugates we obtain

$$\overline{W(\bar{z})} = I + \bar{C}(zI - \bar{A}_\pi)^{-1}\bar{\Gamma}^{-1}\bar{B}.$$

But equalities (3.4)-(3.6) hold. Using these equalities, it is easy to see that $\overline{W(\bar{z})} = W(z)$, i.e., $W(z)$ is real.

For the part (i), without loss of generality we can assume that $\Omega \subseteq \mathbf{C}$. We can (and will) further assume that $\lambda_0 \in \mathbf{R}$ (otherwise, apply a Moebius transformation $z \hookrightarrow f(z)$ which maps $\lambda_0$ to $\mathbf{R}$ and none of the eigenvalues of $A_\pi$ and $A_\zeta$ are mapped to infinity). Since $\tau$ is similar to its complex conjugate, we have

$$C = \bar{C}S, \quad A_\pi = S^{-1}\bar{A}_\pi S,$$

$$B = T^{-1}\bar{B}, \quad A_\zeta = T^{-1}\bar{A}_\zeta T, \quad \Gamma = T^{-1}\bar{\Gamma}S, \tag{3.10}$$

for some invertible complex matrices $S$ and $T$. Taking complex conjugates in these equalities we see that the same equalities are valid with $S$ replaced by $\bar{S}^{-1}$ and $T$ replaced by $\bar{T}^{-1}$. Since the pair $(C, A_\pi)$ is a null kernel pair, the invertible matrix $S$ satisfying (3.10) is unique, and therefore we must have $\bar{S}^{-1} = S$. Analogously, $\bar{T}^{-1} = T$. Factor: $S = S_1\bar{S}_1$, $T = T_1\bar{T}^{-1}$. A straightforward verification shows that all the matrices $CS_1, S_1^{-1}A_\pi S_1, T_1^{-1}A_\zeta T_1, T_1^{-1}B, T_1^{-1}\Gamma S$, are real and the left Sylvester data set $\tau' = (CS_1, S_1^{-1}A_\pi S_1; T_1^{-1}A_\zeta T_1, T_1^{-1}B; T_1^{-1}\Gamma S_1)$ is similar to $\tau$. By Proposition 3.2 we may use $\tau'$ rather than $\tau$. Let $\tau''$ be a completion of $\tau'$ (with respect to $\lambda_0$). By the already proved part (ii) there exists a rational matrix function $W(z)$ having $\tau''$ as its left null-pole triple with respect to $\Omega$. Clearly $W(z)$ satisfies the requirements of part (i). □

Results analogous to Theorem 3.6 are valid for right Sylvester data, and for each of the left and right null and pole pairs. For example:

THEOREM 3.8. *Let $(A_\zeta, B)$ be a full range pair, and let $\Omega \subseteq \mathbf{C} \cup \{\infty\}$ be a symmetric set containing $\sigma(A_\zeta)$. Then $(A_\zeta, B)$ is a left null pair of a real rational matrix function with respect to $\Omega$ if and only if $(A_\zeta, B)$ is similar to $(\bar{A}_\zeta, \bar{B})$. Moreover, in this case a real rational matrix function $W(z)$ exists such that $(A_\zeta, B)$ is a left null pair of $W(z)$ with respect to $\mathbf{C} \cup \{\infty\}$ and $W(z)$ has all its poles in one preselected real point (or infinity) $\lambda_0 \notin \sigma(A_\zeta)$.*

*Proof.* The "only if" part follows from Theorem 3.6. For the "if" part we can assume, arguing as in the proof of Theorem 3.6, that $A_\zeta$ and $B$ are in fact real matrices.

Using, if necessary, a Moebius change of variables, we can further assume that $\lambda_0 = \infty$. Then a construction given in [GLeR] produces a matrix polynomial with left null pair $(A_\zeta, B)$ (with respect to $\mathbf{C}$) and no zeros at infinity. (A complete description of such polynomials is given in [GKR2]). □

## 4. TWO-SIDED TANGENTIAL INTERPOLATION: EXISTENCE OF REAL INTERPOLANTS

In this section we study real solutions (interpolants) of the two-sided tangential interpolation problem studied in [BGR2,BGR3] for (complex) rational matrix functions. We present here this problem in the standard form involving residues.

The statement of the Two-Sided Contour Integral Interpolation problem for (complex) rational matrix functions, as stated and studied in [BGR2,BGR3], is the following:

(TSCII) *Given (complex) matrices $A_\zeta, B_+, B_-, C_+, C_-, A_\pi, S$ of sizes $n_\zeta \times n_\zeta$, $n_\zeta \times M$, $n_\zeta \times N$, $M \times n_\pi$, $N \times n_\pi$, $n_\pi \times n_\pi$, and $n_\zeta \times n_\pi$, respectively, and given a set $\sigma \subseteq \mathbf{C}$ such that $\sigma(A_\zeta) \cup \sigma(A_\pi) \subseteq \sigma$. Find a rational $M \times N$ matrix function $W(z)$ having no poles in $\sigma$ and which satisfies the interpolation conditions*

$$\sum_{z_0 \in \sigma} \operatorname{Res}_{z=z_0} (zI - A_\zeta)^{-1} B_+ W(z) = -B_-; \tag{4.1}$$

$$\sum_{z_0 \in \sigma} \operatorname{Res}_{z=z_0} W(z) C_-(zI - A_\pi)^{-1} = C_+; \tag{4.2}$$

$$\sum_{z_0 \in \sigma} \operatorname{Res}_{z=z_0} (zI - A_\zeta)^{-1} B_+ W(z) C_-(zI - A_\pi)^{-1} = S. \tag{4.3}$$

Alternatively, the conditions (4.1)-(4.3) can be formulated in terms of contour integrals (hence the appearance of Contour Integral in TSCII); for example, (4.1) can be written in the form

$$\frac{1}{2\pi i}\int_{\Gamma}(zI-A_{\zeta})^{-1}B_{+}W(z)dz=-B_{-},$$

where $\Gamma$ is a suitable contour symmetric relative to the real axis and such that all the poles of $W(z)$ are outside $\Gamma$ and all the eigenvalues of $A_\pi$ and $A_\zeta$ are inside $\Gamma$.

As shown in [BGR2], we can (and will) assume without loss of generality that the given data set $(C_+, C_-, A_\pi; A_\zeta, B_+, B_-; S)$ is *$\sigma$-admissible Sylvester*, i.e.:

(i) $(C_-, A_\pi)$ is a null kernel pair;

(ii) $(A_\zeta, B_+)$ is a full range pair;

(iii) $\sigma(A_\pi)\cup\sigma(A_\zeta)\subseteq\sigma$;

(iv) $SA_\pi - A_\zeta S = B_+C_+ + B_-C_-$.

First, we characterize the TSCII problems that admit an interpolant $W(z)$ which is a *real* rational function (on top of the other conditions on $W(z)$ specified in (TSCII)).

Assume that $W(z)$ is analytic on $\sigma$, satisfies $\overline{W(\bar{z}))} = W(z)$ and satisfies (4.1)-(4.3). Using the obvious equality $\mathrm{Res}_{z=\bar{z}_0}\overline{f(\bar{z})} = \overline{\mathrm{Res}_{z=z_0} f(z)}$ for any rational matrix function $f(z)$, we easily obtain that $W(z)$ satisfies also the equalities

$$\sum_{z_0\in\bar{\sigma}}\mathrm{Res}_{z=z_0}(zI-\bar{A}_\zeta)^{-1}\bar{B}_+W(z) = -\bar{B}_-; \tag{4.4}$$

$$\sum_{z_0\in\bar{\sigma}}\mathrm{Res}_{z=z_0}W(z)\bar{C}_-(zI-\bar{A}_\pi)^{-1} = \bar{C}_+; \tag{4.5}$$

$$\sum_{z_0\in\bar{\sigma}}\mathrm{Res}_{z=z_0}(zI-\bar{A}_\zeta)^{-1}\bar{B}_+W(z)\bar{C}_-(zI-\bar{A}_\pi)^{-1} = \bar{S}, \tag{4.6}$$

where $\bar{\sigma} = \{\bar{z} : z\in\sigma\}$.

Observe that the interpolation data set $(\bar{C}_+, \bar{C}_-, \bar{A}_\pi; \bar{A}_\zeta, \bar{B}_+, \bar{B}_-; \bar{S})$ is $\bar{\sigma}$-admissible Sylvester. An obvious necessary condition, then, for existence of a real interpolant $W(z)$ for (TSCII) is that all six interpolation inequalities (4.1)-(4.6) be consistent. This condition turns out to be also sufficient, as expressed in the next theorem.

Given two left Sylvester data sets $\tau_i = (C^{(i)}, A_\pi^{(i)}; A_\zeta^{(i)}, B^{(i)}; \Gamma^{(i)})$ $(i = 1, 2)$ having the same base dimension, we say that $\tau_1$ is a *coextension* of $\tau_2$ (or $\tau_2$ is a *corestriction* of $\tau_1$) if there exist an injective linear map $\Phi$ and a surjective linear map $\Psi$ such that the following equalities hold:

$$C^{(1)}\Phi = C^{(2)}; \quad A_{\pi}^{(1)}\Phi = \Phi A_{\pi}^{(2)}; \tag{4.7}$$

$$A_{\zeta}^{(2)}\Psi = \Psi A_{\zeta}^{(1)}; \quad B^{(2)} = \Psi B^{(1)}; \tag{4.8}$$

$$\Gamma^{(2)} = \Psi\Gamma^{(1)}\Phi.$$

The concept of a *common coextension* of left Sylvester data sets $\tau_1, \ldots, \tau_p$ is now defined naturally. A common coextension $\tau$ of $\tau_1, \ldots, \tau_p$ is called a *least common coextension* of $\tau_1, \ldots, \tau_p$ if any other common coextension of $\tau_1, \ldots, \tau_p$ is in turn a coextension of $\tau$. See [BGR4,BGR5] for more information and applications of common coextensions and corestrictions of left Sylvester data sets.

If equalities (4.7) are satisfied (where $\Phi$, as before, is an injective linear map), we say that the pair $(C^{(1)}, A_{\pi}^{(1)})$ is an *extension* of $(C^{(2)}, A_{\pi}^{(2)})$ with the associated map $\Phi$, while the equalities (4.8) mean that $(A_{\zeta}^{(1)}, B^{(1)})$ is a *coextension* of $(A_{\zeta}^{(2)}, B^{(2)})$, with the associated map $\Psi$. In contrast with left Sylvester data sets, a least common extension of null kernel pairs $(C^{(i)}, A^{(i)})$ $(i = 1, \ldots, p)$ having the same base dimension exists always and is unique up to similarity (see [GKRo], [GLR2]). Analogous statement is valid for coextensions of full range pairs.

THEOREM 4.1. *Let $\sigma \subseteq \mathbf{C}$ and a $\sigma$-admissible Sylvester interpolation data set*

$$w = (C_+, C_-, A_{\pi}; A_{\zeta}, B_+, B_-; S)$$

*be given. Then a necessary and sufficient condition for existence of a real $M \times N$ rational matrix function $W(z)$ without poles in $\sigma$ and satisfying (4.1)-(4.3) is that there exists a common coextension*

$$\left( \begin{bmatrix} C_{e+} \\ C_{e-} \end{bmatrix}, A_{e\pi}; A_{e\zeta}, [B_{e+} B_{e-}]; S_e \right) \tag{4.9}$$

*of the left Sylvester data sets*

$$\left( \begin{bmatrix} C_+ \\ C_- \end{bmatrix}, A_{\pi}; A_{\zeta}, [B_+ B_-]; S \right) \tag{4.10}$$

$$\left(\begin{bmatrix} \bar{C}_+ \\ \bar{C}_- \end{bmatrix}, \bar{A}_\pi; \bar{A}_\zeta, [\bar{B}_+ \bar{B}_-]; \bar{S}\right) \tag{4.11}$$

*with the additional properties that* $(C_{e-}, A_{e\pi})$ *is a null kernel pair and* $(A_{e\zeta}, B_{e+})$ *is a full range pair.*

*Proof.* The necessity of this condition follows from the remark made before Theorem 4.1 and from Theorem 4.3 of [BGR4].

To prove sufficiency, observe that by Theorem 4.3 of [BGR4] there is a (complex) rational matrix function $W(z)$ without poles in $\sigma \cup \bar{\sigma}$ and satisfying the equalities (4.1)-(4.6). It is easy to see that $W_1(z) = \overline{W(\bar{z})}$ satisfies these conditions as well. The rational matrix function $\frac{1}{2}\left(W(z) + \overline{W(\bar{z})}\right)$ is real and also satisfies the interpolation conditions (4.1)-(4.6). □

Using Theorem 5.2 in [BGR5], the conditions in Theorem 4.1 can be expressed in more detail:

THEOREM 4.1′. *Let* $\sigma$ *and* $\omega$ *be as in Theorem* 4.1. *Let* $\left(\begin{bmatrix} C_{+0} \\ C_{-0} \end{bmatrix}, A_{\pi 0}\right)$ *be a least common extension of* $\left(\begin{bmatrix} C_+ \\ C_- \end{bmatrix}, A_\pi\right)$ *and of* $\left(\begin{bmatrix} \bar{C}_+ \\ \bar{C}_- \end{bmatrix}, \bar{A}_\pi\right)$, *with the corresponding associated injective maps* $\Phi_1$ *and* $\Phi_2$, *respectively, and let* $(A_{\zeta 0}, [B_{+0}, B_{-0}])$ *be a least common coextension of* $(A_\zeta, [B_+ B_-])$ *and of* $(\bar{A}_\zeta, [\bar{B}_+ \bar{B}_-])$ *with the corresponding associated surjective maps* $\Psi_1$ *and* $\Psi_2$, *respectively. Then there exists a real rational matrix function without poles in* $\sigma$ *and satisfying* (4.1)-(4.3) *if and only if the following conditions are satisfied:*

(i) $(C_{-0}, A_{\pi 0})$ *is a null kernel pair;*

(ii) $(A_{\zeta 0}, B_{+0})$ *is a full range pair;*

(iii) *there exists a matrix* $T$ *of appropriate size such that the equalities*

$$\Psi_1 T \Phi_1 = S, \quad \Psi_2 T \Phi_2 = \bar{S}, \tag{4.12}$$

$$\Psi_1 T \Phi_2 \bar{A}_\pi - A_\zeta \Psi_1 T \Phi_2 = [B_+ \quad B_-] \begin{bmatrix} \bar{C}_+ \\ \bar{C}_- \end{bmatrix}, \tag{4.13}$$

$$\Psi_2 T\Phi_1 A_\pi - \bar{A}_\zeta \Psi_2 T \Phi_1 = \begin{bmatrix}\bar{B}_+ & \bar{B}_-\end{bmatrix}\begin{bmatrix} C_+ \\ C_- \end{bmatrix}, \tag{4.14}$$

*are valid.*

## 5. TWO-SIDED TANGENTIAL INTERPOLATION WITH REAL-VALUED DATA: DESCRIPTION OF INTERPOLANTS

We study here the (TSCII) interpolation problem with real data:

(RTSCII) *Given real matrices,* $A_\zeta, B_+, B_-, C_+, C_-, A_\pi, S$ *of sizes* $n_\zeta \times n_\zeta$, $n_\zeta \times M$, $n_\zeta \times N$, $M \times n_\pi$, $N \times n_\pi$, $n_\pi \times n_\pi$, *and* $n_\zeta \times n_\pi$, *respectively, and given a set* $\sigma \subseteq \mathbf{C}$ *which is symmetric relative to the real line and such that* $\sigma(A_\zeta) \cup \sigma(A_\pi) \subseteq \sigma$. *Find a real rational* $M \times N$ *matrix function* $W(z)$ *having no poles in* $\sigma$ *and which satisfies the interpolation conditions*

$$\sum_{z_0 \in \sigma} Res_{z=z_0}(zI - A_\zeta)^{-1} B_+ W(z) = -B_-; \tag{5.1}$$

$$\sum_{z_0 \in \sigma} Res_{z=z_0} W(z) C_-(zI - A_\pi)^{-1} = C_+; \tag{5.2}$$

$$\sum_{z_0 \in \sigma} Res_{z=z_0}(zI - A_\zeta)^{-1} B_+ W(z) C_-(zI - A_\pi)^{-1} = S. \tag{5.3}$$

As in the previous section, we assume throughout that the interpolation data set $w = (C_+, C_-, A_\pi; A_\zeta, B_+, B_-; S)$ is $\sigma$-admissible Sylvester. Theorem 4.1 guarantees the existence of real interpolants $W(z)$.

The solution of (RTSCII) is analogous to the complex case:

THEOREM 5.1. *Let* $\sigma \subseteq \mathbf{C}$ *be a set symmetric relative to the real axis, and let* $w = (C_+, C_-, A_\pi; A_\zeta, B_-; S)$ *be a* $\sigma$*-admissible Sylvester TSCII data set consisting of real matrices. Then there exist real rational matrix functions* $W$ *in* $\mathcal{R}_{M\times N}(\sigma)$ *which are solutions of the associated (RTSCII) problem* (5.1)-(5.3). *Moreover, let* $\Theta = \begin{bmatrix} \Theta_{11} & \Theta_{12} \\ \Theta_{21} & \Theta_{22} \end{bmatrix}$ *is any real rational* $(M+N)\times(M+N)$ *matrix function having the set*

$$\tilde{w} = \left(\begin{bmatrix} C_+ \\ C_- \end{bmatrix}, A_\pi; A_\zeta, [B_+, B_-]; S\right) \tag{5.4}$$

*as a (left) null-pole triple over* $\sigma$ *and let* $\phi$ *be a real rational* $M \times M$ *matrix function having no poles in* $\sigma$ *and such that* $(C_-, A_\pi)$ *is a right pole pair of* $\phi^{-1}$ *with respect to* $\sigma$. *Then* $W \in \mathcal{R}_{M\times N}(\sigma)$ *satisfies the (RTSCII) problem* (5.1)-(5.3) *if and only if* $W(z)$ *has the following form: There exist real rational matrix functions* $Q_1 \in \mathcal{R}_{M\times N}(\sigma)$ *and* $Q_2 \in \mathcal{R}_{N\times N}(\sigma)$ *for which the function*

$$\phi(\Theta_{21}Q_1 + \Theta_{22}Q_2)$$

*has no zeros and poles in* $\sigma$, *such that*

$$W = (\Theta_{11}Q_1 + \Theta_{12}Q_2)(\Theta_{21}Q_1 + \Theta_{22}Q_2)^{-1}. \tag{5.5}$$

Observe that the existence of $\Theta$ and $\phi$ with the requisite properties is ensured by Theorems 3.6 and 3.8.

*Proof.* One direction follows immediately from the corresponding result in the complex case (Theorem 16.10.1 in [BGR3]): namely, any function given by (5.6) in a solution of the (RTSCII) problem.

We now prove that every solution of the (RTSCII) problem has the form (5.6). To this end, observe (cf. the proof of Theorem 16.10.1 in [BGR3]) that for every interpolant $W(z)$ one choice of $Q_1(z)$ and $Q_2(z)$ is given by the formula

$$\Theta^{-1}\begin{bmatrix} W \\ I \end{bmatrix}\phi^{-1} = \begin{bmatrix} Q_1 \\ Q_2 \end{bmatrix}.$$

(Observe that, given $W(z)$, the pair $Q_1(z), Q_2(z)$ is defined uniquely up to multiplication on the right by a rational matrix function without poles and zeros in $\sigma$.) This formula shows that if $W(z)$ is real, then so are $Q_1(z)$ and $Q_2(z)$. □

## 6. DEGREES OF INTERPOLANTS

We continue to study the (RTSCII) problem (5.1)-(5.3). In this section we determine the McMillan degrees $\delta(W)$ of the (real) interpolants $W(z)$. We follow the general approach developed in [K], [ABKW] adapting it to the real case.

Several preliminary results are needed, which are also of independent interest. We start with a refinement of Theorem 3.6 concerning column reduced rational matrix functions.

Let us recall the notion of column reducedness of a rational matrix function. Assume a $p \times m$ rational matrix function $W(z)$ has column rank $m$, and make a Laurent expansion at infinity:

$$W(z) = R_{-\ell} z^{\ell} + \cdots R_0 + R_1 z^{-1} + \cdots = [W_0 + W_1 z^{-1} + \cdots] D(z),$$

where

$$D(z) = \text{ diag } \left( z^{\beta_1}, \ldots, z^{\beta_m} \right),$$

and $\beta_i$ is the highest degree of $z$ occurring in the $i$-th column of $W(z)$ $(i = 1, \ldots, m)$. If the $p \times m$ matrix $W_0$ has rank $m$, we say that $W(z)$ is *column reduced* (at infinity); in this case $\beta_i$ is called the *$i$-th column index* (at infinity) of $W(z)$. The concept of column reduced matrix polynomials is well known and used in systems theory (see, e.g., [Kai]); it can be expressed in terms of Wiener-Hopf factorizations (see [GLeR, GKrS]). For rational matrix functions this concept was introduced in [ABKW].

Now let

$$\tau = (C, A_\pi; A_\zeta, B; \Gamma) \tag{6.1}$$

be a left Sylvester data set, where the matrices have sizes $M \times n_\pi$, $n_\pi \times n_\pi$, $n_\zeta \times n_\zeta$, $n_\zeta \times M$, $n_\zeta \times n_\pi$, respectively. We assume that there exists a real $M \times M$ rational matrix function for which $\tau$ is a left null-pole triple (with respect to $\mathbf{C}$). According to Theorem 3.4(a), without loss of generality we can assume that the matrices in $\tau$ are all real matrices. Choose a direct complement $\hat{X}(X)$ of Ker $\Gamma$(Im $\Gamma$) in $\mathbf{R}^{n_\pi}$ ($\mathbf{R}^{n_\zeta}$), and let

$$\rho_\pi \ (\rho_\zeta) \text{ be a projection onto Ker } \Gamma(X) \text{ along } \hat{X}(\text{Im } \Gamma) \tag{6.2}$$

The controllability indices of the pair $(\rho_\zeta A_\zeta|_X, \ \rho_\zeta B)$ can be defined in many ways. Here we introduce them through the following incoming subspaces:

$$H_0 \ := \ \text{Im } \Gamma, \tag{6.3}$$

$$H_j \ := \ \text{Im } \Gamma + \text{Im } B + \cdots + \text{Im } A_\zeta^{j-1} B, \quad j = 1, 2, \ldots. \tag{6.4}$$

We define $\kappa_1 \geq \cdots \geq \kappa_s$ by $s = \dim(H_1/H_0)$ and

$$\kappa_j = \#\{k | \dim(H_k/H_{k-1}) \geq j\} \text{ for } j = 1, \ldots, s. \tag{6.5}$$

Then $\kappa_1 \geq \cdots \geq \kappa_s$ are the *controllability indices* of $(\rho_\zeta A_\zeta|_X,\ \rho_\zeta B)$. Similarly, the observability indices of the pair $\left(C|_{\text{Ker }\Gamma}, \rho_\pi A_\pi|_{\text{Ker }\Gamma}\right)$ are defined through outgoing subspaces:

$$K_0 \ := \ \text{Ker }\Gamma, \tag{6.6}$$

$$K_j \ := \ \text{Ker }\Gamma \cap \text{Ker } C \cap \cdots \cap \text{Ker } CA_\pi^{j-1}, \quad j = 1, 2, \ldots. \tag{6.7}$$

We define $\nu_1 \geq \cdots \geq \nu_t$ by $t = \dim(K_0/K_1)$ and

$$\nu_j = \#\{\ell |\dim(K_{t-1}/K_t) \geq j\} \text{ for } j = 1, \ldots, t. \tag{6.8}$$

Then $\nu_1 \geq \nu_2 \geq \cdots \nu_t$ are the *observability indices* of the pair $\left(C|_{\text{Ker }\Gamma}, \rho_\pi A_\pi|_{\text{Ker }\Gamma}\right)$.

The following result is a refinement of Theorem 3.6 and in the complex case has been proved in [GKR1] (see also [ABKW]).

THEOREM 6.1. *Let $\tau = (C, A_\pi; A_\zeta, B; \Gamma)$ be a left Sylvester data set consisting of real matrices. Then there exists a real $M \times M$ rational matrix function $\Theta(z)$ with the following properties:*

(i) *$\tau$ is a left null-pole triple of $\Theta(z)$ with respect to $\mathbf{C}$;*

(ii) *$\Theta(z)$ is column reduced;*

(iii) *the $j$-th column index $\beta_j$ of $\Theta(z)$ is given by the formula*

$$\beta_j = \begin{cases} -\nu_j, & 1 \leq j \leq t, \\ 0, & t+1 \leq j \leq M-s, \\ \kappa_{M-j+1}, & M-s+1 \leq j \leq M, \end{cases} \tag{6.9}$$

*where $\nu_1 \geq \cdots \geq \nu_t$ are the nonzero observability indices of $(C|_{Ker\ \Gamma}, \rho_\pi A_\pi|_{Ker\ \Gamma})$, and $\kappa_1 \geq \cdots \kappa_s$ are the nonzero controllability indices of $(\rho_\zeta A_\zeta|_X, \rho_\zeta B)$.*

Here $\rho_\pi$ and $\rho_\zeta$ are defined by (6.2).

For the proof of this theorem we need a lemma. Given a nondecreasing sequence of integers $\beta = \{\beta_1 \leq \cdots \leq \beta_M\}$ we define the set $S_\beta = \{P(z)|\ P(z) = [p_{ij}(z)]$ is an $M \times M$ matrix polynomial such that det $P(z)$ is a nonzero constant, and $p_{ij}(z) \equiv 0$ if $\beta_i > \beta_j$, $p_{ij}(z)$ is a scalar polynomial of degree at most $\beta_j - \beta_i$ if $\beta_i \leq \beta_j\}$. The set $S_\beta$ is a multiplicative group. It has been introduced in the theory of Wiener-Hopf factorizations to describe the measure of uniqueness of such factorizations (see [GK]); in [FW] $S_\beta$ is referred to as a factorization group. This group also plays an important role in certain interpolation problems for matrix polynomials [GKrS]. The following proposition is well known (and is not difficult to verify):

PROPOSITION 6.2. *Let $P(z)$ be an $M \times M$ matrix polynomial without constant nonzero determinant. Then the function*

$$diag(z^{\beta_1}, \ldots, z^{\beta_M})P(z)\ diag(z^{-\beta_1}, \ldots, z^{-\beta_M})$$

*is analytic and invertible at infinity if and only if $P(z) \in S_\beta$.*

In the next lemma we use the notation $\bar{V}(z)$ (see the end of Section 1).

LEMMA 6.3. *Let $\beta = \{\beta_1 \le \cdots \le \beta_M\}$ be a nondecreasing sequence of integers, and let $P(z) \in S_\beta$. Then $\bar{P}(z) = (P(z))^{-1}$ if and only if $P(z)$ admits a factorization*

$$P(z) = \bar{Q}(z)(Q(z))^{-1} \tag{6.10}$$

*for some $Q(z) \in S_\beta$.*

*Proof.* The "if" part is clear. We prove the "only if" part by induction on $M$. Thus, assume first $M = 1$. In this case $P(z) \in S_\beta$ if and only if $P(z)$ is a nonzero scalar, call this scalar $\alpha$. The condition $\bar{\alpha} = \alpha^{-1}$ means that $|\alpha| = 1$, and then $\alpha = \bar{q}q^{-1}$ for any $q \in \mathbf{C}$ such that $q^{-2} = \alpha$.

Consider now the case of general $M$, and assume that Lemma 6.3 is proved already for all smaller values of $M$. Suppose first that not all the $\beta_j$'s are equal. Then $P(z)$ has the block triangular form

$$P(z) = \begin{bmatrix} P_1(z) & P_{12}(z) \\ 0 & P_2(z) \end{bmatrix},$$

where $P_1(z) = S_{\beta'}$, $P_2(z) = S_{\beta''}$; here

$$\beta' = \{\beta_1 \le \cdots \le \beta_{i_0}\}; \quad \beta'' = \{\beta_{i_0} \le \cdots \beta_m\}$$

for a suitable $i_0$. Since $\bar{P}(z) = (P(z))^{-1}$, we clearly have $\bar{P}_j(z) = (P_j(z))^{-1}$ $(j = 1, 2)$, and therefore by the induction hypothesis

$$P_j(z) = \bar{Q}_j(z)(Q_j(z)^{-1} \quad (j = 1, 2) \tag{6.11}$$

for some $Q_1(z) \in S'_\beta$, $Q_2(z) = S_{\beta''}$. We seek $Q(z) \in S_\beta$ satisfying (6.10) in the form

$$Q(z) = \begin{bmatrix} Q_1(z) & Q_{12}(z) \\ 0 & Q_2(z) \end{bmatrix},$$

where $Q_{12}(z)$ has to be found. Using (6.11), we obtain (omitting the variable $z$)

$$\bar{Q}Q^{-1} = \begin{bmatrix} P_1 & -\bar{Q}Q_1^{-1}Q_{12}Q_2^{-1} + \bar{Q}_{12}Q_2^{-1} \\ 0 & P_2 \end{bmatrix}.$$

Thus, (6.10) will be satisfied if

$$-\bar{Q}_1Q_1^{-1}Q_{12} + \bar{Q}_{12} = P_{12}Q_2,$$

or

$$R - \bar{R} = -\bar{Q}_1^{-1}P_{12}Q_2, \tag{6.12}$$

where $R = Q_1^{-1}Q_{12}$. The equality $\bar{P}P = I$, together with (6.11), easily implies

$$\bar{Q}_1^{-1}P_{12}Q_2 = -Q_1^{-1}\bar{P}_{12}\bar{Q}_2.$$

So we take $R = -\frac{1}{2}\bar{Q}_1^{-1}P_{12}Q_2$ to satisfy (6.12). This leads to the formula

$$Q_{12} = -\frac{1}{2}\bar{P}_1P_{12}Q_2. \tag{6.13}$$

Now clearly

$$\begin{bmatrix} \bar{P}_1^{-1} & \frac{1}{2}P_{12} \\ 0 & Q_2^{-1} \end{bmatrix} \in S_\beta,$$

therefore, since $S_\beta$ is a group,

$$\begin{bmatrix} \bar{P}_1 & -\frac{1}{2}\bar{P}_1P_{12}Q_2 \\ 0 & Q_2 \end{bmatrix}^{-1} = \begin{bmatrix} \bar{P}_1^{-1} & \frac{1}{2}P_{12} \\ 0 & Q_2^{-1} \end{bmatrix} \in S_\beta.$$

This shows that the degrees of the entries of $Q_{12}$ (defined by (6.13)) satisfy the requisite inequalities to ensure that

$$Q = \begin{bmatrix} Q_1 & Q_{12} \\ 0 & Q_2 \end{bmatrix} \in S_\beta.$$

It remains to consider the case when all the $\beta_j$'s are equal. Then $S_\beta$ consists of constant invertible $M \times M$ matrices. The result then follows from a well-known fact (see, e.g., Section

1a in [BH] or Lemma 4.6.9 in [HJ]) that every invertible matrix $P$ satisfying $\bar{P} = P^{-1}$ can be written in the form $P = \bar{Q}Q^{-1}$ for some matrix $Q$. □

*Proof of Theorem 6.1.* Let $\tau$ be a real left Sylvester data set. By Theorem 2.1 of [GKR1] (or Theorem 6.1 of [ABKW]) there exists a complex $M \times M$ rational matrix function $V(z)$ satisfying the properties (i)-(iii) of Theorem 6.1. In particular,

$$V(z) = Q(z) \operatorname{diag}(z^{\beta_1}, \ldots, z^{\beta_M}), \tag{6.14}$$

where $Q(z)$ is analytic and invertible at infinity. The rational matrix function with complex conjugate coefficients $\bar{V}(z)$ also has $\tau$ as its left null-pole triple with respect to $\mathbf{C}$ (cf. Proposition 3.1 and formula (3.2)). By the divisibility theorem (see [BGR6] or Theorem 4.5.8 in [BGR3]), we have $\bar{V}(z) = V(z)R(z)$, where the rational matrix function $R(z)$ has no poles and zeros in $\mathbf{C}$ (in other words, $R(z)$ is a matrix polynomial with constant nonzero determinant). In view of (6.14),

$$\operatorname{diag}(z^{\beta_1}, \ldots, z^{\beta_M})R(z) \operatorname{diag}(z^{-\beta_1}, \ldots, z^{-\beta_M}) = Q(z)^{-1}\bar{Q}(z), \tag{6.15}$$

and so by Proposition 6.2, $R \in S_\beta$. The equality (6.15) shows that $\bar{R}(z) = (R(z))^{-1}$. By Lemma 6.3 write

$$R(z) = \bar{T}(z)T(z)^{-1}$$

for some $T \in S_\beta$. Now the rational matrix function $\Theta(z) := V(z)\bar{T}(z)$ is real, and has $\tau$ as its left null-pole triple with respect to $\mathbf{C}$ (by [BGR6] or Theorem 4.5.8 in [BGR3]). Furthermore, we have

$$\begin{aligned} \Theta(z) &= Q(z) \operatorname{diag}(z^{\beta_1}, \ldots, z^{\beta_M})\bar{T}(z) \\ &= Q(z)Q_1(z) \operatorname{diag}(z^{\beta_1}, \ldots, z^{\beta_M}), \end{aligned}$$

where $Q_1(z)$ is analytic and invertible at infinity by Proposition 6.2. This equality shows that $\Theta(z)$ is column reduced and has column indices $\beta_1 \leq \cdots \leq \beta_M$. □

Using Theorem 6.1, we can now state and prove an alternative description of solutions to the (RTSCII) problem which will allow us to identify the McMillan degrees of the interpolants. Recall that the McMillan degree $\delta(W)$ of an $M \times N$ rational matrix function is defined as its total number of poles (counted with multiplicities) in $\mathbf{C} \cup \{\infty\}$.

THEOREM 6.4. *Consider the (RTSCII) problem* (5.1)-(5.3) *with $\sigma$-admissible Sylvester data* $(C_+, C_-, A_\pi; A_\rho, B_+, B_-; S)$. *Let* $\Theta(z)$ *be the real* $(M+N)\times(M+N)$ *rational matrix function satisfying* (i)-(iii) *of Theorem* 6.1, *where*

$$C = \begin{bmatrix} C_+ \\ C_- \end{bmatrix}, \quad B = [B_+ \quad B_-], \Gamma = S.$$

*Then a real rational matrix function* $W \in \mathcal{R}_{M\times N}(\sigma)$ *is a solution of this (RTSCII) problem if and only if there exist real matrix polynomials* $P(z)$ *and* $Q(z)$ *of sizes* $M\times N$ *and* $N\times N$ *respectively with the following properties:*

(i)

$$W = (\Theta_{11}P + \Theta_{12}Q)(\Theta_{21}P + \Theta_{22}Q)^{-1}, \tag{6.16}$$

*where* $\begin{bmatrix} \Theta_{11} & \Theta_{22} \\ \Theta_{21} & \Theta_{22} \end{bmatrix}$ *is the partition of* $\Theta$ *with* $\Theta_{11}$ *and* $\Theta_{22}$ *of sizes* $M\times M$ *and* $N\times N$, *respectively;*

(ii) $\{(C_-, A_\pi), (0,0); 0\}$ *is a left* $\sigma$*-null-pole triple for* $\Theta_{21}P + \Theta_{22}Q$ *(in particular,* $\Theta_{21}P + \Theta_{22}Q$ *has no zeros in* $\sigma$*);*

(iii) $P$ *and* $Q$ *are right coprime, i.e.,*

$$P_1(z)P(z) + Q_1(z)Q(z) \equiv I$$

*for some real matrix polynomials* $P_1(z)$ *and* $Q_1(z)$ *of suitable sizes;*

(iv) *the rational matrix function* $\Theta\begin{bmatrix} P \\ Q \end{bmatrix}$ *is column reduced.*

*If* (i)-(iv) *are satisfied, then*

$$\delta(W) = n_\pi + \sum_{i=1}^{N} \gamma_i, \tag{6.17}$$

*where* $\gamma_i$ *is the i-th column index of* $\Theta\begin{bmatrix} P \\ Q \end{bmatrix}$.

*Proof.* By Theorem 6.2 in [ABKW], every $W \in \mathcal{R}_{M\times N}(\sigma)$ of the form (6.16) which is obviously real, solves the (RTSCII) problem and its McMillan degree is given by (6.17).

Let now real $W(z) \in \mathcal{R}_{M\times N}(\sigma)$ be a solution of the RTSCII) problem. By Theorem 6.2 in [ABKW], $W(z)$ is given by (6.16), where $P(z)$ and $Q(z)$ are right coprime complex matrix polynomials having the properties (ii) and (iv). Since $\Theta(z)$ and $W(z)$ are real, we have also

$$W = (\Theta_{11}\bar{P} + \Theta_{12}\bar{Q})(\Theta_{21}\bar{P} + \Theta_{22}\bar{Q})^{-1}.$$

Moreover, $\bar{P}$ and $\bar{Q}$ are right coprime, $\Theta\begin{bmatrix}\bar{P}\\ \bar{Q}\end{bmatrix}$ is column reduced, and (by Proposition 3.3) $\{(C_-, A_\pi), (0,0); 0\}$ is a left $\sigma$-null-pole triple for $\Theta_{21}\bar{P}+\Theta_{22}\bar{Q}$. But the block column $\begin{bmatrix}P\\ Q\end{bmatrix}$ with the above properties is uniquely determined by $W$ (for the fixed $\Theta$) up to multiplication on the right by a rational matrix function without poles and zeros in $\sigma$. Thus,

$$\begin{bmatrix}\bar{P}\\ \bar{Q}\end{bmatrix} = \begin{bmatrix}P\\ Q\end{bmatrix} Z, \tag{6.18}$$

where $Z(z)$ has no poles and zeros in $\sigma$. Since $P$ and $Q$ are right coprime, in fact $Z$ is a polynomial, and since $\bar{P}$ and $\bar{Q}$ are right coprime, $Z^{-1}$ is a polynomial as well. So $Z$ is a matrix polynomial with constant nonzero determinant. Using the column reducedness of $\Theta\begin{bmatrix}P\\ Q\end{bmatrix}$, write

$$\Theta(z)\begin{bmatrix}P(z)\\ Q(z)\end{bmatrix} = A(z)\ \mathrm{diag}(z^{\beta_1}, \ldots, \ldots, z^{\beta_N})E, \tag{6.19}$$

where $\beta_1 \le \cdots \le \beta_N$ are integers, $A(z)$ is analytic and invertible at infinity, and $E$ is a permutation matrix. (The matrix $E$ is needed in case the column indices of $\Theta\begin{bmatrix}P\\ Q\end{bmatrix}$ do not appear in the nondecreasing order.) Clearly,

$$\Theta(z)\begin{bmatrix}\bar{P}(z)\\ \bar{Q}(z)\end{bmatrix} = \bar{A}(z)\ \mathrm{diag}(z^{\beta_1}, \ldots, z^{\beta_N})E, \tag{6.20}$$

and comparing with (6.18) we see (in view of Proposition 6.2) that

$$Z_0(z) := EZ(z)E^{-1} \in S_\beta,$$

where $\beta = \{\beta_1 \leq \cdots \leq \beta_N\}$. From (6.18)-(6.20) it follows also that $\bar{Z}_0(z) = (Z_0(z))^{-1}$. By Lemma 6.3 we can factor:

$$Z_0(z) = \bar{Y}(z)(Y(z))^{-1},$$

where $Y(z) \in S_\beta$. Let

$$\begin{bmatrix} P_1(z) \\ Q_1(z) \end{bmatrix} = \begin{bmatrix} P(z) \\ Q(z) \end{bmatrix} E^{-1}\bar{Y}(z)E;$$

then the polynomials $P_1(z)$ and $Q_1(z)$ are real and satisfy all the requirements of Theorem 6.4. □

## 7. GENERALIZED NEVANLINNA-PICK INTERPOLATION FOR REAL RATIONAL MATRIX FUNCTIONS

Many results concerning generalized Nevanlinna-Pick interpolation and its many variants and related interpolation problems for *real* rational matrix functions (such as Takagi, Nudelman, Nehari, Caratheodory-Toeplitz, boundary Nevanlinna-Pick) can be extended without difficulties from the known results for *complex* matrix functions (see [BGR3, BGR7, BGR8, BGR9, BGR10]). Indeed, the results (in the complex case) are obtained using the complex case version of Theorem 5.1, where the function $\Theta$ is selected in a particular way; in fact, formulas for a suitable $\Theta$ are available. An inspection of these formulas reveals immediately that if all interpolation data are given in terms of real matrices, then the formulas produce a real $\Theta$. Thus, Theorem 5.1 is applicable.

We state here the real analogue of only one result (Theorem 18.5.1 in [BGR3]) on generalized Nevanlinna-Pick interpolation for complex rational matrix functions.

THEOREM 7.1. *Let $\sigma$ be the open right half-plane, and let $w = (C_+, C_-, A_\pi; A_\zeta, B_+, B_-; S)$ be a $\sigma$-admissible Sylvester TSCII data set consisting of real matrices. Let $S_1$ and $S_2$ be the unique (necessarily real symmetric) solutions of the Lyapunov equations*

$$\begin{aligned} S_1 A_\pi + A_\pi^T S_1 &= C_+^T C_+; \\ S_2 A_\zeta^T + A_\zeta S_2 &= B_+ B_+^T - B_- B_-^T. \end{aligned}$$

*Let $\Lambda = \begin{bmatrix} S_1 & S^T \\ S & S_2 \end{bmatrix}$. Then there exists a real rational $M \times N$ matrix function $F$ satisfying* *the interpolation conditions*

$$\sum_{z_0 \in \sigma} Res_{z=z_0} F(z)C_-(zI - A_\pi)^{-1} = C_+;$$
$$\sum_{z_0 \in \sigma} Res_{z=z_0} (zI - A_\pi)^{-1}B_+F(z) = -B_-;$$
$$\sum_{z_0 \in \sigma} Res_{z=z_0} (zI - A_\pi)^{-1}B_+F(z)C_-(zI - A_\pi)^{-1} = S;$$

*together with the metric constraint* $\sup_{z\in\sigma} \|F(z)\| < 1$ *if and only if* $\Lambda$ *is positive definite. In this case, the set of all such interpolants* $F$ *is given by*

$$F = (\Theta_{11}G + \Theta_{12})(\Theta_{21}G + \Theta_{22})^{-1},$$

*where* $G$ *is an arbitrary real rational* $M \times N$ *matrix function analytic on* $\sigma$ *with* $\sup_{z\in\sigma} \|G(z)\| <$ 1, *and*

$$\Theta(z) = \begin{bmatrix} \Theta_{11}(z) & \Theta_{12}(z) \\ \Theta_{21}(z) & \Theta_{22}(z) \end{bmatrix}$$

*is the real rational matrix function determined from the set* $w$ *by the formula*

$$\Theta(z) = I + \begin{bmatrix} C_+ & -B_+^T \\ C_- & B_-^T \end{bmatrix} \begin{bmatrix} (zI - A_\pi)^{-1} & 0 \\ 0 & (zI + A_\zeta^T)^{-1} \end{bmatrix} \Lambda^{-1} \begin{bmatrix} -C_+^T & C_-^T \\ B_+ & B_- \end{bmatrix}.$$

## References


[ABGR] D. Alpay, J. A. Ball, I. Gohberg, and L. Rodman. The two-sided residue interpolation in the Stiltjes class for matrix functions, preprint.

[ABKW] A. C. Antoulas, J. A. Ball, J. Kang, and J. C. Willems. On the solution of the minimal rational interpolation problem. *Linear Algebra Appl.* **137/138** (1990), 511–573.

[BH] J. A. Ball and J. W. Helton. Beurling-Lax representation using classical Lie groups with many applications, IV. $GL(n, R)$, $U^*(2n)$, $SL(n, C)$ and a solvable group. *J. Functional Analysis* **69** (1986), 178–206.

[BGR1] J. A. Ball, I. Gohberg, and L. Rodman. Tangential interpolation problems for rational matrix functions. *Proc. of Symposia in Applied Math.* **40** (1990), 59–86.

[BGR2] J. A. Ball, I. Gohberg, and L. Rodman.Two-sided Lagrange-Sylvester interpolation problems for rational matrix functions. *Proc. Sympos. Pure Math.* **51** (1990), 17–83.

[BGR3] J. A. Ball, I. Gohberg, and L. Rodman. *Interpolation of Rational Matrix Functions*, OT 45, Birkhäuser, Basel, 1990.

[BGR4] J. A. Ball, I. Gohberg, and L. Rodman. Simultaneous residue interpolation problems for rational matrix functions. *Integral Equations Operator Theory* **13** (1990), 611–637.

[BGR5] J. A. Ball, I. Gohberg, and L. Rodman. Common minimal multiples and divisors for rational matrix functions. *Linear Algebra Appl.* **137/138** (1990), 621–662.

[BGR6] J. A. Ball, I. Gohberg, and L. Rodman. Minimal factorization of meromorphic matrix functions in terms of local data. *Integral Equations Operator Theory* (1987), 309–348.

[BGR7] J. A. Ball, I. Gohberg, and L. Rodman. Sensitivity minimization and tangential Nevanlinna-Pick interpolation in contour integral form, in: *Signal Processing Part II: Control Theory and Applications* (eds. F. A. Grünbaum, J. W. Helton, P. Khargonekar), *IMA Vol. in Math. and Appl.* **23** (1990), 3–25, Springer Verlag, New York.

[BGR8] J. A. Ball, I. Gohberg, and L. Rodman. Boundary Nevanlinna-Pick interpolation for rational matrix functions. *J. Math. Systems, Estimation and Control* **1** (1991), 131–164.

[BGR9] J. A. Ball, I. Gohberg, and L. Rodman. Two-sided Nudelman interpolation problem for rational matrix functions. *Analysis and Partial Differential Equations* (ed. C. Sadony), 371-416 (1990), Marcel Dekker.

[BGR10] J. A. Ball, I. Gohberg, and L. Rodman. Nehari interpolation problem for rational matrix functions: The generic case, in: C. Foias, et al., *$H^\infty$-Control Theory* (E. Mosca, L. Pandolfi, eds.), *Lecture Notes in Mathematics* **1496** (1991), 277-308, Springer-Verlag.

[BGR11] J. A. Ball, I. Gohberg, and L. Rodman. The structure of flat gain rational matrices that satisfy two-sided interpolation requirements. *Systems and Control Letters*, to appear.

[BKRV] J. A. Ball, J. Kim, L. Rodman, and M. Verma. Minimal degree coprime factorizations of rational matrix functions. *Linear Algebra Appl.*, to appear.

[BRa2] J. A. Ball and M. Rakowski. Interpolation by rational matrix functions and stability of feedback systems: the 4-block case. *Operator Theory: Advances and Applications* **59** (1992), 96-142.

[BRa1] J. A. Ball and M. Rakowski. Interpolation by rational matrix functions and stability of feedback systems: the 2-block case. *Journal of Mathematical Systems, Estimation and Control*, to appear.

[FW] P. A. Fuhrmann and J. C. Willems. Factorization indices at infinity for rational matrix functions. *Integral Equation Operator Theory* **2/3** (1979), 287–301.

[Gl1] K. Glover. Robust stabilization of linear multivariable systems: relations to approximation. *Intern. J. Control* **43** (1986), 741–766.

[Gl2] K. Glover. All optimal Hankel-norm approximations of linear multivariable systems and their $L^\infty$ error bounds. *Intern. J. Control* **39** (1984), 1115–1193.

[GKLR] I. Gohberg, M. A. Kaashoek, L. Lerer, and L. Rodman. Minimal divisors of rational matrix functions with prescribed zero and pole structure, in: *Topics in Operator Thoery, Systems and Networks* (H. Dym and I. Gohberg, eds.), OT 12, Birkhäuser, Basel (1984), 241–275.

[GKR1] I. Gohberg, M. A. Kaashoek, and A. C. M. Ran. Regular rational matrix functions with prescribed null and pole data except at infinity. *Linear Algebra Appl.* **137/138** (1990), 387-412.

[GKR2] I. Gohberg, M. A. Kaashoek, and A. C. M. Ran. Matrix polynomials with prescribed zero structure in the finite complex plane, to appear.

[GKRo] I. Gohberg, M. A. Kaashoek, and L. Rodman. Spectral analysis of families of operator polynomials and a generalized Vandermonde matrix, I. The finite dimensional case, in: *Topics in Functional Analysis* (I. Gohberg, M. Kac, eds.) (1978), 91–128, Academic Press.

[GKrS] I. Gohberg, M. A. Kaashoek, and F. van Schagen. Rational matrix and operator functions with prescribed singularities. *Integral Equations Operator Theory* **5** (1982), 673–717.

[GK] I. Gohberg and M. G. Krein. Systems of integral equations on a half line with kernels depending on the differences of arguments. *Amer. Math. Soc. Trans.* **14** (1960), 217–282 (English translation).

[GLR1] I. Gohberg, P. Lancaster, and L. Rodman. *Invariant Subspaces of Matrices with Applications.* Wiley, 1986.

[GLR2] I. Gohberg, P. Lancaster, and L. Rodman. *Matrix Polynomials.* Academic, 1982.

[GLeR] I. Gohberg, L. Lerer, and L. Rodman. On factorization, indices and completely decomposable matrix polynomials. Technical Report 80-47 (1980), Dept. of Mathematical Sciences, Tel-Aviv University.

[HJ] R. A. Horn, C. R. Johnson. *Matrix Analysis.* Cambridge University Press, Cambridge, 1985.

[Kai] T. Kailath. *Linear Systems.* Prentice Hall, 1980.

[K] J. Kang. *Interpolation by Rational Matrix Functions with Minimal McMillan Degree*, Ph.D. Thesis, Virginia Tech, 1990.

[Ki] H. Kimura. Directional interpolation approach to $H^\infty$-optimization and robust stabilization. *IEEE Trans. Autom. Control*, **AC-32** (1987), 1085–1093.

[VK] M. Vidyasagar and K. Kimura. Robust controllers for uncertain linear multivariable systems. *Automatica* **22** (1986), 85-94.

J. A. Ball
Department of Mathematics
Virginia Tech
Blacksburg, VA 24061
USA

I. Gohberg
School of Mathematical Sciences
Raymond and Beverly Sackler
Faculty of Exact Sciences
Tel-Aviv University
Tel-Aviv, Ramat Aviv 69978
ISRAEL

L. Rodman
Department of Mathematics
College of William and Mary
Williamsburg, VA 23187-8795
USA

MSC: 15A54, 47A56, 30E05.

Operator Theory:
Advances and Applications, Vol. 64



# ON THE SPECTRA OF OPERATOR COMPLETION PROBLEMS

HONG-KE DU CAIXING GU

For an $2 \times 2$ operator matrix $M_X = \begin{pmatrix} A & C \\ X & B \end{pmatrix}$ on the Hilbert space $H \oplus K$, if $A, B$ and $C$ are given, we study the intersection and the union of the spectra of all $M_X$ when $X$ is taken over $B(H, K)$.

## 1. INTRODUCTION

Let $H$ and $K$ be two complex separable Hilbert spaces. Let $B(H)$, $B(H, K)$ denote the set of all linear bounded operators from $H$ into $H$, from $H$ into $K$, respectively. Let $M_X(A, B, C)$ (for short, $M_X$) be the operator from $H \oplus K$ into $H \oplus K$,

$$M_X(A,B,C) = \begin{pmatrix} A & C \\ X & B \end{pmatrix} \tag{1.1}$$

where $A \in B(H)$, $B \in B(K)$, $C \in B(K, H)$ are given and $X \in B(H, K)$ is arbitrary.

In this paper we will study the spectra for this $2 \times 2$ operator matrix completion problem. The main purpose of this paper is to compute the intersection and the union of the spectra of all $M_X$ when $X$ is taken over $B(H, K)$ under some appropriate conditions. Partial results have been obtained for the following spectrum assignment problem: Given a compact subset $D$ of complex plane, find the necessary and sufficient conditions in terms of operators $A$, $B$, $C$ and compact set $D$ such that there exists a completion of $M_X$ (a choice of $X$) with its spectrum equal to the compact set $D$.

Our problem is a special case of the following general operator completion problem: Given part of entries of a matrix, find completions of the matrix ( choices of other unspecified entries) with various properties. This problem has been studied in a variety of directions by a number of authors. Here are some recent papers. J. Ball, I. Gohberg, L. Rodman and T. Shalom in [1] describe the possible eigenvalues of matrices with given triangular part. C. Davis, W. M. Kahan and H. F. Weinberger [2] and S. Parrott [9] find the minimal norm completions for $M_X$. L. Gurvits, L. Rodman and I. Spitkovsky in [8] give a description of the possible spectra of operators of the form $A - BF$ for given Hilbert space operators $A$ and $B$, under suitable hypotheses. See also I. Gohberg and S. Rubinstein [6], L. Rodman and T. Shalom [10] and H. J. Woerdman [11].

Now let us introduce some notations and terminologies. Throughout this paper, an operator always means a linear bounded operator. For an operator $S$ on a Hilbert space, $\sigma(S)$ denotes the spectrum of $S$. In the sequel, for convenience, we call $(A, B, C)$ a 3-tuple. For a given 3-tuple $(A, B, C)$, if for any $X \in B(H, K)$, we always have $\lambda \in \sigma(M_X)$, then $\lambda$ is called a stability point of spectrum for the 3-tuple $(A, B, C)$. The set of all stability points of spectrum for a 3-tuple $(A, B, C)$ is denoted by $\sigma(A, B, C)$, that is, the set $\sigma(A, B, C)$ is the intersection of the spectra of all $M_X$ when $X$ is taken over $B(H, K)$. If for any $X \in B(H, K)$, we always have $\lambda \notin \sigma(M_X)$, then $\lambda$ is called a stable regular point for the 3-tuple $(A, B, C)$. The set of all stable regular points for a 3-tuple $(A, B, C)$ is denoted by $\rho(A, B, C)$, that is, the set $\rho(A, B, C)$ is the complement of the union of the spectra of all $M_X$ when $X$ is taken over $B(H, K)$. Note that the set $\sigma(A, B, C)$ and $\rho(A, B, C)$ are disjoint, but their union may be all of $\mathbf{C}$.

In Section 2 we will characterize the set $\sigma(A, B, C)$ and the set $\rho(A, B, C)$ when $H$ and $K$ are finite dimensional. Also we discuss the spectrum assignment problem of $M_X$ for a given 3-tuple $(A, B, C)$ when $X$ runs over $B(H, K)$. Namely we proved that if $rank\, C \geq k$ then for any given complex numbers $\{\lambda_1, \cdots, \lambda_k\}$ there exists an operator $X \in B(H, K)$ such that $\{\lambda_1, \cdots, \lambda_k\} \subset \sigma(M_X)$ (see Theorem 2.6 below). Section 3 is devoted to the infinite dimensional case. As it will be seen, the problems mentioned above have different answers in the infinite dimensional case. For example, in the finite dimensional case, for any given 3-tuple $(A, B, C)$, if $C \neq 0$, then the set $\rho(A, B, C)$ is the empty set. But in the infinite dimensional case the set $\rho(A, B, C)$ is always an open set and there exists a 3-tuple $(A, B, C)$ with $C \neq 0$ such that this set is not empty (see Example 3.1 below). Some examples are presented to illustrate the complicated nature of the spectrum assignment problem and of the identification of the set $\sigma(A, B, C)$ in the infinite dimensional case.

## 2. Case of finite dimensional spaces

For a vector space $M$, we shall denote by $dim M$ the dimension of $M$. Throughout this section both $dim H$ and $dim K$ are finite.

We begin by noting the following simple fact.

**Proposition 2.1.** *For a 3-tuple* $(A, B, 0)$, *we have* $\sigma(A, B, 0) = \sigma(A) \cup \sigma(B)$ *and* $\rho(A, B, 0) = \rho(A) \cap \rho(B)$.

Here and throughout $\rho(S) = \mathbf{C} \setminus \sigma(S)$ denotes the resolvent of an operator $S$.

In general, it is clear that

$$\sigma(A,B,C) := \cap_{X\in B(H,K)}\sigma(M_X) \subseteq \sigma(A)\cup\sigma(B) \tag{2.1}$$

The following theorem shows that the inclusion in (2.1) may be proper.

**Theorem 2.2.** *For a given 3-tuple $(A,B,C)$, a complex number $\lambda$ is in $\sigma(A,B,C)$ if and only if $R(A-\lambda)+R(C)\neq H$ or $R(B^*-\bar{\lambda})+R(C^*)\neq K$.*

Here above, as well as in the sequel, we denote by $R(S)$ the range of an operator $S$.

**Proof.** The proof of the "if" part is clear. In fact, if $R(A-\lambda)+R(C)\neq H$, then for any $X\in B(H,K)$, $R(M_X-\lambda)\neq H\oplus K$, so that $\lambda\in\sigma(M_X)$. Similarly, if $R(B^*-\bar{\lambda})+R(C^*)\neq K$, then for any $X\in B(H,K)$, $R(M_X^*-\bar{\lambda})\neq H\oplus K$, so that $\lambda\notin\sigma(M_X)$ too.

For the proof of the "only if" part, consider a $\lambda\in\mathbb{C}$ with $R(A-\lambda)+R(C)=H$ and $R(B^*-\bar{\lambda})+R(C^*)=K$. We shall choose an operator $X\in B(H,K)$ such that $M_X-\lambda$ is invertible.

Without loss of generality, assume that $\lambda=0$. First we introduce some notations. For a matrix $D$, let $rank\, D$, $N(D)$ and $n(D)$ denote the rank of $D$, the kernel of $D$ and the dimension of the kernel of $D$, respectively. Let $m := dim H$, $n = dim K$.

Let $\{e_i\}_{i=1}^{n(A)}$ be an orthonormal basis of $N(A)$. Let $\{x_i\}_{i=1}^{t}\subset N(A)^\perp$ be t orthonormal unit vectors such that there exist $\{y_i\}_{i=1}^{t}\subset N(C)^\perp$ with

$$Ax_i + Cy_i = 0,$$

where $t = dim\,\{R(A)\cap R(C)\} = rank\, A + rank\, C - n = rank\, C - n(A)$, since $R(A)+R(C)=H$. Let $\{f_i\}_{i=1}^{k}$ be an orthonormal basis of $\{BN(C)\}^\perp$, where $BN(C)$ is the subspace spanned by $\{By, y\in N(C)\}$. By assumption $N(B)\cap N(C)=\{0\}$, so $k=n-n(C)=rank\, C$. Note that $t+n(A)=rank\, C=k$. Let

$$X=\sum_{i=1}^{t}(-By_i+f_i)\otimes x_i+\sum_{j=1}^{n(A)} f_{j+t}\otimes e_j,$$

where for $f\in H, g\in K$, $g\otimes f$ denote the rank one operator defined by $(g\otimes f)u=(u,f)g$, for any $u\in H$. It is easy to check that with this choice of $X$, $M_X$ is invertible. Indeed for $x\in H$ and $y\in K$ if we have

$$\begin{pmatrix} A & C\\ X & B\end{pmatrix}\begin{pmatrix} x\\ y\end{pmatrix}=0, \tag{3.4}$$

then $x=\sum_{j=1}^{n(A)}\beta_j e_j+\sum_{i=1}^{t}\alpha_i x_i$ and $y=y_0+\sum_{i=1}^{t}\alpha_i y_i$ for some $y_0\in N(C)$. But by the definition of $X$,

$$\sum_{j=1}^{n(A)} \beta_j f_{j+t} + \sum_{i=1}^{t} \alpha_i f_i + B y_0$$
$$= X(\sum_{j=1}^{n(A)} \beta_j e_j + \sum_{i=1}^{t} \alpha_i x_i) + B(y_0 + \sum_{i=1}^{t} \alpha_i y_i) = 0$$

implies that $\alpha_i = 0, i = 1, \cdots, t$, $\beta_j = 0, j = 1, \cdots, n(A)$, and $By_0 = 0$; but $y_0 = 0$ since $N(B) \cap N(C) = \{0\}$. Therefore $x = 0$ and $y = 0$. The proof is complete.

The following corollary is an immediate consequence of Theorem 2.2.

**Corollary 2.3.** *For a 3-tuple $(A, B, C, )$ with $C$ an invertible operator from $K$ onto $H$, the set $\sigma(A, B, C)$ is empty.*

Proposition 2.1 and the next proposition give a complete description of $\rho(A, B, C)$. Namely, $\rho(A, B, 0) = \rho(A) \cap \rho(B)$; and the set $\rho(A, B, C)$ is empty for $C \neq 0$.

**Theorem 2.4.** *For any 3-tuple $(A, B, C)$ with $C \neq 0$, the set*

$$\rho(A, B, C) := \cap_{X \in B(H,K)} \rho(M_X)$$

*is empty.*

**Proof.** Assume on the contrary that there exists a 3-tuple $(A, B, C)$ with $C \neq 0$, but $\rho(A, B, C) \neq \emptyset$ ($\emptyset$ is the empty set). Let us take a $\lambda \in \rho(A, B, C)$. This means that for any $X \in B(H, K)$, the operator

$$\begin{pmatrix} A - \lambda & C \\ X & B - \lambda \end{pmatrix}$$

is invertible. Put $X = 0$, then $\begin{pmatrix} A - \lambda & C \\ 0 & B - \lambda \end{pmatrix}$ is invertible. Hence $A - \lambda$ is also invertible on $H$. Since $C \neq 0$, let $y \in K$ be a unit vector such that $Cy \neq 0$. Since $A - \lambda$ is invertible, there must be a vector $x \in H$ such that

$$(A - \lambda)x + Cy = 0.$$

Then for

$$X_0 = -\frac{1}{\|x\|}(B - \lambda)y \otimes x,$$

we have

$$(M_{X_0} - \lambda)\begin{pmatrix} x \\ y \end{pmatrix} = \begin{pmatrix} A - \lambda & C \\ X_0 & B - \lambda \end{pmatrix}\begin{pmatrix} x \\ y \end{pmatrix} = 0.$$

That is $\lambda \in \sigma(M_{X_0})$, a contradiction. This completes the proof.

In other words, we have that if $C \neq 0$, then $\cup_{X \in B(H,K)} \sigma(M_X) = \mathbf{C}$ .

Next we turn to the assignment of the spectrum of $M_X$ for a given 3-tuple $(A, B, C)$.

**Lemma 2.5.** *For a given 3-tuple $(A, B, C)$ with $dimR(C) \geq k$, and for $m$ distinct complex numbers $\lambda_i$ $(1 \leq i \leq m)$ and positive integers $n_i$ $(1 \leq i \leq m)$ satisfying $dim(R(A-\lambda_i) \cap R(C)) \geq n_1 + \cdots + n_i, 1 \leq i \leq m$, and $n_1 + \cdots + n_m = k$, there exists an operator $X \in B(H, K)$ such that $\{\lambda_1, \lambda_2, \cdots \lambda_k\} \subset \sigma(M_X)$; moreover the dimension of the eigenspace of $M_X$ corresponding to $\lambda_i$ is at least $n_i$ $(i = 1, \cdots, m)$.*

**Proof.** For $\lambda_1$, consider the space $(A-\lambda_1)^{-1}(R(A-\lambda_1) \cap R(C))$ which is the preimage of space $(R(A-\lambda_1) \cap R(C))$ under operator $(A-\lambda_1)$. Since $dim(R(A-\lambda_1) \cap R(C)) \geq n_1$, there are $n_1$ orthogonal unit vectors $\{x_1^{j_1}\}_{j_1=1}^{n_1} \subset (A-\lambda_1)^{-1}$ $(R(A-\lambda_1) \cap R(C))$. For $\lambda_2$, since $dim(A-\lambda_2)^{-1}(R(A-\lambda_2) \cap R(C)) \geq dim(R(A-\lambda_2) \cap R(C)) \geq n_1 + n_2$, there are $n_2$ unit vectors $\{x_2^{j_2}\}_{j_2=1}^{n_2} \subset (A-\lambda_2)^{-1}$ $(R(A-\lambda_2) \cap R(C))$ and $x_1^{j_1}, x_2^{j_2}$ $(j_1 = 1, \cdots, n_1, j_2 = 1, \cdots, n_2)$ are orthogonal.

By induction on $l$, assume we have taken $n_1 + \cdots + n_l$ orthogonal unit vectors $x_i^{j_i}$ $(1 \leq i \leq l, 1 \leq j_i \leq n_i)$ where $x_i^{j_i} \in (A-\lambda_i)^{-1}(R(A-\lambda_i) \cap R(C))$, $1 \leq i \leq l, 1 \leq j_i \leq n_i$. As before we choose $n_{l+1}$ unit vectors $x_{l+1}^{j_{l+1}} \in (A-\lambda_{l+1})^{-1}(R(A-\lambda_{l+1}) \cap R(C))$ such that $x_i^{j_i}$ $(1 \leq i \leq l+1, 1 \leq j_i \leq n_i)$ are orthogonal. Continue the procedure and stop at $l = m$. We get $k$ orthogonal unit vectors $x_i^{j_i}$ $(1 \leq i \leq m, 1 \leq j_i \leq n_i)$ where $x_i^{j_i} \in (A-\lambda_i)^{-1}(R(A-\lambda_i) \cap R(C))$, $1 \leq i \leq l, 1 \leq j_i \leq n_i$.

Clearly, for each $x_i^{j_i}$ $(1 \leq i \leq m, 1 \leq j_i < n_i)$, we can choose a vector $y_i^{j_i}$ such that $(A-\lambda_i)x_i^{j_i} + Cy_i^{j_i} = 0$. Put

$$X = \sum_{i=1}^{m} \sum_{j_i=1}^{n_i} -(B-\lambda_i)y_i^{j_i} \otimes x_i^{j_i};$$

by a straightforward computation we have $\{\lambda_1, \lambda_2, \cdots \lambda_m\} \subset \sigma(M_X)$, and the dimension of the eigenspace of $M_X$ corresponding to $\lambda_i$ is at least $n_i$ $(i = 1, \cdots, m)$. This completes the proof.

**Theorem 2.6.** *The following two statements are equivalent.*

(1). *$rank\, C \geq k$.*

(2). *For any complex numbers $\{\lambda_1, \cdots, \lambda_m\}$ and positive integers $\{n_1, \cdots, n_m\}$ satisfying $n_1 + \cdots + n_m = k$, there exists an operator $X \in B(H, K)$ such that $\{\lambda_1, \cdots, \lambda_k\} \subset \sigma(M_X)$ and the multiplicity of $\lambda_i$ is at least $n_i$ for all $i$; moreover the eigenspace of $M_X$ corresponding to $\lambda_i$ has a dimension at least $n_i$ for those $i$ such that $\lambda_i \in \rho(A)$.*

**Proof.** We first prove that statement (1) implies statement (2). Note that for $V$ and $S$ invertible operators on $H$ and $K$, respectively, the condition and the conclusion in the theorem are invariant under the following similarity transformation

$$\begin{pmatrix} V & 0 \\ 0 & S \end{pmatrix} \begin{pmatrix} A & C \\ X & B \end{pmatrix} \begin{pmatrix} V^{-1} & 0 \\ 0 & S^{-1} \end{pmatrix} = \begin{pmatrix} VAV^{-1} & VCS^{-1} \\ SXV^{-1} & SBS^{-1} \end{pmatrix}.$$

Therefore we may assume $A$ is in its Jordan form. Let $k_i$ be the multiplicity of $\lambda_i$ in $\sigma(A)$ ($i = 1, \cdots, m$, $k_i$ might be zero). We write $A$ in the following form

$$A = \begin{pmatrix} A_{11} & A_{12} \\ 0 & A_{22} \end{pmatrix}, \quad A_{11} = \begin{pmatrix} \lambda_1 & & & & & \\ & \ddots & & & & \\ & & \lambda_1 & & * & \\ & & & \ddots & & \\ & & & & \lambda_m & \\ & & & & & \ddots \\ & & & & & & \lambda_m \end{pmatrix}.$$

where $A_{11}$ is a upper triangular matrix and the characteristic polynomial of $A_{11}$ is

$$|\lambda I - A_{11}| = (A - \lambda_1)^{min\{n_1,k_1\}} \cdots (A - \lambda_m)^{min\{n_m,k_m\}}. \tag{2.2}$$

Then $M_X$ take the following form

$$\begin{pmatrix} A_{11} & A_{12} & C_1 \\ 0 & A_{22} & C_2 \\ X_1 & X_2 & B \end{pmatrix}.$$

Let $X_1 = 0$, then we have

$$|\lambda I - M_X| = |\lambda I - A_{11}||\lambda I - M_{X_2}(A_{22}, B, C_2)|. \tag{2.3}$$

Note that

$$rank\, C_2 \geq (n_1 - min\{n_1, k_1\}) + \cdots + (n_m - min\{n_m, k_m\});$$

and if $n_i - min\{n_i, k_i\} > 0$, then $\lambda_i \in \rho(A_{22})$. If this is the case we have that $dim(R(A_{22} - \lambda_i) \cap R(C_2)) = dim R(C_2)$. So by Lemma 2.5 with $C = C_2$ and all the complex numbers $\lambda_i$ $(1 \leq i \leq m)$ such that $n_i - min\{n_i, k_i\} > 0$ we can choose $X_2$ such that

$$|\lambda I - M_{X_2}(A_{22}, B, C_2)| = (\lambda - \lambda_1)^{n_1 - min\{n_1,k_1\}} \cdots (\lambda - \lambda_m)^{n_m - min\{n_m,k_m\}} p(\lambda) \tag{2.4}$$

for some polynomial $p(\lambda)$. Combining (2.2) and (2.4) we prove the promised result.

Now we prove that statement (2) implies statement (1). Take $\lambda_0 \in \rho(A) \cap \rho(B)$. If there is an operator $X_0$ such that $\lambda_0 \in \sigma(M_{X_0})$ and the eigenspace of $M_{X_0}$ corresponding to $\lambda_0$ has a dimension at least $k$, then there exist $k$ linearly independent vectors $z_i = x_i + y_i$ $(i = 1, 2, \cdots, k)$, where $x_i \in H, y_i \in K$, such that

$$\begin{cases} (A - \lambda_0)x_i + Cy_i = 0 \\ X_0 x_i + (B - \lambda_0)y_i = 0 \end{cases}$$

$i = 1, 2, \cdots, k$. Note that $x_i$ $(i = 1, 2, \cdots, k)$ are linearly independent.

Indeed since $\lambda_0 \in \rho(A) \cap \rho(B)$, we have

$$y_i = -(B - \lambda_0)^{-1} X_0 x_i, \qquad i = 1, 2, \cdots .k.$$

If $\{x_i\}_{i=1}^k$ is linearly dependent, that is, there exist $\alpha_i$ $(i = 1, \cdots, k)$ with at least one of them nonzero such that $\sum_{i=1}^k \alpha_i x_i = 0$, then $\sum_{i=1}^k \alpha_i y_i = 0$, so $\sum_{i=1}^k \alpha_i z_i = 0$. This is a contradiction. Again, since $\lambda_0 \in \rho(A)$, we have $Cy_i = -(A - \lambda_0)x_i$ $( i = 1, 2, \cdots, k)$ are linearly independent, thus $rank\, C \geq k$.

We conclude this section to remark that Theorem 2.6 implies Theorem 2.4.

## 3. Case of Infinite Dimensional Spaces

In this section we assume $dim H = dim K = \infty$ unless stated otherwise. As expected in the case of infinite dimensional spaces some questions discussed in the previous section will be more complicated.

We first discuss the set $\rho(A, B, C)$ for a 3-tuple $(A, B, C)$. In the infinite dimensional case, unlike in the finite dimensional case (see Theorem 2.4), the set $\rho(A, B, C)$ may not be empty even if $C \neq 0$.

**Example 3.1.** Let $H = K$ and $\{e_i\}_{i=0}^\infty$ be orthonormal basis of $H$. Let $A$ be a unilateral shift, $Ae_i = e_{i+1}$, $i = 0, 1, \cdots$, $B = A^*$ and $C = I_H - AA^*$, where $I_H$ is the identity operator on $H$. For this 3-tuple $(A, B, C)$, we have that the operator $M_X$ is invertible for any $X \in B(H, K)$. To see this, let $u, v \in H$, we claim that there are unique $x, y \in H$ such that

$$\begin{pmatrix} A & C \\ X & B \end{pmatrix} \begin{pmatrix} x \\ y \end{pmatrix} = \begin{pmatrix} u \\ v \end{pmatrix}.$$

Indeed, by applying $A^*$ to both sides of $Ax + Cy = u$, we get $x = A^*u$ and $(I_H - AA^*)y = (I_H - AA^*)u$; and by applying $A$ to both sides of $Xx + By = v$, we obtain $AA^*y = Av - AXx$. Hence $y = (I_H - AA^*)y + AA^*y = Av - AXA^*u + (I_H - AA^*)u$. In fact the next theorem shows that for this example the set $\rho(A, B, C) = \{z, |z| < 1\}$.

**Theorem 3.2.** *$\lambda$ is a stable regular point for a 3-tuple $(A, B, C)$ with $C \neq 0$ if and only if following statements (1)-(5) hold:*

(1). *Both $A - \lambda$ and $B - \lambda$ are non-invertible.*

(2). *There is a positive number $\delta > 0$ such that $(A^* - \bar{\lambda})(A - \lambda) \geq \delta I_H$ and $(B - \lambda)(B^* - \bar{\lambda}) \geq \delta I_K$.*

(3). *There is a positive number $\xi > 0$ such that $(A - \lambda)(A^* - \bar{\lambda}) + CC^* \geq \xi I_H$ and $(B^* - \bar{\lambda})(B - \lambda) + C^*C \geq \xi I_K$.*

(4). *$R(A - \lambda) \cap R(C) = \{0\}$ and $R(B^* - \bar{\lambda}) \cap R(C^*) = \{0\}$.*

(5). *$R(C)$ is closed (and thus $R(C^*)$ is also closed).*

**Proof.** Without loss of generality, assume $\lambda = 0$.

The proof of the "only if" part.

(4). If $R(A) \cap R(C) \neq \{0\}$, let $x \in H$ and $y \in K$ be two nonzero vectors such that $Ax + Cy = 0$. Define $X_1 \in B(H, K)$ by

$$X_1 = -\frac{1}{\|x\|^2} By \otimes x,$$

then $0 \in \sigma_p(M_{X_1})$, where $\sigma_p(S)$ is the point spectrum of an operator $S$. This contradicts the assumption that 0 is a stable regular point for the 3-tuple $(A, B, C)$. By a similar argument, we obtain $R(B^*) \cap R(C^*) = \{0\}$.

(1). It follows from (4) proved above.

(2) and (3). Take $X = 0$; since $\begin{pmatrix} A & C \\ 0 & B \end{pmatrix}$ and its adjoint $\begin{pmatrix} A^* & 0 \\ C^* & B^* \end{pmatrix}$ are invertible; so $B$, $A^*$ , $(\, A \quad C\,)$ and $(\, C^* \quad B^*\,)$ are surjective; thus there are positive numbers $\delta, \xi$ such that $A^*A \geq \delta I_H$, $BB^* \geq \delta I_K$, $AA^* + CC^* \geq \xi I_H$ and $B^*B + C^*C \geq \xi I_K$.

(5). We see that statement (3) is equivalent to that $R(A) + R(C) = H$ and $R(B^*) + R(C^*) = K$. So statements (3) and (4) show us that $R(A)$ and $R(C)$ ($R(B^*)$ and $R(C^*)$) are complemented, and one is other's complement. By Theorem 2.4 in [4], $R(A)$ and $R(C)$ ($R(B^*)$ and $R(C^*)$) are closed subspaces.

Now we will prove the "if" part.

We start by first noticing that if statements (1)-(5) hold, then as above statements (3) and (4) show us that $R(A)$ and $R(C)$ ($R(B^*)$ and $R(C^*)$) are complemented, and one is other's complement. Thus there exist invertible operators $V$ on $H$ and $S$ on $K$ such that $R(VA) \perp R(VC)$ and $R(S^*B^*) \perp R(S^*C^*)$.

On the other hand, statements (1), (2), (3), (4) and (5) are obviously equivalent to the following statments (1)', (2)', (3)', (4)' and (5)', respectively,

(1)'. Both $VA$ and $BS$ are non-invertible.

(2)'. There is a positive number $\delta' > 0$ such that $(VA)^*(VA) \geq \delta' I_H$ and $(BS)(BS)^* \geq \delta' I_K$.

(3)'. There is a positive number $\xi' > 0$ such that $(VA)(VA)^* + (VCS)(VCS)^* \geq \xi' I_H$ and $(BS)(BS)^* + (VCS)(VCS)^* \geq \xi' I_K$.

(4)'. $R(VA) \cap R(VC) = \{0\}$ and $R(S^*B^*) \cap R(S^*C^*) = \{0\}$.

(5)'. $R(VCS)$ is closed.

Therefore, without loss of generality, we can assume that $R(A) \perp R(C)$ and $R(B^*) \perp$

$R(C^*)$. Thus $M_X$ becomes the following operator matrix

$$M_X = \begin{pmatrix} A_1 & A_2 & 0 & 0 \\ 0 & 0 & 0 & C_1 \\ X_{11} & X_{12} & B_1 & 0 \\ X_{21} & X_{22} & B_2 & 0 \end{pmatrix} \tag{3.1}$$

according to the space decomposition $H = R(A) \oplus N(A^*)$ and $K = R(B^*) \oplus N(B)$, where $N(S)$ denotes the nullspace of an operator $S$.

To prove that $M_X$ is invertible, it is enough to prove that $M_X$ and its adjoint $M_X^*$ are bounded below. We will only show that $M_X$ is bounded below since the proof for $M_X^*$ is similar. Suppose that there is a sequence of unit vectors $\{z_n = x_{1n} + x_{2n} + y_{1n} + y_{2n} : x_{1n} \in R(A), x_{2n} \in N(A), y_{1n} \in R(B^*), y_{2n} \in N(B)\}$ such that $M_X z_n \to 0$ (as $n \to \infty$), then from (3.1)

$$\begin{cases} A_1 x_{1n} + A_2 x_{2n} \to 0, \\ C_1 y_{2n} \to 0, \\ X_{11} x_{1n} + X_{12} x_{2n} + B_1 y_{1n} \to 0, \\ X_{21} x_{1n} + X_{22} x_{2n} + B_2 y_{1n} \to 0. \end{cases}$$

Since $A$ is bounded below on $H$, so $x_{1n} + x_{2n} \to 0$. Thus $B_1 y_{1n} \to 0$ and $B_2 y_{1n} \to 0$. But $B$ is bounded below on $R(B^*)$, hence $y_{1n} \to 0$. Moreover, since statements (3)'and (4)' imply that $R(C_1)$ is equal to $N(A^*)$ and $R(C_1^*)$ is equal to $N(B)$, we conclude that $C_1$ is invertible. Hence $y_{2n} \to 0$. In conclusion, $z_n = x_{1n} + x_{2n} + y_{1n} + y_{2n} \to 0$, a contradiction. This completes the proof.

It is clear from the proof of Theorem 3.2 that Theorem 3.2 holds without any restriction on the dimensions of $H$ and $K$. But note that if at least one of the spaces $H$ and $K$ is finite dimensional, then the properties (1) and (2) in Theorem 3.2 are not compatible. This gives the following generalization of Theorem 2.4.

**Corollary 3.3** *If at least one of the spaces $H$ and $K$ is finite dimensional, then for a given 3-tuple $(A, B, C)$ with $C \neq 0$ the set $\rho(A, B, C)$ is empty.*

Next note that if $\lambda \in \rho(A, B, C)$, then by property (2) in Theorem 3.2 both $A - \lambda$ and $B - \lambda$ are semi-Fredholm operators. As we all know that the semi-Fredholm domain of an operator is an open set. This motivates the following theorem.

**Theorem 3.4.** *For a 3-tuple $(A, B, C)$, the set $\rho(A, B, C)$ is an open set.*

In the proof of this theorem we need the following well-known fact.

**Lemma 3.5.** *Let $H_1$ and $H_2$ be two closed subspaces of a Hilbert space $H$. Let $P_i$ be the orthogonal projection from $H$ onto $H_i, i = 1, 2$. If $\|P_1 P_2\| < 1$ (or $\|P_2 P_1\| < 1$), then $H_1 \cap H_2 = \{0\}$*

**Proof of Theorem 3.4:** By Theorem 3.2, if $\lambda_0 \in \rho(A, B, C)$ then the statements (1), (2), (3) and (4) in Theorem 3.2 are satisfied; moreover, by (4)' in the proof of Theorem 3.2, without loss of generality, we can assume that $R(A-\lambda_0) \perp R(C)$, $R(B^*-\bar{\lambda}_0) \perp R(C^*)$ and that $R(C)$ (thus $R(C^*)$) is closed. It is easy to see that there exists an open neighborhood $D_1(\lambda_0)$ of $\lambda_0$ such that (1), (2), (3) are still satisfied for any $\lambda \in D_1(\lambda_0)$. To prove that $\rho(A,B,C)$ is open, it is enough to show that there is an open neighborhood $D_0(\lambda_0) \subset D_1(\lambda_0)$ of $\lambda_0$ such that the statement (4) in Theorem 3.2 holds for $\lambda \in D_0(\lambda_0)$. First we show that there is an open neighborhood $D'(\lambda_0)$ of $\lambda_0$ such that $R(A-\lambda) \cap R(C) = \{0\}$ holds for any $\lambda \in D'(\lambda_0)$. For $\lambda \in D_1(\lambda_0)$, by the statement (2) in Theorem 3.2 , the generalized inverse (Moore- Penrose inverse [7]) $(A-\lambda)^+$ of $(A-\lambda)$ is given by

$$(A-\lambda)^+ = (A^*-\bar{\lambda})((A^*-\bar{\lambda})(A-\lambda))^{-1};$$

and it is easy to see that $(A-\lambda)^+$ is continous at $\lambda_0$. Note that $(A-\lambda_0)(A-\lambda_0)^+$ and $(A-\lambda)(A-\lambda)^+$ are orthogonal projections on $R(A-\lambda_0)$ and $R(A-\lambda)$, respectively. By $R(A-\lambda_0) \perp R(C)$, we have $\|(A-\lambda_0)(A-\lambda_0)^+P\| = 0$, where $P$ denotes the orthogonal projection on $R(C)$. The continuity of $(A-\lambda)^+$ at $\lambda_0$ implies that there is an open neighborhood $D'(\lambda_0) \subset D_1(\lambda_0)$ such that

$$\|(A-\lambda)(A-\lambda)^+P\| < 1$$

for any $\lambda \in D'(\lambda_0)$. By Lemma 3.5, we see that $R(A-\lambda) \cap R(C) = \{0\}$ holds for any $\lambda \in D'(\lambda_0)$. Similarly, there exists an open neighborhood $D''(\lambda_0) \subset D_1(\lambda_0)$ of $\lambda_0$ such that $R(B^*-\bar{\lambda}) \cap R(C^*) = \{0\}$ for any $\lambda \in D''(\lambda_0)$. Take $D_0(\lambda_0) = D'(\lambda_0) \cap D''(\lambda_0)$; then for any $\lambda \in D_0(\lambda_0)$ the four statements (1), (2), (3) and (4) in Theorem 3.2 are satisfied. That is, $D_0(\lambda_0) \subset \rho(A,B,C)$, so the set $\rho(A,B,C)$ is an open set.

Next we will discuss the set $\sigma(A,B,C)$. In the infinite dimensional case, unlike in the finite dimensional case (see Proposition 2.1), the computation of the set $\sigma(A,B,C)$ is not trivial even in the case $C=0$. This follows from a result of [3] which we will recall below, just for the sake of completeness.

**Theorem 3.6.** (see [3]) *For a given 3-tuple $(A,B,C)$ with $C=0$, we have*

$$\sigma(A,B,C) := \bigcap_{X \in B(H,K)} \sigma(M_X(A,B,0)) = \sigma_\delta(A) \cup \sigma_\pi(B) \cup \{\lambda : n(A-\lambda) \neq n(B^*-\bar{\lambda})\},$$

*where $N(A-\lambda)$ is the kernel of operator $N-\lambda$, $n(A-\lambda) = dim N(A-\lambda)$, $\sigma_\pi(B)$ is the approximate point spectrum of $B$ and $\sigma_\delta(A)$ $(= \{\bar{\lambda} : \lambda \in \sigma_\pi(A^*)\})$ is the defect spectrum of $A$.*

It is easy to see that for a given 3-tuple $(A,B,C)$, if $R(A-\lambda)+R(C) \neq H$ or $R(B^*-\bar{\lambda})+R(C^*) \neq K$, then $\lambda \in \sigma(A,B,C)$. But in the infinite dimensional case, unlike in the finite dimensional case (see Theorem 2.2), the condition that $R(A-\lambda)+R(C) \neq H$ or

$R(B^* - \bar{\lambda}) + R(C^*) \neq K$ is not a necessary condition for $\lambda \in \sigma(A, B, C)$. By applying Theorem 3.6, we illustrate this point in the following example.

**Example 3.7.** Let $S$ be a unilateral shift on a Hilbert space $H$ and $K = H \oplus H$. Let $A = S^*$, $B = S \oplus S$ and $C = 0$. Then it is clear that $R(A) = H$ and $R(B^*) = K$. But by Theorem 3.6 above, $0 \in \sigma(A, B, C)$, since $1 = n(A) \neq n(B^*) = 2$. That is for any $X \in B(H, K)$ the operator

$$M_X = \begin{pmatrix} S^* & 0 & 0 \\ X_1 & S & 0 \\ X_2 & 0 & S \end{pmatrix}$$

is not invertible. In fact, by Theorem 3.6,

$$\sigma(M_X(S^*, S \oplus S, 0)) = \{z : |z| \leq 1\} \text{ for any } X \in B(H, H \oplus H)$$

Next we introduce some notations. For an operator $D$, let $\sigma_e(D)$, $\sigma_{e\pi}(D)$, and $\sigma_{e\delta}(D)$ $(= \{\bar{\lambda} : \lambda \in \sigma_{e\pi}(D^*)\})$ be the essential spectrum of $D$, the essential approximate point spectrum of $D$, and the essential defect spectrum of $D$, respectively; and by $D \in SF$ we mean that $D$ is semi-Fredholm. If $D \in SF$ we define the index of $D$ to be $ind\, D = n(D) - n(D^*)$.

To identify the set $\sigma(A, B, C)$ for a 3-tuple $(A, B, C)$ with $C$ a finite rank operator, we begin with a result concerning essential spectra.

**Proposition 3.8.** *For a given 3-tuple $(A, B, C)$ with $C$ a compact operator, we have*

$$\bigcap_{X \in B(H,K)} \sigma_e(M_X) = \sigma_{e\delta}(A) \cup \sigma_{e\pi}(B) \cup G(A, B) \cup G(B, A) \tag{3.2}$$

*where*

$$G(A, B) = \{\lambda : A - \lambda, B - \lambda \in SF,\ ind(A - \lambda) = \infty \text{ and } |ind(B - \lambda)| < \infty\},$$

*and $G(B, A)$ is defined analogously.*

**Proof.** Since $C$ is compact, $M_X$ is essentially invertible if and only if $M_X(A, B, 0)$ is essentially invertible; to prove (3.2), we can assume $C = 0$. We first show that the right hand side of (3.2) is included in the left hand side of (3.2). Without loss of generality, assume $\lambda = 0$. If $0 \in \sigma_{e\delta}(A) \cup \sigma_{e\pi}(B)$, it is clear that $M_X$ is not Fredhlom for any $X$, so 0 is in the left hand side of (3.2). If $0 \in G(A, B) \cup G(B, A)$, then it is easy to see that $M_X$ is semi-Fredholm for any $X$. By the basic propeties of index function, we get that $|ind\, M_X| = |ind\, M_0| = |ind\, A + ind\, B| = \infty$. That is, $M_X$ is not Fredholm for any $X$. This proves that the right hand side of (3.2) is included in the left hand side of (3.2).

On the other hand, if 0 is not in the right hand side of (3.2), then either both $A$ and $B$ are Fredholm or $A, B \in SF$ and $ind\, A = \infty, ind\, B = -\infty$. In the first case, it is clear that $M_X$ is Fredholm for any $X$. In the second case, we claim that there exists $X_0$ such that $M_{X_0}$ is Fredholm. Indeed let $\{x_i\}_{i=1}^{\infty} \subset H$ be an orthonormal basis of $N(A)$ and $\{y_i\}_{i=1}^{\infty} \subset K$ be an orthonormal basis of $R(B)^{\perp}$. Let

$$X_0 = \sum_{i=1}^{\infty} y_i \otimes x_i.$$

Let $A^+$, $B^+$ be the Moore-Penrose inverses of $A$ and $B$, respectively. Then it is easy to verify that

$$\begin{pmatrix} A & 0 \\ X_0 & B \end{pmatrix} \begin{pmatrix} A^+ & X_0^* \\ 0 & B^+ \end{pmatrix} = \begin{pmatrix} P_{R(A)} & 0 \\ 0 & I \end{pmatrix}, \begin{pmatrix} A^+ & X_0^* \\ 0 & B^+ \end{pmatrix} \begin{pmatrix} A & 0 \\ X_0 & B \end{pmatrix} = \begin{pmatrix} I & 0 \\ 0 & P_{R(B^*)} \end{pmatrix},$$

where $P_{R(A)}$, $P_{R(B^*)}$ are orthogonal projections onto $R(A)$ and $R(B)$, respectively. Hence $M_{X_0}$ is Fredholm. The proof is complete.

The following theorem is a natural generalization of Theorem 3.6.

**Theorem 3.9.** *For a given* 3 *-tuple* $(A, B, C)$ *with* $C$ *a finite rank operator, we have*

$$\begin{aligned} \sigma(A, B, C) = \{\lambda : R(A-\lambda) + R(C) \neq H\} & \\ \cup \{\lambda : R(B^* - \bar{\lambda}) + R(C^*) \neq K\} \cup G(A, B) \cup G(B, A) & \\ \cup \{\lambda : \text{ both } A - \lambda \text{ and } B - \lambda \text{ are Fredholm, and } ind\, A + ind\, B \neq 0\}. & \quad (3.3) \end{aligned}$$

**Proof.** We first show that the right hand side of (3.3) is included in the left hand side of (3.3). Without loss of generality, assume $\lambda = 0$. If $R(A) + R(C) \neq H$ or $R(B^*) + R(C^*) \neq K$, it is clear that $M_X$ is not invertible for any $X$, so $0 \in \sigma(A, B, C)$. If $R(A) + R(C) = H$, $R(B^*) + R(C^*) = K$ (thus $A, B \in SF$ because $C$ is a finite rank operator) and $ind\, A + ind\, B \neq 0$, it is easy to see that $M_X$ is semi-Fredholm for all $X$, thus by the basic properties of the index function, we have that

$$ind\, M_X = ind \begin{pmatrix} A & 0 \\ 0 & B \end{pmatrix} = ind\, A + ind\, B \neq 0.$$

Hence $M_X$ is not invertible for any $X$, that is $0 \in \sigma(A, B, C)$.

Now we prove that the left hand side of (3.3) is included in the right hand side of (3.3). Assume 0 is not in the right hand side of (3.3), then there are two possible cases: in the first case, $R(A) + R(C) = H$, $R(B^*) + R(C^*) = K$, both $A$ and $B$ are Fredholm, and $ind\, A + ind\, B = 0$; in the second case, $R(A) + R(C) = H$, $R(B^*) + R(C^*) = K$ (thus $A, B \in SF$), and $ind\, A = \infty$, $ind\, B = -\infty$. We claim that in both cases there exists $X$ such that $M_X$ is invertible.

First case. By a similar argument as above, we see that $M_X$ is Fredholm and $ind\, M_X = 0$ for any $X$. Thus it is enough to choose a $X$ such that $N(M_X) = \{0\}$. Let $\{e_i\}_{i=1}^{n(A)}$ be an orthonormal basis of $N(A)$. Let $\{x_i\}_{i=1}^{t} \subset N(A)^\perp$ be an orthonormal set such that there exist $\{y_i\}_{i=1}^{t} \subset N(C)^\perp$ with

$$Ax_i + Cy_i = 0,$$

where $t = dim\,\{R(A) \cap R(C)\} = rank\, C - n(A^*)$, since $R(A) + R(C) = H$. Let $\{f_i\}_{i=1}^{k}$ be an orthonormal basis of $\{BN(C)\}^\perp$, where $BN(C)$ is the subspace spanned by $\{By, y \in N(C)\}$. $BN(C)$ is closed because $B$ is Fredholm and $N(C)$ is closed. Since

$$\{BN(C)\}^\perp = N(B^*)\dot{+}\{f : f \in R(B), B^* f \in R(C^*)\}$$

and $R(B^*) + R(C^*) = K$, we have that

$$k = n(B^*) + dim\,\{R(B^*) \cap R(C^*)\} = n(B^*) + rank\, C^* - n(B) = rank\, C - ind\, B.$$

By the assumption that $ind\, A + ind\, B = 0$, we get that

$$t + n(A) = rank\, C + ind\, A = rank\, C - ind\, B = k.$$

Let

$$X = \sum_{i=1}^{t}(-By_i + f_i) \otimes x_i + \sum_{j=1}^{n(A)} f_{j+t} \otimes e_j.$$

It is easy to check that for this $X$, $N(M_X) = \{0\}$; see the proof of Theorem 2.2 for details.

Second case. Let $\{e_i\}_{i=1}^{\infty}$ be an orthonormal basis of $N(A)$, and $\{x_i\}_{i=1}^{t} \subset N(A)^\perp$ be an orthonormal set such that there exist $\{y_i\}_{i=1}^{t} \subset N(C)^\perp$ with

$$Ax_i + Cy_i = 0,$$

where $t = dim\,\{R(A) \cap R(C)\} = rank\, C - n(A^*)$, since $R(A) + R(C) = H$. Let $\{f_i\}_{i=1}^{\infty}$ be an orthonormal basis of $\{BN(C)\}^\perp$. We define

$$X = \sum_{i=1}^{t}(-By_i + f_i) \otimes x_i + \sum_{j=1}^{\infty} f_{j+t} \otimes e_j.$$

Then $M_X$ is Fredholm because

$$\begin{pmatrix} A & C \\ X_0 & B \end{pmatrix}\begin{pmatrix} A^+ & X_0^* \\ 0 & B^+ \end{pmatrix} = \begin{pmatrix} P_{R(A)} & D_1 \\ D_2 & I + D_3 \end{pmatrix}$$

and

$$\begin{pmatrix} A^+ & X_0^* \\ 0 & B^+ \end{pmatrix} \begin{pmatrix} A & C \\ X_0 & B \end{pmatrix} = \begin{pmatrix} I + D_4 & D_5 \\ D_6 & P_{R(B^*)} \end{pmatrix}$$

for some finite rank operators $D_i, i = 1, \cdots, 6$, where $P_{R(A)}$, $P_{R(B^*)}$ are orthogonal projections onto $R(A)$ and $R(B)$, respectively. We see that $N(M_X) = \{0\}$ by repeating almost verbatim the proof of Theorem 2.2. Now we prove that $M_X$ is onto. Let $z_1 \in H$ and $z_2 \in K$. By $R(A) + R(C) = H$, there exist $x_0$ and $y_0$ such that

$$Ax_0 + Cy_0 = z_1.$$

Since $\{BN(C)\}^\perp + BN(C) = K$, there exist $\alpha_i, \beta_j$ $(1 \le i \le t, 1 \le j \le \infty)$, and $z \in N(C)$ such that

$$\sum_{i=1}^{t} \alpha_i f_i + \sum_{j=1}^{\infty} \beta_j f_{j+t} + Bz = -Xx_0 - By_0 + z_2.$$

Now let

$$x = \sum_{i=1}^{t} \alpha_i x_i + \sum_{j=1}^{\infty} \beta_j e_j + x_0, \quad y = \sum_{i=1}^{t} \alpha_i y_i + z + y_0.$$

By a straightforward computation, we obtain

$$\begin{pmatrix} A & C \\ X & B \end{pmatrix} \begin{pmatrix} x \\ y \end{pmatrix} = \begin{pmatrix} z_1 \\ z_2 \end{pmatrix}.$$

This completes the proof.

**Remark.** The proof of Theorem 3.9 also shows that the right hand side of (3.3) is included in the left hand side of (3.3) in the case $C$ is assumed only to be a compact operator. But our proof for the other inclusion does not extend to this more general case. Nevertheless we conjecture that (3.3) holds for any compact operator $C$.

**Acknowledgements.** The authors thank Professors C. Foias and I. Gohberg for their helpful comments. The first named author would like to thank Department of Mathematics of Indiana University at Bloomington for their hospitality.

## References

[1]. J. A. Ball, I. Gohberg, L. Rodman and T. Shalom, *On the eigenvalues of matrices with given upper triangular part*, Integral Equations and Operator Theory **13** (1990), 488-497.

[2]. C. Davis, W. M. Kahan and H. F. Weinberger, *Norm-preserving dilations and the application to optimal error bounds*, SIAM J. Num. Anal. **19** (1982), 445-469.

[3]. Hong-ke Du and Jin Pan, *Perturbation of spectra of* $2\times 2$ *operator matrices*, preprint.

[4]. P. A. Fillmore and J. P. Williams, *On operator ranges*, Adv. Math. **7** (1971), 254-281.

[5]. C. Foias and A. Tannenbaum, *A strong Parrott's theorem*, Proc. Amer. Math. Soc. **106** (1989), 777-784.

[6]. I. Gohberg and S. Rubinstein, *A classification of upper equivalent matrices the generic case*, Integral Equations and Operator Theory **14** (1991), 533-544.

[7]. C. W. Groetsch, *Generalized Inverses of Linear Operators*, Marcel Dekker Inc. New York and Basel. 1977.

[8]. L. Gurvits, L. Rodman and I. Spitkovsky, *Spectral assignment for Hilbert space operators*, Houston Journal of Mathematics **17** (1991), 501-523.

[9]. S. Parrott, *On the quotient norn and Sz. Nagy-Foias lifting theorem*, J. Funct. Anal. **30** (1978), 311-328.

[10]. L. Rodman and T. Shalom, *Jordan form of completion of partial upper triangular matrices*, Linear Algebra Appl **168** (1992), 221-249.

[11]. H. J. Woerdeman, *Matrix and Operator Extentions*, CWT, Trant **68** (1989).

Hong-ke Du
Department of Mathematics
Shaanxi Normal University
Xi'an, 710062.
P. R. China

Caixing Gu
Department of Mathematics
Indiana University
Bloomington, Indiana 47405
U. S. A.

1991 Mathematics Subject Classification. 47A55, 47A10, 15A09, 15A18

Operator Theory:
Advances and Applications, Vol. 64



# The Exact $H^2$ Estimate for the Central $H^\infty$ Interpolant

C. Foias, A. E. Frazho and W. S. Li

The central intertwining lifting for the commutant lifting theorem is used to generalize and tighten the $H^2$–$H^\infty$ bounds of Kaftal-Larson-Weiss. Several explicit formulas for the central intertwining lifting and the outer factor of its corresponding defect operator are given. Some applications to a mixed $H^2$–$H^\infty$ Carathéodory and four-block interpolation problems are also presented.

## Introduction

Kaftal-Larson-Weiss [34] discovered that given any g in $L^\infty$ and $\delta > 1$, there exists an h in $H^\infty$ satisfying the following mixed $H^2 - H^\infty$ bounds

$$\|g + h\|_\infty \leq \delta\, d_\infty \quad \text{and} \quad \|g + h\|_2 \leq \frac{\delta\, d_2}{\sqrt{\delta^2 - 1}}$$

where $d_\infty$ is the distance from g to $H^\infty$ in the $L^\infty$ norm $\|\cdot\|_\infty$, and $d_2$ is the distance from g to $H^2$ in the $L^2$ norm $\|\cdot\|_2$. The Kaftal-Larson-Weiss result was generalized to the two-sided block Nehari setting by using the central solution for the commutant lifting in [19], where state space formulas to compute a solution in the rational Nehari setting were also given. It is worth mentioning here other work inspired by the Kaftal-Larson-Weiss result. A proof of the Kaftal-Larson-Weiss result based on the Dym-Gohberg maximal entropy theory [15] is presented in [24]. By applying skew-Toeplitz techniques, an algorithm to find an h in $H^\infty$ satisfying the Kaftal-Larson-Weiss bounds for certain nonrational g is given in [20]. For some other results

This research was supported in part by the Research Fund of Indiana University and grants from the National Science Foundation Nos. DMS-9024769, DMS-9007802, and DMS-8921369.

concerning the Kaftal-Larson-Weiss bounds for the four-block problem, tangential interpolation and certain $H^2 - H^\infty$ control problems see [32], [23] and [41], respectively. Some nice generalizations of the Kaftal-Larson-Weiss bounds to $H^p$ spaces are presented in [33], [36], [40].

In this paper, we will use the central solution for the commutant lifting theorem to generalize and tighten the Kaftal-Larson-Weiss bounds. We will present several explicit formulas to compute the central intertwining lifting and apply some of our results to a mixed $H^2 - H^\infty$ Carathéodory and four-block interpolation problems. Finally, we note that both $H^2$ and $H^\infty$ optimization problems have played an important role in control theory [4,5,7,8,9,12,13,14,21,26,27,28,35,37,38,39,42]. For this reason, it is believed that some of our results may be useful in control theory.

## 1. An Improved Kaftal-Larson-Weiss Estimate

Throughout this paper, we will follow the standard notation for Hilbert spaces presented in [18], [45]. For example, the orthogonal projection onto a subspace $\mathcal{M}$ is denoted by $P_{\mathcal{M}}$, and $H^2(\mathcal{E})$ is the Hardy space of all analytic functions in the open unit disc with values in $\mathcal{E}$ whose Fourier coefficients are square summable. To begin, let T be an isometry on $\mathcal{H}$ and $T'$ a contraction on $\mathcal{H}'$. Let A be an operator mapping $\mathcal{H}$ into $\mathcal{H}'$ satisfying $T'A = AT$. Recall that $U'$ on $\mathcal{K}'(\supseteq \mathcal{H}')$ is an isometric lifting of $T'$ if $U'$ is an isometry satisfying $U'^* \mid \mathcal{H}' = T'^*$. An operator B mapping $\mathcal{H}$ into $\mathcal{K}'$ is called *an intertwining lifting* of A, if

$$U'B = BT \quad \text{and} \quad P_{\mathcal{H}'}B = A\,. \tag{1.1}$$

The commutant lifting theorem ([18], [44], [45]) states that there exists an intertwining lifting B of A satisfying $\|B\| = \|A\|$. In this paper we will present some new results on the central intertwining lifting $B_\gamma$ of A presented in [18], [19] and used in [19], [23], [32] to solve certain $H^2 - H^\infty$ interpolation problems. For example, we will show that one can use the central intertwining lifting $B_\gamma$ to obtain some stronger $H^2 - H^\infty$ bounds for the Nehari problem, than the bounds presented in [19], [24], [34]. Moreover, we will give some explicit formulas to compute $B_\gamma$ and the "outer spectral factor" for $\gamma^2 I - B_\gamma^* B_\gamma$. Some of our formulas are the commutant lifting generalizations of formulas introduced in [3], [30], for the block Nehari case. We mention that our approach is quite different in its geometrical framework and also leads to some new mixed $H^2$–$H^\infty$ bounds for the four-block Nehari problem.

Since any isometric lifting $U'$ of $T'$ admits a reducing decomposition of the form $U' = U_m \oplus U_1$ where $U_m$ is the minimal isometric dilation of $T'$ (see Remark VI.3.3 in [18]), we assume without loss of generality $U'$ is the minimal isometric dilation of $T'$. (Recall that $U'$ on $\mathcal{K}'$ is a *minimal isometric dilation* of $T'$ if $U'$ is an isometric lifting of $T'$ and $\mathcal{H}'$ is cyclic for $U'$; see [18], [45].) Furthermore, because all minimal isometric dilations of $T'$ are unitarily equivalent, we always assume that $U'$ on $\mathcal{K}'$ is the Sz.-Nagy-Schäffer minimal isometric dilation of $T'$. To be precise, we assume that $U'$ is the isometry on $\mathcal{K}' = \mathcal{H}' \oplus H^2(\mathcal{D}')$ given by

$$U' = \begin{bmatrix} T' & 0 \\ D' & S' \end{bmatrix} \quad \text{on} \quad \begin{bmatrix} \mathcal{H}' \\ H^2(\mathcal{D}') \end{bmatrix} \tag{1.2}$$

where $S'$ is the unilateral shift on $H^2(\mathcal{D}')$. Here $D'$ is the positive square root of $I - T'^*T'$, while $\mathcal{D}'$ is the closed range of $D'$. Notice that the operator $D'$ in (1.2) maps $\mathcal{H}'$ into the constant functions of $H^2(\mathcal{D}')$. In other words,

$$U'(h \oplus g) = T'h \oplus (D'h + zg) \qquad (h \oplus g \in \mathcal{H}' \oplus H^2(\mathcal{D}')) . \tag{1.3}$$

We say that an operator C is *bounded by* $\gamma$ if $\|C\| \le \gamma$ and *strictly bounded by* $\gamma$ if $\|C\| < \gamma$. If C is an operator mapping $\mathcal{H}$ into $\mathcal{H}_1$ bounded by a *specified* $\gamma$, then $D_C$ is the positive square root of $\gamma^2 I - C^*C$ and $\mathcal{D}_C$ is the closed range of $D_C$. Obviously if $\|C\| < \gamma$, then $D_C$ is invertible and $\mathcal{D}_C = \mathcal{H}$. Throughout this paper A is an operator mapping $\mathcal{H}$ into $\mathcal{H}'$ bounded by $\gamma$ satisfying $T'A = AT$ where T on $\mathcal{H}$ is an isometry and $T'$ on $\mathcal{H}'$ is contractive. In order to introduce the central intertwining lifting $B_\gamma$ of A, let $\omega$ be the isometry from $\mathcal{F} = \overline{D_A T \mathcal{H}}$ into $\mathcal{D}' \oplus \mathcal{D}_A$ defined by

$$\omega D_A Th = D'Ah \oplus D_A h \qquad (h \in \mathcal{H}) . \tag{1.4}$$

(Using $T'A = AT$ it is easy to verify that $\omega$ is an isometry; see Section V.1 in [18] for further details.) Let $\Pi'$ be the operator from $\mathcal{D}' \oplus \mathcal{D}_A$ to $\mathcal{D}'$ which picks out the first component of $\mathcal{D}' \oplus \mathcal{D}_A$ and $\Pi_A$ the operator from $\mathcal{D}' \oplus \mathcal{D}_A$ into $\mathcal{D}_A$ which picks out the second component of $\mathcal{D}' \oplus \mathcal{D}_A$. To be precise,

$$\Pi'(d' \oplus d_A) = d' \quad \text{and} \quad \Pi_A(d' \oplus d_A) = d_A \qquad (d' \oplus d_A \in \mathcal{D}' \oplus \mathcal{D}_A) . \tag{1.5}$$

Finally, by modifying equation (4.10) in Chapter XIV of [18] we define the operator $B_\gamma$ mapping $\mathcal{H}$ into $\mathcal{K}'$ by

$$B_\gamma = \begin{bmatrix} A \\ \Pi' \omega P_{\mathcal{F}} (I - z\Pi_A \omega P_{\mathcal{F}})^{-1} D_A \end{bmatrix} : \mathcal{H} \to \begin{bmatrix} \mathcal{H}' \\ H^2(\mathcal{D}') \end{bmatrix} \tag{1.6}$$

where $P_{\mathcal{F}}$ is the orthogonal projection onto $\mathcal{F}$. It is easy to verify that $B_\gamma$ is indeed an intertwining lifting of A bounded by $\gamma$, that is, $\|B_\gamma\| \le \gamma$ and $B_\gamma$ satisfies (1.1) where $U'$ is the Sz.-Nagy-Schäffer minimal isometric dilation of $T'$ in (1.2). The intertwining lifting $B_\gamma$ of A in (1.6) is called the *central intertwining lifting* of A, because the set of all intertwining lifting B of A satisfying $\|B\| \le \gamma$ is parameterized by the closed unit ball in $H^\infty(\mathcal{G}, \mathcal{G}')$, and the central intertwining lifting $B_\gamma$ of A in (1.6) is the one obtained by choosing the free parameter in the closed unit ball of $H^\infty(\mathcal{G}, \mathcal{G}')$ to be trivial function zero; see Chapter XIV in [18] for further details.

Throughout this paper $\Pi_o$ is the operator from $\mathcal{H}$ onto $\mathcal{L}$ defined by $\Pi_o = P_{\mathcal{L}}$ where $\mathcal{L} = \ker T^*$. Notice that $\Pi_o^*$ is precisely the identity operator embedding $\mathcal{L}$ into $\mathcal{H}$. Now assume that $\|A\| < \gamma$. Obviously $D_A^2$ is invertible. Moreover, since $D_A^2 \le \gamma^2 I$, we have, $\gamma^{-2} I \le D_A^{-2}$. (Recall that if X and Y are two strictly positive operators satisfying $X \le Y$, then $Y^{-1} \le X^{-1}$.) This implies that $\gamma^{-2} I \le \Pi_o D_A^{-2} \Pi_o^*$. Therefore $(\Pi_o D_A^{-2} \Pi_o^*)^{-1} \le \gamma^2 I$. Throughout, $N_A$ is the strictly positive operator bounded by $\gamma$ on $\mathcal{L}$ defined by

$$N_A = (\Pi_o D_A^{-2} \Pi_o^*)^{-1/2} \,. \tag{1.7}$$

The following result uses the operator $N_A$ to give some bounds on the central intertwining lifting $B_\gamma$ of A.

**THEOREM 1.1.** *Let* T *be an isometry on* $\mathcal{H}$ *and* A *be an operator bounded by* $\gamma$ *mapping* $\mathcal{H}$ *into* $\mathcal{H}'$ *satisfying* $T'A = AT$. *Then* $B_\gamma$, *defined in (1.6), is an intertwining lifting of* A *satisfying* $\|B_\gamma\| \le \gamma$. *Moreover, if* A *is strictly bounded by* $\gamma$, *then the central intertwining lifting* $B_\gamma$ *of* A *satisfies the following inequality:*

$$\|B_\gamma a\|^2 \le \|D_{N_A} a\|^2 \le \frac{\gamma^2 \|Aa\|^2}{\gamma^2 - \|A\|^2 + \|Aa\|^2} \qquad (a \in \ker T^* \text{ and } \|a\| = 1) \tag{1.8}$$

*for all unit vectors a in the kernel of* $T^*$.

**PROOF.** Obviously we have $P_{\mathcal{H}'} B_\gamma = A$. To verify that $U' B_\gamma = B_\gamma T$ where $U'$ is the Sz.-Nagy-Schäffer minimal isometric of $T'$ defined in (1.2), first notice that

$$D' A = \Pi' \omega P_{\mathcal{F}} D_A T \quad \text{and} \quad D_A = \Pi_A \omega P_{\mathcal{F}} D_A T \,. \tag{1.9}$$

So using this along with the definition of $B_\gamma$ in (1.6) we have

$$U'B_\gamma = \begin{bmatrix} T'A \\ D'A + z\Pi'\omega P_{\mathcal{F}}(I - z\Pi_A\omega P_{\mathcal{F}})^{-1}D_A \end{bmatrix} =$$

$$\begin{bmatrix} AT \\ \Pi'\omega P_{\mathcal{F}}D_AT + z\Pi'\omega P_{\mathcal{F}}(I - z\Pi_A\omega P_{\mathcal{F}})^{-1}\Pi_A\omega P_{\mathcal{F}}D_AT \end{bmatrix} =$$

$$\begin{bmatrix} AT \\ \Pi'\omega P_{\mathcal{F}}(I - z\Pi_A\omega P_{\mathcal{F}})^{-1}D_AT \end{bmatrix} = B_\gamma T .$$

Therefore $B_\gamma$ is an intertwining lifting of A.

To show that $B_\gamma$ is bounded by $\gamma$ let h be in $\mathcal{H}$. Then the definition of the central contractive intertwining lifting $B_\gamma$ of A in (1.6), along with $\|\Pi' f\|^2 = \|f\|^2 - \|\Pi_A f\|^2$ for any f in $\mathcal{D}' \oplus \mathcal{D}_A$ gives

$$\|B_\gamma h\|^2 = \|Ah\|^2 + \sum_{j=0}^{\infty} \|\Pi'\omega P_{\mathcal{F}}(\Pi_A\omega P_{\mathcal{F}})^j D_A h\|^2 =$$

$$\|Ah\|^2 + \sum_{j=0}^{\infty} (\|\omega P_{\mathcal{F}}(\Pi_A\omega P_{\mathcal{F}})^j D_A h\|^2 - \|\Pi_A\omega P_{\mathcal{F}}(\Pi_A\omega P_{\mathcal{F}})^j D_A h\|^2) \le$$

$$\|Ah\|^2 + \lim_{n\to\infty} \sum_{j=0}^{n} (\|(\Pi_A\omega P_{\mathcal{F}})^j P_{\mathcal{F}} D_A h\|^2 - \|(\Pi_A\omega P_{\mathcal{F}})^{j+1} D_A h\|^2) =$$

$$\|Ah\|^2 + \|P_{\mathcal{F}} D_A h\|^2 - \lim_{n\to\infty} \| (\Pi_A\omega P_{\mathcal{F}})^{n+1} D_A h\|^2 \le \|Ah\|^2 + \|P_{\mathcal{F}} D_A h\|^2 .$$

So, we obtain the following inequality

$$\|B_\gamma h\|^2 \le \|Ah\|^2 + \|P_{\mathcal{F}} D_A h\|^2 \qquad (h \in \mathcal{H}) . \tag{1.10}$$

In particular, $\|B_\gamma h\|^2 \le \|Ah\|^2 + \|D_A h\|^2 = \gamma^2\|h\|^2$. Therefore $B_\gamma$ is is an intertwining lifting of A bounded by $\gamma$.

To complete the proof we assume that A is strictly bounded by $\gamma$. To establish (1.8) we need an explicit expression for the orthogonal projection $P_{\mathcal{F}}$. Since $\|A\| < \gamma$, the operator $D_A$ is invertible. Because T is an isometry, the operator $X = D_AT$ on $\mathcal{H}$ has closed range. In fact, the range of X is $\mathcal{F}$. Therefore, $X^*X$ is invertible. Moreover, using $T^*T = I$ we have

$$X^*X = T^*(\gamma^2 I - A^*A)T = \gamma^2 I - T^*A^*AT = D_{AT}^2 .$$

From the fact that the operator $P = X(X^*X)^{-1}X^*$ is onto $\mathcal{F}$ and $P^2 = P = P^*$, we conclude that the orthogonal projection $P_{\mathcal{F}}$ onto $\mathcal{F}$ is given by

$$P_{\mathcal{F}} = X(X^*X)^{-1}X^* = D_A T D_{AT}^{-2} T^* D_A \,. \tag{1.11}$$

Let a be any unit vector in $\mathcal{L} = \ker T^*$. Using the previous expression for $P_{\mathcal{F}}$ in (1.10) and $T^*a = 0$ we have

$$\begin{aligned} &\|B_\gamma a\|^2 \le \|Aa\|^2 + (P_{\mathcal{F}} D_A a, D_A a) = \|Aa\|^2 + (D_{AT}^{-2} T^* D_A^2 a, T^* D_A^2 a) = \\ &\|Aa\|^2 + (D_{AT}^{-2} T^* A^* Aa, T^* A^* Aa) = \\ &(a, A^*Aa) + (T(\gamma^2 I - T^* A^* AT)^{-1} T^* A^* Aa, A^* Aa) = \\ &(a, A^*Aa) + ((\gamma^2 I - TT^* A^* A)^{-1} TT^* A^* Aa, A^* Aa) = \\ &\gamma^2 (A^* A(\gamma^2 I - TT^* A^* A)^{-1} a, a) \,. \end{aligned} \tag{1.12}$$

We claim that

$$\gamma^2(\gamma^2 I - TT^* A^* A)^{-1} a = D_A^{-2} \Pi_o^* N_A^2 a \qquad (a \in \ker T^*) \tag{1.13}$$

where $N_A$ on $\mathcal{L}$ is defined in (1.7). To prove this let

$$f = (\gamma^2 I - TT^* A^* A) D_A^{-2} \Pi_o^* N_A^2 a \,. \tag{1.14}$$

Then using $TT^* = I - P_{\mathcal{L}}$ we have

$$\begin{aligned} (I - P_{\mathcal{L}})f = TT^* f = \gamma^2 TT^* D_A^{-2} \Pi_o^* N_A^2 a - TT^* A^* A D_A^{-2} \Pi_o^* N_A^2 a = \\ TT^*(\gamma^2 I - A^* A) D_A^{-2} \Pi_o^* N_A^2 a = TT^* \Pi_o^* N_A^2 a = 0 \,. \end{aligned}$$

Thus $P_{\mathcal{L}} f = f$. So according to (1.14) we have

$$f = P_{\mathcal{L}} f = P_{\mathcal{L}} (\gamma^2 I - TT^* A^* A) D_A^{-2} \Pi_o^* N_A^2 a = \gamma^2 P_{\mathcal{L}} D_A^{-2} \Pi_o^* N_A^2 a = \gamma^2 a \,.$$

Therefore $f = \gamma^2 a$ and equation (1.13) follows from (1.14). Now using (1.13), we have

$$\begin{aligned} \gamma^2 (A^* A(\gamma^2 I - TT^* A^* A)^{-1} a, a) = (A^* A D_A^{-2} \Pi_o^* N_A^2 a, a) = \\ ((\gamma^2 D_A^{-2} - I) \Pi_o^* N_A^2 a, a) = \gamma^2 (a, a) - (N_A^2 a, a) = \|D_{N_A} a\|^2 \,. \end{aligned}$$

This along with (1.12) readily yields the first inequality for $\|B_\gamma a\|^2$ in (1.8).

To complete the proof notice that

$$\begin{aligned} &\gamma^2 I - N_A^2 = \gamma^2 I - (\Pi_o D_A^{-2} \Pi_o^*)^{-1} = \\ &\gamma^2 I - \gamma^2 (\Pi_o (\gamma^2 I - A^* A + A^* A) D_A^{-2} \Pi_o^*)^{-1} = \\ &\gamma^2 I - \gamma^2 (I + \Pi_o A^* D_{A^*}^{-2} A \Pi_o^*)^{-1} \,. \end{aligned} \tag{1.15}$$

Now let Y be any operator mapping $\mathcal{H}_1$ into $\mathcal{H}_2$. We claim that for any unit vector a in $\mathcal{H}_1$ the following inequality holds:

$$\frac{1}{1+\|Ya\|^2} \le ((I+Y^*Y)^{-1}a, a) \qquad (\|a\|=1)\,. \tag{1.16}$$

This result follows from the Cauchy-Schwartz inequality, that is,

$$\begin{aligned}&1 = \|a\|^4 = ((I+Y^*Y)^{1/2}a, (I+Y^*Y)^{-1/2}a)^2 \le \\ &\|(I+Y^*Y)^{1/2}a\|^2\|(I+Y^*Y)^{-1/2}a\|^2 = \\ &((I+Y^*Y)a, a)((I+Y^*Y)^{-1}a, a) = (1+\|Ya\|^2)((I+Y^*Y)^{-1}a, a)\,.\end{aligned} \tag{1.17}$$

Dividing by $1+\|Ya\|^2$ readily produces the inequality in (1.16).

Now let $Y = D_{A^*}^{-1}A\Pi_o^*$. Using (1.16) in (1.15) along with $\|B_\gamma a\|^2 \le \|D_{N_A}a\|^2$ we obtain

$$\begin{aligned}&\|B_\gamma a\|^2 \le \|D_{N_A}a\|^2 = ((\gamma^2 I - N_A^2)a, a) = \\ &\gamma^2 - \gamma^2((I+\Pi_o A^* D_{A^*}^{-2} A\Pi_o^*)^{-1}a, a) = \\ &\gamma^2 - \gamma^2((I+Y^*Y)^{-1}a, a) \le \gamma^2 - \frac{\gamma^2}{1+\|Ya\|^2} = \\ &\frac{\gamma^2\|Ya\|^2}{1+\|Ya\|^2} = \frac{\gamma^2\|D_{A^*}^{-1}Aa\|^2}{1+\|D_{A^*}^{-1}Aa\|^2}\,.\end{aligned} \tag{1.18}$$

Because $\|A\| < \gamma$, we have $(\gamma^2 - \|A\|^2)I \le D_{A^*}^2$. Thus

$$\|D_{A^*}^{-1}Aa\|^2 = (D_{A^*}^{-2}Aa, Aa) \le \frac{\|Aa\|^2}{\gamma^2 - \|A\|^2}\,. \tag{1.19}$$

Using (1.19) in the last term of (1.18), along with the fact that the function $\tau(1+\tau)^{-1}$ is increasing, equation (1.18) now becomes

$$\begin{aligned}&\|B_\gamma a\|^2 \le \|D_{N_A}a\|^2 \le \frac{\gamma^2\|D_{A^*}^{-1}Aa\|^2}{1+\|D_{A^*}^{-1}Aa\|^2} \le \\ &\frac{\gamma^2\|Aa\|^2(\gamma^2-\|A\|^2)^{-1}}{1+\|Aa\|^2(\gamma^2-\|A\|^2)^{-1}} = \frac{\gamma^2\|Aa\|^2}{\gamma^2-\|A\|^2+\|Aa\|^2}\,.\end{aligned}$$

This gives (1.8) and completes the proof.

Let $\mathcal{X}$ and $\mathcal{Y}$ be Hilbert spaces and L be an operator mapping $\mathcal{X}$ into $\mathcal{Y}$. Recall that the Hilbert-Schmidt norm of L is defined by

$$\|L\|_2^2 = \sum_i \|L\phi_i\|^2 \tag{1.20}$$

where $\{\phi_i\}$ is an orthonormal basis for $\mathcal{X}$. The Hilbert-Schmidt norm of L is independent of the choice for the orthonormal basis. The following result, which is essentially folklore, will be useful in generalizing the Kaftal-Larson-Weiss $H^2 - H^\infty$ Nehari interpolation result [34] to the multidimensional setting.

**LEMMA 1.2.** *Let* L *be a linear operator mapping a n dimensional space* $\mathcal{X}$ *into* $\mathcal{Y}$. *Then*

$$\frac{\|L\|_2^2}{n} = \max \min \|L\phi_i\|^2 \tag{1.21}$$

*where the minimum is taken over a specified orthonormal basis* $\{\phi_i\}_1^n$ *for* $\mathcal{X}$ *and the maximum is taken over all orthonormal basis for* $\mathcal{X}$.

**PROOF.** For completeness a self contained proof is given. Let $\{\phi_i\}_1^n$ be any orthonormal basis for $\mathcal{X}$ arranged in the following way $\|L\phi_1\|^2 \geq \|L\phi_2\|^2 \geq \cdots \geq \|L\phi_n\|^2$. According to the definition of the Hilbert-Schmidt norm

$$\|L\|_2^2 = \sum_{i=1}^n \|L\phi_i\|^2 \geq \sum_{i=1}^n \|L\phi_n\|^2 = n\|L\phi_n\|^2 .$$

This readily implies that the mean m of $\{\|L\phi_i\|^2\}$ satisfies the following inequality

$$m = \frac{\|L\|_2^2}{n} \geq \max \min \|L\phi_i\|^2 . \tag{1.22}$$

To complete the proof we must show that there is equality in (1.22). To this end, let r be a scalar in [0,1] and t the positive square root of $1 - r^2$. Let $\psi$ be the unit vector defined by $\psi_1 = r\phi_1 + t\phi_n$, and f(r) the continuous function defined by

$$\begin{gathered} f(r) = \|L\psi_1\|^2 = \|Lr\phi_1 + Lt\phi_n\|^2 = \\ r^2\|L\phi_1\|^2 + 2\,r\,t\,\mathrm{Re}\,(L\phi_1, L\phi_n) + t^2\|L\phi_n\|^2 . \end{gathered} \tag{1.23}$$

Notice that because m is the mean of $\{\|L\phi_i\|^2\}$, we have

$$f(0) = \|L\phi_n\|^2 \leq m \leq \|L\phi_1\|^2 = f(1) .$$

Since f(r) is a positive continuous function for $0 \leq r \leq 1$, the previous inequality shows that there exists an r such that

$$\|L\psi_1\|^2 = f(r) = m = \frac{\|L\|_2^2}{n}. \tag{1.24}$$

Let $\mathcal{X}_{n-1}$ be the orthogonal complement of $\psi_1$. (In fact, $\{t\phi_1 - r\phi_n, \phi_2, \phi_3, ..., \phi_{n-1}\}$ is an orthonormal basis for $\mathcal{X}_{n-1}$.) By using the definition of the Hilbert-Schmidt norm and equation (1.24) we have

$$\|L\|_2^2 = \|L\psi_1\|^2 + \|L \,|\, \mathcal{X}_{n-1}\|_2^2 = \frac{\|L\|_2^2}{n} + \|L \,|\, \mathcal{X}_{n-1}\|_2^2$$

This readily implies that

$$\|L \,|\, \mathcal{X}_{n-1}\|_2^2 = \frac{\|L\|_2^2(n-1)}{n}.$$

By applying equation (1.24) to the $n-1$ dimensional space $\mathcal{X}_{n-1}$, we see that there exists a unit vector $\psi_2$ in $\mathcal{X}_{n-1}$ satisfying

$$\|L\psi_2\|^2 = \frac{\|L \,|\, \mathcal{X}_{n-1}\|_2^2}{n-1} = \frac{\|L\|_2^2}{n} = m.$$

By continuing in this fashion there exist an orthonormal basis $\{\psi_i\}_1^n$ for $\mathcal{X}$ satisfying $\|L\psi_i\|^2 = m$ for all $i = 1, , ..., n$. (Moreover, by applying the previous procedure to the orthonormal basis $\{t\phi_1 - r\phi_n, \phi_2, ..., \phi_{n-1}\}$ for $\mathcal{X}_{n-1}$ and continuing on, one can easily construct a recursive procedure to compute the orthonormal basis $\{\psi_i\}_1^n$.) Using $\phi_i = \psi_i$ yields equality in (1.22). This completes the proof.

The following result is a slightly improved and abstract generalization of the Kaftal-Larson-Weiss solution to an $H^2 - H^\infty$ Nehari interpolation problem [34].

**COROLLARY 1.3.** *Let* T *be an isometry on* $\mathcal{H}$ *and assume that* $\mathcal{L} = \ker T^*$ *is an* n *dimensional vector space. Let* A *be an operator mapping* $\mathcal{H}$ *into* $\mathcal{H}'$ *satisfying* $T'A = AT$. *Finally, let* $\delta > 1$ *and* $\gamma$ *the constant defined by* $\gamma = \delta\|A\|$. *Then the central intertwining* $B_\gamma$ *of* A *satisfies the following inequalities*

$$\text{(i)} \quad \|B_\gamma\| \le \delta\|A\| \tag{1.25}$$

$$\text{(ii)} \quad \|B_\gamma \mid \mathcal{L}\|_2^2 \le \frac{\delta^2\|A \mid \mathcal{L}\|_2^2}{\delta^2 - 1 + \dfrac{\|A \mid \mathcal{L}\|_2^2}{n\|A\|^2}} . \tag{1.26}$$

*In particular, by choosing* $\delta^2 = 2 - \|A \mid \mathcal{L}\|_2^2/n\|A\|^2$, *the corresponding central intertwining lifting* $B_\gamma$ *of* A *satisfies the following bounds*

$$\text{(iii)} \quad \|B_\gamma\| \le \|A\| \sqrt{2 - \frac{\|A \mid \mathcal{L}\|_2^2}{n\|A\|^2}}$$

$$\text{(iv)} \quad \|B_\gamma \mid \mathcal{L}\|_2 \le \|A \mid \mathcal{L}\|_2 \sqrt{2 - \frac{\|A \mid \mathcal{L}\|_2^2}{n\|A\|^2}} .$$

**PROOF**. The inequality in part (i) follows from $\|B_\gamma\| \le \gamma$. To obtain part (ii) let $\{\phi_i\}_1^n$ be an orthonormal basis for $\mathcal{L}$. By consulting (1.8) we see that

$$\|B_\gamma\phi_i\|^2 \le \frac{\delta^2\|A\|^2\|A\phi_i\|^2}{(\delta^2 - 1)\|A\|^2 + \|A\phi_i\|^2} =$$

$$\frac{\delta^2\|A\phi_i\|^2}{\delta^2 - 1 + \dfrac{\|A\phi_i\|^2}{\|A\|^2}} \le \frac{\delta^2\|A\phi_i\|^2}{\delta^2 - 1 + \dfrac{\|A\phi_{min}\|^2}{\|A\|^2}} \tag{1.27}$$

where $\phi_{min}$ is any unit vector from $\{\phi_i\}$ satisfying $\|A\phi_{min}\| \le \|A\phi_i\|$ for all i. According to the previous lemma we can choose an orthonormal basis $\{\phi_i\}$ such that $\|A\phi_{min}\|^2 = \|A\phi_i\|^2 = \|A \mid \mathcal{L}\|_2^2/n$. Notice that we are choosing an orthonormal basis $\{\phi_i\}$ which solves the optimization problem max min $\|A\phi_i\|^2$, because we want to use a basis $\{\phi_i\}$ to make the coefficient $\delta^2(\delta^2 - 1 + \|A\phi_{min}\|^2/\|A\|^2)^{-1}$ of $\|A\phi_i\|^2$ as small as possible. So by choosing our basis such that $\|A\phi_{min}\|^2 = \|A \mid \mathcal{L}\|_2^2/n$, the equation (1.27) yields

$$\|B_\gamma\phi_i\|^2 \le \frac{\delta\|A\phi_i\|^2}{\delta^2 - 1 + \dfrac{\|A \mid \mathcal{L}\|_2^2}{n\|A\|^2}} .$$

By using the definition of the Hilbert-Schmidt norm we obtain part (ii). This completes the proof.

To complete this section we will present a more explicit formula for the central intertwining lifting; see Chapter XIV in [18] for the formula parameterizing the set of all contractive intertwining liftings B of A. To this end, assume that A is strictly bounded by $\gamma$ and let $T_A$ be the operator on $\mathcal{H}$ defined by

$$T_A = D_A^2 T D_{AT}^{-2} = (\gamma^2 I - A^* A)T(\gamma^2 I - T^* A^* AT)^{-1} . \tag{1.28}$$

We can now state the following useful result.

**PROPOSITION 1.4.** *Let* T *be an isometry on* $\mathcal{H}$ *and* A *an operator strictly bounded by* $\gamma$ *mapping* $\mathcal{H}$ *into* $\mathcal{H}'$ *satisfying* $T'A = AT$. *Then* $T_A^*$ *is similar to the contraction* $\Pi_A \omega P_{\mathcal{F}}$. *Moreover, the central intertwining lifting* $B_\gamma$ *of* A *can also be computed by*

$$B_\gamma = \begin{bmatrix} A \\ D'AT_A^*(I - zT_A^*)^{-1} \end{bmatrix} . \tag{1.29}$$

*In particular, for all* a *in* $\mathcal{L} = \ker T^*$ *we have*

$$B_\gamma D_A^{-2} a = AD_A^{-2} a \qquad (a \in \ker T^*) . \tag{1.30}$$

**PROOF.** Using the form of the orthogonal projection $P_{\mathcal{F}}$ in (1.11), along with (1.9) we have

$$\Pi_A \omega P_{\mathcal{F}} = \Pi_A \omega D_A T D_{AT}^{-2} T^* D_A = D_A D_{AT}^{-2} T^* D_A =$$

$$D_A D_{AT}^{-2} T^* D_A^2 D_A^{-1} = D_A T_A^* D_A^{-1} .$$

Therefore $\Pi_A \omega P_{\mathcal{F}}$ is similar to $T_A^*$. Substituting

$$D_A T_A^* D_A^{-1} = \Pi_A \omega P_{\mathcal{F}} \tag{1.31}$$

into the second component of the central intertwining lifting $B_\gamma$ of A in (1.6) and using (1.9) we have

$$\Pi' \omega P_{\mathcal{F}} (I - z\Pi_A \omega P_{\mathcal{F}})^{-1} D_A = \Pi' \omega P_{\mathcal{F}} (I - zD_A T_A^* D_A^{-1})^{-1} D_A =$$

$$\Pi' \omega P_{\mathcal{F}} D_A (I - zT_A^*)^{-1} = \Pi' \omega D_A T D_{AT}^{-2} T^* D_A^2 (I - zT_A^*)^{-1} =$$

$$D'AT_A^*(I - zT_A^*)^{-1} .$$

This yields the form of $B_\gamma$ in (1.29). Now notice that if a is in $\ker T^*$, then

$T_A^* D_A^{-2} a = D_{AT}^{-2} T^* a = 0$. This along with the formula for $B_\gamma$ in (1.29) gives (1.30). The proof is now complete.

## 2. Some Formulas for $D_{B_\gamma}$

In this section we will present some explicit formulas to compute $D_{B_\gamma}$ where $B_\gamma$ is the central intertwining lifting of A. As before, A is an operator bounded by (a specified) $\gamma$ satisfying $T'A = AT$ where T on $\mathcal{H}$ is an isometry and $T'$ on $\mathcal{H}'$ is a contraction. Let $\mathcal{G}$ be the orthogonal complement of $\mathcal{F}$ in $\mathcal{D}_A$ and $B_\gamma$ the central intertwining lifting of A given by (1.6). For h in $\mathcal{H}$ we have

$$\|D_{B_\gamma} h\|^2 = \gamma^2 \|h\|^2 - \|B_\gamma h\|^2 = \gamma^2 \|h\|^2 - \|Ah\|^2 - \|\Pi' \omega P_{\mathcal{F}} (I - z\Pi_A \omega P_{\mathcal{F}})^{-1} D_A h\|^2 =$$

$$\|D_A h\|^2 - \lim_{n\to\infty} \sum_{i=0}^{n} \|\Pi' \omega P_{\mathcal{F}} (\Pi_A \omega P_{\mathcal{F}})^i D_A h\|^2 =$$

$$\|D_A h\|^2 + \lim_{n\to\infty} \sum_{i=0}^{n} (- \|\omega P_{\mathcal{F}} (\Pi_A \omega P_{\mathcal{F}})^i D_A h\|^2 + \|\Pi_A \omega P_{\mathcal{F}} (\Pi_A \omega P_{\mathcal{F}})^i D_A h\|^2) =$$

$$\|D_A h\|^2 - \|\omega P_{\mathcal{F}} D_A h\|^2 + \lim_{n\to\infty} \left[ \sum_{i=0}^{n} \|(\Pi_A \omega P_{\mathcal{F}})^{i+1} D_A h\|^2 - \sum_{i=1}^{n} \|P_{\mathcal{F}} (\Pi_A \omega P_{\mathcal{F}})^i D_A h\|^2 \right] =$$

$$\|P_{\mathcal{G}} D_A h\|^2 + \lim_{n\to\infty} \left[ \|(\Pi_A \omega P_{\mathcal{F}})^{n+1} D_A h\|^2 + \sum_{i=1}^{n} (\|(\Pi_A \omega P_{\mathcal{F}})^i D_A h\|^2 - \|P_{\mathcal{F}} (\Pi_A \omega P_{\mathcal{F}})^i D_A h\|^2) \right] =$$

$$\|P_{\mathcal{G}} D_A h\|^2 + \sum_{i=1}^{\infty} \|P_{\mathcal{G}} (\Pi_A \omega P_{\mathcal{F}})^i D_A h\|^2 + \lim_{n\to\infty} \|(\Pi_A \omega P_{\mathcal{F}})^n D_A h\|^2 .$$

Therefore we see that

$$\|D_{B_\gamma} h\|^2 = \|P_{\mathcal{G}} (I - z\Pi_A \omega P_{\mathcal{F}})^{-1} D_A h\|^2 + \lim_{n\to\infty} \|(\Pi_A \omega P_{\mathcal{F}})^n D_A h\|^2 , \qquad (2.1)$$

where the first norm on the right of the equality is the $H^2(\mathcal{G})$-norm of the $\mathcal{G}-$ valued function $P_{\mathcal{G}} (I - z\Pi_A \omega P_{\mathcal{F}})^{-1} D_A h$ for $|z| < 1$.

Obviously $\Pi_A \omega P_{\mathcal{F}} = C$ is a contraction. So $C^{*n} C^n$ is a sequence of decreasing operators. This implies that

$$\text{strong-}\lim_{n\to\infty} (P_{\mathcal{F}} \omega^* \Pi_A^*)^n (\Pi_A \omega P_{\mathcal{F}})^n = R_2 \qquad (2.2)$$

defines a positive operator $R_2$ on $\mathcal{D}_A$. Let R be the positive square root of $R_2$ and $\mathcal{R}$ be the closed range of R. Obviously $R^2 = R_2$. According to (2.2) we have

$$\|RD_A h\|^2 = \lim_{n\to\infty} \|(\Pi_A \omega P_{\mathcal{F}})^n D_A h\|^2 \qquad (h \in \mathcal{H}). \tag{2.3}$$

This along with the second equation in (1.9) gives

$$\|RD_A Th\|^2 = \lim_{n\to\infty} \|(\Pi_A \omega P_{\mathcal{F}})^n D_A Th\|^2 = \lim_{n\to\infty} \|(\Pi_A \omega P_{\mathcal{F}})^{n-1} D_A h\|^2 = \|RD_A h\|^2 . \tag{2.4}$$

Therefore there exists an isometry W on $\mathcal{R}$ satisfying $RD_A T = WRD_A$. However, because $\mathcal{F} = \overline{D_A T\mathcal{H}}$ and the range of R is contained in $\mathcal{F}$, we see that $R\{D_A T\mathcal{H}\}^-$ equals $R\mathcal{D}_A$ and so $RD_A T\mathcal{H}$ is dense in $\mathcal{R}$. Thus W is onto. In other words, there is a unitary operator W on $\mathcal{R}$ satisfying

$$RD_A T = WRD_A . \tag{2.5}$$

Now let $\Theta$ be the operator from $\mathcal{H}$ to $H^2(\mathcal{G}) \oplus \mathcal{R}$ defined by

$$\Theta = \begin{bmatrix} P_{\mathcal{G}}(I - z\Pi_A \omega P_{\mathcal{F}})^{-1} D_A \\ RD_A \end{bmatrix} . \tag{2.6}$$

Equations (2.1) and (2.3) show that $\Theta$ is a spectral factor of $D^2_{B_\gamma}$. In the sense that, $D^2_{B_\gamma} = \Theta^* \Theta$. Let S be the unilateral shift on $H^2(\mathcal{G})$. We claim that

$$(S \oplus W)\Theta = \Theta T . \tag{2.7}$$

This follows from (2.5), (1.9) and the following calculation

$$P_{\mathcal{G}}(I - z\Pi_A \omega P_{\mathcal{F}})^{-1} D_A T = P_{\mathcal{G}} D_A T + zP_{\mathcal{G}}(I - z\Pi_A \omega P_{\mathcal{F}})^{-1} \Pi_A \omega P_{\mathcal{F}} D_A T =$$

$$SP_{\mathcal{G}}(I - z\Pi_A \omega P_{\mathcal{F}})^{-1} D_A .$$

Thus (2.7) holds.

We say that an operator $\Psi$ mapping $\mathcal{H}$ into $\mathcal{H}_1$ is an *outer spectral factor* for $D^2_{B_\gamma}$ if $D^2_{B_\gamma} = \Psi^* \Psi$ and the range of $\Psi$ is dense in $\mathcal{H}_1$. Notice that if $\Psi$ is an outer spectral factor for $D^2_{B_\gamma}$, then for h in $\mathcal{H}$ we have

$$\|\Psi Th\|^2 = \|D_{B_\gamma} Th\|^2 = \gamma^2 \|Th\|^2 - \|B_\gamma Th\|^2 =$$

$$\gamma^2 \|h\|^2 - \|U' B_\gamma h\|^2 = \gamma^2 \|h\|^2 - \|B_\gamma h\|^2 = \|D_{B_\gamma} h\|^2 = \|\Psi h\|^2 .$$

This implies that there exists an isometry $U_1$ on $\mathcal{H}_1$ satisfying $U_1 \Psi = \Psi T$. This sets the stage for the following result.

**THEOREM 2.1.** *Let* T *be an isometry on* $\mathcal{H}$ *and* A *an operator bounded by* $\gamma$ *satisfying* $T'A = AT$. *Let* $B_\gamma$ *be the central intertwining lifting of* A *defined in (1.6). Then* $\Theta$ *is an outer spectral factor of* $D^2_{B_\gamma}$ *satisfying* $(S \oplus W)\Theta = \Theta T$.

**PROOF.** Since $D^2_{B_\gamma} = \Theta^*\Theta$, it remains to show that the range of $\Theta$ is dense in $H^2(\mathcal{G}) \oplus \mathcal{R}$. To this end, first notice that equation (2.2) shows that $\mathcal{R} \subseteq \mathcal{F}$. Let $g \oplus f$ be any function in $H^2(\mathcal{G}) \oplus \mathcal{R}$ orthogonal to $\Theta\mathcal{H}$ and let $g = \Sigma g_n z^n$ be the power series expansion of g. According to the definition of $\Theta$ in (2.6) we have for all h in $\mathcal{H}$

$$0 = (g \oplus f, \Theta h) = (f, RD_A h) + \sum_{n=0}^{\infty} (g_n, P_{\mathcal{G}} (\Pi_A \omega P_{\mathcal{F}})^n D_A h) =$$

$$(D_A Rf, h) + \sum_{n=0}^{\infty} (D_A (P_{\mathcal{F}} \omega^* \Pi_A^*)^n g_n, h) .$$

This implies that

$$D_A[Rf + g_o + P_{\mathcal{F}} \omega^* \Pi_A^* g_1 + (P_{\mathcal{F}} \omega^* \Pi_A^*)^2 g_2 + \cdots ] = 0 . \tag{2.8}$$

(The sum in (2.8) is finite because it is obtained by applying the bounded operator $\Theta^*$ to $g \oplus f$.) Since $D_A$ is one to one on $\mathcal{D}_A = \mathcal{F} \oplus \mathcal{G}$ and the range of R is contained in $\mathcal{F}$, we see that $g_o = 0$. This implies that

$$Rf + P_{\mathcal{F}} \omega^* \Pi_A^* g_1 + (P_{\mathcal{F}} \omega^* \Pi_A^*)^2 g_2 + \cdots = 0 . \tag{2.9}$$

Now notice that

$$\|R\Pi_A \omega P_{\mathcal{F}} x\|^2 = \lim_{n\to\infty} \|(\Pi_A \omega P_{\mathcal{F}})^n \Pi_A \omega P_{\mathcal{F}} x\|^2 = \|Rx\|^2 .$$

This implies that there exists an isometry $W_1$ on $\mathcal{R}$ satisfying

$$W_1 R = R\Pi_A \omega P_{\mathcal{F}} . \tag{2.10}$$

Using (1.9) we have

$$W_1 RD_A T^2 = R\Pi_A \omega P_{\mathcal{F}} D_A T^2 = RD_A T .$$

Since $\mathcal{F} = \overline{D_A T \mathcal{H}}$, this implies that $W_1$ is onto. Therefore $W_1$ is a unitary operator satisfying (2.10). By applying the adjoint of (2.10) in (2.9) we have

$$P_{\mathcal{F}} \omega^* \Pi_A^* [RW_1 f + g_1 + (P_{\mathcal{F}} \omega^* \Pi_A^*) g_2 + (P_{\mathcal{F}} \omega^* \Pi_A^*)^2 g_3 + \cdots ] = 0 . \tag{2.11}$$

Notice that $\Pi_A \omega P_{\mathcal{F}}$ is onto a dense set in $\mathcal{D}_A$. This follows because $\Pi_A \omega P_{\mathcal{F}} D_A T = D_A$.

Therefore $P_{\mathcal{F}}\omega^*\Pi_A^*$ is one to one. Since the range of R is contained in $\mathcal{F}$ and $g_n$ is in $\mathcal{G}$, equation (2.11) implies that $g_1 = 0$. By continuing in a similar fashion $g_n = 0$ for all $n \geq 0$. Hence $g = 0$.

To complete the proof it remains to show that $f = 0$. According to equation (2.8) we now have $D_A Rf = 0$. Therefore $Rf = 0$. Since f is in $\mathcal{R}$ we have $f = 0$. This completes the proof.

The following result provides us with another outer spectral factor $\Theta_{\mathcal{L}}$ for $D_{B_\gamma}^2$ involving the operators $N_A$ and $T_A$ defined in (1.7) and (1.28), respectively.

**PROPOSITION 2.2.** *Let* T *be an isometry on* $\mathcal{H}$ *and* $\mathcal{L}$ *be the kernel of* $T^*$. *Let* A *be an operator strictly bounded by* $\gamma$ *satisfying* $T'A = AT$ *and* $B_\gamma$ *the central intertwining lifting of* A. *Let* $\Theta_{\mathcal{L}}$ *be the operator mapping* $\mathcal{H}$ *into* $H^2(\mathcal{L}) \oplus \mathcal{R}$ *defined by*

$$\Theta_{\mathcal{L}} = \begin{bmatrix} N_A\Pi_o(I - zT_A^*)^{-1} \\ RD_A \end{bmatrix}. \tag{2.12}$$

*Then* $\Theta_{\mathcal{L}}$ *is an outer spectral factor for* $D_{B_\gamma}^2$. *Moreover,* $(S \oplus W)\Theta_{\mathcal{L}} = \Theta_{\mathcal{L}} T$ *where* S *is the unilateral shift on* $H^2(\mathcal{L})$.

**PROOF.** According to equation (1.31) we have

$$P_{\mathcal{G}}(I - z\Pi_A\omega P_{\mathcal{F}})^{-1}D_A = P_{\mathcal{G}}D_A(I - zT_A^*)^{-1}. \tag{2.13}$$

We claim that $\mathcal{G} = D_A^{-1}\mathcal{L}$. To see this notice that g is in $\mathcal{G}$ if and only if g is orthogonal to $D_A T\mathcal{H}$. Therefore g is in $\mathcal{G}$ if and only if $D_A g$ is orthogonal to $T\mathcal{H}$, or equivalently, $D_A g$ is in $\mathcal{L}$. Thus g is in $\mathcal{G}$ if and only if g is in $D_A^{-1}\mathcal{L}$. This implies that $\mathcal{G} = D_A^{-1}\mathcal{L}$. Now let X be the operator mapping $\mathcal{L}$ into $\mathcal{H}$ defined by $X = D_A^{-1}\Pi_o^*$. Because $\mathcal{G} = D_A^{-1}\mathcal{L}$, the range of X is $\mathcal{G}$. So the orthogonal projection $P_{\mathcal{G}}$ onto $\mathcal{G}$ is given by

$$P_{\mathcal{G}} = X(X^*X)^{-1}X^* = D_A^{-1}\Pi_o^*(\Pi_o D_A^{-2}\Pi_o^*)^{-1}\Pi_o D_A^{-1} = D_A^{-1}\Pi_o^* N_A^2 \Pi_o D_A^{-1}. \tag{2.14}$$

By using (2.14) in (2.13) we obtain

$$P_{\mathcal{G}}(I - z\Pi_A\omega P_{\mathcal{F}})^{-1} = D_A^{-1}\Pi_o^* N_A^2 \Pi_o (I - zT_A^*)^{-1}. \tag{2.15}$$

Notice that

$$\Omega = X(X^*X)^{-1/2} = D_A^{-1}\Pi_o^*(\Pi_o D_A^{-2}\Pi_o^*)^{-1/2} = D_A^{-1}\Pi_o^* N_A$$

is a unitary operator mapping $\mathcal{L}$ onto $\mathcal{G}$. Obviously $\Omega$ can be trivially extended to a unitary operator $\Omega_+$ mapping $H^2(\mathcal{L})$ onto $H^2(\mathcal{G})$ by

$$\Omega_+ f = \Omega_+ \sum_0^\infty f_n z^n = \sum_0^\infty z^n \Omega f_n = \sum_0^\infty z^n D_A^{-1} \Pi_0^* N_A f_n \,.$$

Moreover, $\Omega_+$ commutes with the unilateral shifts on $H^2(\mathcal{L})$ and $H^2(\mathcal{G})$. By using $\Omega_+$ in (2.15) we have

$$P_{\mathcal{G}} (I - z\Pi_A \omega P_{\mathcal{F}})^{-1} = \Omega_+ N_A \Pi_0 (I - zT_A^*)^{-1} \,. \tag{2.16}$$

This along with Theorem 2.1 produces the form of $\Theta_{\mathcal{L}}$ in (2.12) and completes the proof.

**REMARK 2.3.** Notice that the operator R is not necessarily zero. For example, if $||A|| < \gamma$, then there exists an $\varepsilon > 0$ satisfying $0 < \varepsilon^2 I \le D_A^2 \le \gamma^2 I$. If in addition T is not a unilateral shift, then there exists a nonzero h in $\mathcal{H}_u = \bigcap_0^\infty T^n \mathcal{H}$. Recall that from the Wold decomposition of an isometry $T \,|\, \mathcal{H}_u$ is unitary. So using (1.9) for this h we have

$$||(\Pi_A \omega P_{\mathcal{F}})^n D_A h||^2 = ||(\Pi_A \omega P_{\mathcal{F}})^n D_A T^n T^{*n} h||^2 =$$

$$||D_A T^{*n} h||^2 \ge \varepsilon^2 ||T^{*n} h||^2 = \varepsilon^2 ||h||^2 \ne 0 \,.$$

According to (2.3) we see that $||RD_A h|| \ge \varepsilon ||h||$. Hence R is nonzero. In the next section, we will show that if T is a unilateral shift and A is strictly bounded by $\gamma$, then the operator $R = 0$.

## 3. The Role of $D_A^{-2}\Pi_0^*$

In this section we will show that the operator $D_A^{-2}\Pi_0^*$ plays a key role in computing the central intertwining lifting $B_\gamma$ for A and the outer spectral factor $\Theta_{\mathcal{L}}$ when T is a unilateral shift. Recall that a subspace $\mathcal{M}$ is *cyclic* for an operator C on $\mathcal{H}$ if $\mathcal{H}$ equals the closed linear span of $\{C^n \mathcal{M} : n \ge 0\}$. We begin with the following useful result.

**THEOREM 3.1.** *Let* T *on* $\mathcal{H}$ *be a unilateral shift,* $\mathcal{L} = \ker T^*$ *and* $T'$ *a contraction on* $\mathcal{H}'$. *If* A *is an operator from* $\mathcal{H}$ *to* $\mathcal{H}'$ *strictly bounded by* $\gamma$ *satisfying* $T'A = AT$, *then* $D_A^{-2}\mathcal{L}$ *is cyclic for* T.

**PROOF.** Assume that there exists a h in $\mathcal{H}$ orthogonal to $T^n D_A^{-2} \mathcal{L}$ for all $n \ge 0$. This implies that $D_A^{-2} T^{*n} h$ is orthogonal to $\mathcal{L}$ for all $n \ge 0$. Since $\mathcal{L} = \ker T^*$ we see that $D_A^{-2} T^{*n} h$ is contained in the range of T. Therefore

$$D_A^{-2} T^{*n} h = T h_n \tag{3.1}$$

for some $h_n$ in $\mathcal{H}$. This readily implies that

$$T^{*n+1}h = T^*T^{*n}h = T^*D_A^2Th_n = D_{AT}^2h_n .$$

Thus $h_n = D_{AT}^{-2}T^{*n+1}h$. Using this and (3.1) we have

$$TD_{AT}^{-2}T^{*n+1}h = Th_n = D_A^{-2}T^{*n}h .$$

By applying $D_A^2$ to both sides and using the definition of $T_A = D_A^2TD_{AT}^{-2}$, we obtain the formula $T^{*n}h = T_AT^{*n+1}h$ for all $n \geq 0$. Thus

$$h = T_AT^*h = T_A^2T^{*2}h = T_A^3T^{*3}h = \cdots = T_A^nT^{*n}h .$$

However, by Proposition 1.4 the operator $T_A$ is similar to a contraction. So $T_A$ is power bounded. In other words, $\|T_A^n\| \leq \beta < \infty$ for some $\beta$ and all $n \geq 0$. Because T is a unilateral shift, $T^{*n}h \to 0$ as $n \to \infty$. Thus $h = T_A^nT^{*n}h$ for all $n \geq 0$ implies $h = 0$. This completes the proof.

**THEOREM 3.2.** *Let* T *be a unilateral shift on* $\mathcal{H}$ *and* A *an operator strictly bounded by* $\gamma$ *satisfying* $T'A = AT$. *Then* $R = 0$, *or equivalently,*

$$\|RD_Ah\|^2 = \lim_{n\to\infty} \|(\Pi_A\omega P_{\mathcal{F}})^nD_Ah\|^2 = 0 \qquad (\textit{for all } h \textit{ in } \mathcal{H}) . \tag{3.2}$$

*Moreover,* $T_A^{*n} \to 0$ *strongly as* $n \to \infty$.

**PROOF.** Let C be the operator on $\mathcal{D}_A$ defined by $C = \Pi_A\omega P_{\mathcal{F}}$. Using (1.9) we see that $CD_AT = D_A$. Thus

$$C^nD_AT^n = D_A \qquad (\text{for all } n \geq 0) . \tag{3.3}$$

Recall that $\mathcal{G} = D_A^{-1}\mathcal{L}$. So equation (3.3) gives

$$C^{n+1}D_AT^nD_A^{-2}\mathcal{L} = CD_A^{-1}\mathcal{L} = \Pi_A\omega P_{\mathcal{F}}D_A^{-1}\mathcal{L} = 0 . \tag{3.4}$$

According to Theorem 3.1, the space $D_A^{-2}\mathcal{L}$ is cyclic for T. Therefore $D_AT^nD_A^{-2}\mathcal{L}$ for $n \geq 0$ spans $\mathcal{H}$. This along with (3.4) shows that the limit in (3.2) is zero. Finally, recall that $T_A^*$ is similar to $\Pi_A\omega P_{\mathcal{F}}$; see (1.31). Thus $T_A^{*n} \to 0$ strongly as $n \to \infty$. This completes the proof.

Let $T = S$ be the unilateral shift on $\mathcal{H} = H^2(\mathcal{L})$. For convenience we shall denote by $H^2(\mathcal{L}, \mathcal{L})$ the set of all analytic functions F(z) in $|z| < 1$, whose values are operators on $\mathcal{L}$ satisfying $F(z)a \in H^2(\mathcal{L})$ for all a in $\mathcal{L}$. A function F(z) in $H^2(\mathcal{L}, \mathcal{L})$ is *outer* if $F(z)\mathcal{L}$ is cyclic for T. If F(z) is in $H^\infty(\mathcal{L}, \mathcal{L})$, then $F_+$ is the multiplication operator on $H^2(\mathcal{L})$ defined by $F_+h = Fh$ for h in $H^2(\mathcal{L})$. In this case F(z) is outer if and only if the range of $F_+$ is dense in $H^2(\mathcal{L})$. Recall that an operator $T_o$ on $\mathcal{H}$ is called *Toeplitz* if $T^*T_oT = T_o$. We say that a function F(z) in $H^\infty(\mathcal{L}, \mathcal{L})$ is an *outer spectral factor* for the Toeplitz operator $T_o$, if F(z) is an outer function and $F_+^* F_+ = T_o$. In this case, by a slight abuse of terminology, we call both the function F(z) and its corresponding operator $F_+$ an outer spectral factor for $T_o$. If $B_\gamma$ is the central intertwining lifting of A, then obviously $D_{B_\gamma}^2$ is Toeplitz. Now assume that A is strictly bounded by $\gamma$. According to Proposition 2.2 the operator $\Theta_{\mathcal{L}}$ defined in (2.12) is an outer spectral factor for the Toeplitz matrix $D_{B_\gamma}^2$. Moreover, by the previous theorem $R = 0$. Therefore the outer spectral factor $\Theta_{\mathcal{L}}$ is an operator on $H^2(\mathcal{L})$ commuting with the unilateral shift S. This implies that there exists an analytic function F in $H^\infty(\mathcal{L}, \mathcal{L})$ satisfying $\Theta_{\mathcal{L}} = F_+$. Obviously F(z) is an outer spectral factor for the Toeplitz operator $D_{B_\gamma}^2$. Furthermore, this function F is given by $F(z)a = \Theta_{\mathcal{L}} a$ where a is in $\mathcal{L}$. Summing up this analysis along with the form of $\Theta_{\mathcal{L}}$ in (2.12) we obtain the following result.

**COROLLARY 3.3.** *Let* $T = S$ *be a unilateral shift on* $H^2(\mathcal{L})$ *and* A *an operator strictly bounded by* $\gamma$ *satisfying* $T'A = AT$. *Let* $B_\gamma$ *be the central intertwining lifting of* A. *Then the function* F *in* $H^\infty(\mathcal{L}, \mathcal{L})$ *defined by*

$$F(z) = N_A \Pi_0 (I - zT_A^*)^{-1} \Pi_0^* \tag{3.5}$$

*is an outer spectral factor for the Toeplitz operator* $D_{B_\gamma}^2$.

The following result gives us an explicit form for the outer spectral factor F(z) for $D_{B_\gamma}^2$.

**THEOREM 3.4.** *Let* $T = S$ *be a unilateral shift on* $H^2(\mathcal{L})$ *and* A *an operator strictly bounded by* $\gamma$ *satisfying* $T'A = AT$. *Let* $B_\gamma$ *be the central intertwining lifting of* A *and* Q(z) *the analytic function whose values are linear operators on* $\mathcal{L}$ *defined by*

$$Q(z) = (D_A^{-2} \Pi_0^*)(z) . \tag{3.6}$$

*Then* Q(z) *is an outer function. Moreover, the function* F(z) *in* $H^\infty(\mathcal{L}, \mathcal{L})$ *defined by*

$$F(z) = N_A^{-1} Q(z)^{-1} = (\Pi_0 D_A^{-2} \Pi_0^*)^{1/2} Q(z)^{-1} = Q(0)^{1/2} Q(z)^{-1} \tag{3.7}$$

*is precisely the outer spectral factor for the Toeplitz operator* $D_{B_\gamma}^2$ *given in (3.5). In particular,* $Q(z)^{-1}$ *is an outer function in* $H^\infty(\mathcal{L}, \mathcal{L})$.

Recall that $\Pi_0$ is the operator from $\mathcal{H}$ onto $\mathcal{L} = \ker T^*$ defined by $\Pi_0 = P_{\mathcal{L}}$. Therefore the function Q(z) in (3.6) is defined by

$$Q(z)a = (D_A^{-2} a)(z) \qquad (\text{for all } |z| < 1 \text{ and } a \in \mathcal{L}) . \tag{3.8}$$

Moreover, it is easy to verify that $\Pi_o D_A^{-2} \Pi_o^* = Q(0)$.

**PROOF.** According to Theorem 3.1 the space $D_A^{-2}\mathcal{L}$ is cyclic for T. Therefore Q(z) is an outer function. Using the fact that $T_A^* T = I$ we see that

$$D_{AT}^{-2} T^* D_A^2 = T_A^* = T_A^* T T^* + T_A^* P_{\mathcal{L}} = T^* + Y\Pi_o \tag{3.9}$$

where Y is an operator from $\mathcal{L}$ to $\mathcal{H}(=\mathcal{D}_A)$ which we are going to determine. By applying $D_A^{-2}\Pi_o^*$ on the right to both sides of (3.9) we have

$$0 = D_{AT}^{-2} T^* \Pi_o^* = T^* D_A^{-2} \Pi_o^* + Y\Pi_o D_A^{-2} \Pi_o^* = T^* D_A^{-2} \Pi_o^* + Y N_A^{-2} .$$

Thus $Y = -T^* D_A^{-2} \Pi_o^* N_A^2$. Substituting this into equation (3.9) yields

$$T_A^* = T^* - T^* D_A^{-2} \Pi_o^* N_A^2 \Pi_o . \tag{3.10}$$

Using (3.10) in (3.5) we see that an outer spectral factor F for the Toeplitz operator $D_{B_\gamma}^2$ is given by

$$F(z) = N_A \Pi_o (I - zT_A^*)^{-1} \Pi_o^* = N_A \Pi_o (I - zT^* + zT^* D_A^{-2} \Pi_o^* N_A^2 \Pi_o)^{-1} \Pi_o^* . \tag{3.11}$$

The inverse in (3.11) is well defined for all $|z| < 1$, because $T_A^*$ is similar to a contraction; see Proposition 1.4. We note that for any operator X, C and D on the appropriate spaces,

$$D(X + CD)^{-1} = (DX^{-1}C + I)^{-1} DX^{-1} \tag{3.12}$$

holds as long as X and (X + CD) are both invertible. In this case, the inverse of $DX^{-1}C + I$ is well defined and given by

$$(DX^{-1}C + I)^{-1} = I - D(X + CD)^{-1} C . \tag{3.13}$$

In our setting $X = I - zT^*$ and $C = zT^* D_A^{-2} \Pi_o^* N_A^2$ and $D = \Pi_o$. Therefore

$$DX^{-1}C + I = \Pi_o (I - zT^*)^{-1} zT^* D_A^{-2} \Pi_o^* N_A^2 + I =$$

$$[\Pi_o (I - zT^*)^{-1} zT^* D_A^{-2} \Pi_o^* + \Pi_o D_A^{-2} \Pi_o^*] N_A^2 = \Pi_o (I - zT^*)^{-1} D_A^{-2} \Pi_o^* N_A^2 = Q(z) N_A^2 \tag{3.14}$$

where Q(z) is the function defined by

$$Q(z) = \Pi_o (I - zT^*)^{-1} D_A^{-2} \Pi_o^* . \tag{3.15}$$

Notice that $Q(z) N_A^2$ is invertible for all $|z| < 1$, because according to equations (3.10), (3.13) and (3.14)

$$I - D(I - zT_A^*)^{-1}C = (DX^{-1}C + I)^{-1} = N_A^{-2}Q(z)^{-1} . \tag{3.16}$$

Now using (3.11), (3.12) and (3.16) we have

$$F(z) = N_A\Pi_0(I - zT^* + zT^*D_A^{-2}\Pi_0^*N_A^2\Pi_0)^{-1}\Pi_0^* =$$

$$N_AD(X + CD)^{-1}\Pi_0^* = N_A(DX^{-1}C + I)^{-1}DX^{-1}\Pi_0^* =$$

$$N_AN_A^{-2}Q(z)^{-1}\Pi_0(I - zT^*)^{-1}\Pi_0^* = N_A^{-1}Q(z)^{-1} .$$

This gives the form of F(z) in (3.7). Since the multiplication operator $F_+$ is an outer spectral factor of $D_{B_\gamma}^2$, it follows that $F_+$ is bounded by $\gamma$. (Recall that $\|F_+\| = \|F\|_\infty$.) Thus F(z) is an outer function in $H^\infty(\mathcal{L}, \mathcal{L})$ satisfying $\|F\|_\infty \le \gamma$. Therefore $Q(z)^{-1} = N_AF(z)$ is an outer function in $H^\infty(\mathcal{L}, \mathcal{L})$.

To complete the proof it remains to show that $Q(z) = (D_A^{-2}\Pi_0^*)(z)$. To this end, notice that if $h = \Sigma h_nz^n$ is in $H^2(\mathcal{L})$, then for $|z| < 1$

$$\Pi_0(I - zT^*)^{-1}h = \sum_{n=0}^{\infty} z^n\Pi_0T^{*n}h = \sum_{n=0}^{\infty} h_nz^n = h(z) \qquad (h \in H^2(\mathcal{L})) . \tag{3.17}$$

So using $h = D_A^{-2}a$ for a in $\mathcal{L}$ in (3.17) we have by (3.15)

$$Q(z)a = \Pi_0(I - zT^*)^{-1}D_A^{-2}a = (D_A^{-2}a)(z) . \tag{3.18}$$

This completes the proof.

In the following two Corollaries, we will use the function Q(z) to obtain an explicit formula for the central intertwining lifting $B_\gamma$ of A when T is a unilateral shift. To this end, let T be the unilateral shift on $\mathcal{H} = H^2(\mathcal{L})$, and assume that the Sz.-Nagy-Schäffer minimal isometric dilation $U'$ on $\mathcal{K}'$ is also a unilateral shift. Recall that $U'$ is a unilateral shift if and only if $T'^*$ is strongly stable, that is, $T'^{*n} \to 0$ strongly as $n \to \infty$; see [18], [45]. In this case there exists a unitary operator W mapping $\mathcal{K}'$ onto $H^2(\mathcal{L}_*)$ satisfying $SW = WU'$ where S is the unilateral shift on $H^2(\mathcal{L}_*)$. In fact, the dimension of $\mathcal{L}_*$ is precisely the rank of $I - T'T'^*$. As before, let $B_\gamma$ be the central intertwining lifting of A. Obviously $SWB_\gamma = WB_\gamma T$. This implies that there exists a function $G_\gamma$ in $H^\infty(\mathcal{L}, \mathcal{L}_*)$ satisfying $WB_\gamma = M_{G_\gamma} | H^2(\mathcal{L})$. Moreover, $\|B_\gamma\| = \|G_\gamma\|_\infty$. Furthermore, let us assume that A is strictly bounded by $\gamma$ and $\mathcal{L}$ is finite dimensional. Now let a be in $\mathcal{L}$. By using our definition of Q in (3.6) along with (1.30) we have

$$G_\gamma Qa = WB_\gamma D_A^{-2}a = WAD_A^{-2}a . \tag{3.19}$$

Let $P(e^{it})$ be the function defined by $P(e^{it}) = (WAD_A^{-2}\Pi_0^*)(e^{it})$. Because $\mathcal{L}$ is finite dimensional, and $Q(z)^{-1}$ is in $H^\infty(\mathcal{L}, \mathcal{L})$, equation (3.19) implies that $G_\gamma = PQ^{-1}$. Summing up this analysis leads to the following formula for computing the central intertwining lifting $B_\gamma$ of A.

**COROLLARY 3.5.** *Let* T *be a unilateral shift on* $H^2(\mathcal{L})$ *where* $\mathcal{L}$ *is finite dimensional. Let* $T'^*$ *on* $\mathcal{H}'$ *be a strongly stable contraction and* A *an operator strictly bounded by* $\gamma$ *satisfying* $T'A = AT$. *Let* W *mapping* $\mathcal{K}'$ *onto* $H^2(\mathcal{L}_*)$ *be a unitary operator satisfying* $SW = WU'$ *where* S *is the unilateral shift on* $H^2(\mathcal{L}_*)$ *and* $U'$ *on* $\mathcal{K}'$ *is the Sz.-Nagy-Schäffer minimal isometric dilation of* $T'$. *Finally, let* P *and* Q *be the function defined by*

$$P(z) = (WAD_A^{-2}\Pi_o^*)(z) \text{ and } Q(z) = (D_A^{-2}\Pi_o^*)(z) . \tag{3.20}$$

*Then the central intertwining lifting* $B_\gamma$ *of* A *is given by* $WB_\gamma = G_{\gamma+}$ *where* $G_\gamma$ *is the function in* $H^\infty(\mathcal{L}, \mathcal{L}_*)$ *computed by* $G_\gamma = PQ^{-1}$.

**EXAMPLE 3.6.** In this example we will use Corollary 3.5 to solve an $H^2$–$H^\infty$ Carathéodory interpolation problem. To this end, let G(z) be the polynomial in $H^\infty(\mathcal{E}, \mathcal{E}')$ defined by

$$G(z) = A_0 + A_1 z + \cdots + A_{m-1} z^{m-1} \tag{3.21}$$

where $\mathcal{E}$ is finite dimensional and the coefficients $A_i$ are operators mapping $\mathcal{E}$ into $\mathcal{E}'$. Now let $H^2(\mathcal{E}, \mathcal{E}')$ be the Hilbert space defined by the Hardy space of all analytic functions $\Psi$ in the open unit disc, whose values are operators mapping $\mathcal{E}$ into $\mathcal{E}'$, and the norm is defined by

$$\|\Psi\|_2^2 = \frac{1}{2\pi}\int_0^{2\pi} \text{trace } (\Psi(e^{it})^*\Psi(e^{it}))dt = \|\Psi_+ \mid \mathcal{E}\|_2^2 .$$

Notice that the $H^2$-norm of $\Psi$ equals the Hilbert-Schmidt norm of $\Psi_+ \mid \mathcal{E}$. Now consider the following two optimization problems.

$$\begin{aligned} d_2 &= \inf \{\|G(z) + z^m H\|_2 \quad : \quad H \in H^2(\mathcal{E}, \mathcal{E}')\} \\ d_\infty &= \inf \{\|G(z) + z^m H\|_\infty \quad : \quad H \in H^2(\mathcal{E}, \mathcal{E}')\} \end{aligned} \tag{3.22}$$

where $d_2$ is the distance from the polynomial G to $H^2(\mathcal{E}, \mathcal{E}')$ in the $H^2$ norm, while $d_\infty$ is the distance from G to $H^\infty(\mathcal{E}, \mathcal{E}')$ in the $H^\infty$ norm. By using the projection theorem, it is easy to show that $d_2 = \|G\|_2$. To obtain an expression for $d_\infty$, let M be the lower triangular m by m block Toeplitz matrix generated by $\{A_i\}$, that is, let

$$M = \begin{bmatrix} A_0 & 0 & 0 & \dots & 0 \\ A_1 & A_0 & 0 & \dots & 0 \\ \vdots & \vdots & \vdots & \vdots & \vdots \\ A_{m-1} & A_{m-2} & A_{m-3} & \dots & A_0 \end{bmatrix} . \tag{3.23}$$

Then it is well known that $d_\infty = \|M\|$; see [5], [18].

Now let A be the operator from $H^2(\mathcal{E})$ into $\mathcal{H}' = H^2(\mathcal{E}') \ominus z^m H^2(\mathcal{E}')$ defined by $A = P_{\mathcal{H}'} G_+$. Let T be the unilateral shift on $H^2(\mathcal{E})$ and S the unilateral shift on $H^2(\mathcal{E}')$. Let $T'$ on $\mathcal{H}'$ be the compression of S to $\mathcal{H}'$. Because $\mathcal{H}'$is an invariant subspace for $S^*$, it follows that $P_{\mathcal{H}'} S = T' P_{\mathcal{H}'}$. Since $\mathcal{H}'$ is cyclic for S, the unilateral shift S is also the minimal isometric dilation of $T'$. Recall that all minimal isometric dilations of $T'$ are isomorphic; see [45] or Theorem VI.3.2 in [18]. Therefore, there exists a unitary W mapping $\mathcal{K}'$ onto $H^2(\mathcal{E}')$ satisfying $SW = WU'$ and $W \mid \mathcal{H}' = I$, where $U'$ on $\mathcal{K}'$ is the Sz.-Nagy-Schäffer minimal isometric dilation of $T'$ defined in (1.2). Corollary 3.5 and $W \mid \mathcal{H}' = I$ imply that the central intertwining lifting $WB_\gamma$ of A is given by $WB_\gamma = G_{\gamma +}$ where $G_\gamma$ is the function in $H^\infty(\mathcal{E}, \mathcal{E}')$ defined by $G_\gamma = P Q^{-1}$ and

$$P(z) = (AD_A^{-2}\Pi_o^*)(z) \qquad \text{and} \qquad Q(z) = (D_A^{-2}\Pi_o^*)(z) . \tag{3.24}$$

Now we can use the lower triangular Toeplitz matrix in (3.23) to compute $G_\gamma$. To this end, let $W_1$ be the unitary operator mapping $\ell^2(\mathcal{E})$ (the Hilbert space formed by the square summable one-sided sequences with values in $\mathcal{E}$) onto $H^2(\mathcal{E})$ defined by

$$W_1 [f_0, f_1, f_2, \dots]^{tr} = \sum_{n=0}^{\infty} f_n z^n$$

where tr denotes the transpose and $f_i \in \mathcal{E}$ for all i. Let $W_2$ be the unitary operator mapping $\ell_m^2(\mathcal{E}')$ (the Hilbert space of m-tuples with components in $\mathcal{E}'$) onto $\mathcal{H}'$ defined by

$$W_2 [g_0, g_1, \dots, g_{m-1}]^{tr} = \sum_{i=0}^{m-1} g_i z^i .$$

Using the unitary operators $W_1$ and $W_2$, it is easy to show that $W_2 [M, 0] = AW_1$. So if $\Pi_{\mathcal{E}}$ is the block row matrix of length m defined by $\Pi_{\mathcal{E}} = [I, 0, 0, \dots, 0]$, then P and Q in (3.24) can be computed by

$$P(z) = [I, zI, z^2I, \dots, z^{m-1}I] M D_M^{-2} \Pi_{\mathcal{E}}^* \quad \text{and} \quad Q(z) = [I, zI, z^2I, \dots, z^{m-1}I] D_M^{-2} \Pi_{\mathcal{E}}^* . \tag{3.25}$$

Therefore, the central intertwining lifting $G_\gamma$ is given by $G_\gamma(z) = P(z)Q(z)^{-1}$ where P(z) and Q(z) are the polynomials computed according to (3.25). Moreover, the polynomial Q(z) is outer. In particular, in the scalar case, all the zeros of Q(z) are outside the closed unit disc. Finally, it is noted that our central solution $G_\gamma$ is precisely the central solution one obtains by applying the Schur algorithm to the Carathéodory data $\{A_i\}$; see [18], [22].

To complete this example, we will show that our central intertwining lifting satisfies some $H^2$–$H^\infty$ bounds. To see this, first recall that $d_2 = \|G\|_2 = \|A \,|\, \mathcal{E}\|_2$ and $\mathcal{E} = \ker T^*$. Then using the fact that $W_2[M, 0] = AW_1$, we have $d_\infty = \|M\| = \|A\|$. If $a$ is in $\mathcal{E}$, then $P_{\mathcal{H}'} G_\gamma a = P_{\mathcal{H}'} W B_\gamma a = Aa = Ga$. Since $\mathcal{H}'$ is formed by polynomials up to degree $m-1$, this implies that there exists a function H in $H^\infty(\mathcal{E}, \mathcal{E}')$ satisfying $G_\gamma = G + z^m H$. In fact, H is computed by $z^m H = PQ^{-1} - G$. Now let $\delta > 1$. Using $d_2 = \|A \,|\, \mathcal{E}\|_2$ and $d_\infty = \|A\|$ along with $WB_\gamma = G_{\gamma+}$ in Corollary 1.3, we see that our central intertwining lifting, $G_\gamma = G + z^m H$ is computed by $G_\gamma = PQ^{-1}$, and satisfies the following $H^2 - H^\infty$ bounds

$$\|G + z^m H\|_\infty \le \delta d_\infty \quad \text{and} \quad \|G + z^m H\|_2 \le \frac{\delta d_2}{\sqrt{\delta^2 - 1 + \dfrac{d_2^2}{n d_\infty^2}}} \tag{3.26}$$

where n is the dimension for $\mathcal{E}$. In particular, by choosing $\delta^2 = 2 - d_2^2/nd_\infty^2$ our central solution $G_\gamma = G + z^m H$ satisfies the following bounds

$$\|G + z^m H\|_\infty \le d_\infty \sqrt{2 - \frac{d_2^2}{n d_\infty^2}} \quad \text{and} \quad \|G + z^m H\|_2 \le d_2 \sqrt{2 - \frac{d_2^2}{n d_\infty^2}}\,.$$

For some other results on mixed $H^2$–$H^\infty$ tangential interpolation problems see [23].

Recall that an isometry U on $\mathcal{K}$ is a bilateral shift if $\mathcal{K}$ contains a cyclic wandering subspace $\mathcal{L}_*$; see [18], [45]. So if U is a bilateral shift, then there exists a unitary operator W mapping $\mathcal{K}$ onto $L^2(\mathcal{L}_*)$ satisfying $VW = WU$ where V is the bilateral shift (multiplication by $e^{it}$) on $L^2(\mathcal{L}_*)$. By an easy adaptation of the proof of Corollary 3.5, we obtain the following result.

**COROLLARY 3.7.** *Let* T *be a unilateral shift on* $H^2(\mathcal{L})$ *where* $\mathcal{L}$ *is finite dimensional. Let* A *be an operator strictly bounded by* $\gamma$ *satisfying* $T'A = AT$ *where* $T'$ *on* $\mathcal{H}'$ *is a contraction. Assume that the Sz.-Nagy-Schäffer minimal isometric dilation* $U'$ *on* $\mathcal{K}'$ *of* $T'$ *is a bilateral shift, and let* W *be the unitary operator mapping* $\mathcal{K}'$ *onto* $L^2(\mathcal{L}_*)$ *satisfying* $VW = WU'$ *where* V *is the bilateral shift on* $L^2(\mathcal{L}_*)$. *Finally, let* P *and* Q *be the functions defined by (3.20). Then the central intertwining lifting* $B_\gamma$ *of* A *is given by* $WB_\gamma = M_{G_\gamma} | H^2(\mathcal{L})$ *where* $G_\gamma$ *is the function in* $L^\infty(\mathcal{L}, \mathcal{L}_*)$ *computed by* $G_\gamma = PQ^{-1}$.

**REMARK 3.8.** The conclusion of the last Corollary can be extended to the case when $T'$ is a completely nonunitary contraction. Indeed in this case $\mathcal{K}'$ can be viewed as an invariant subspace for an adequate bilateral shift of appropriate multiplicity; see Chapter II in [45].

Finally, we will note that the role of the contraction $T'$ in Theorem 3.4 is rather secondary, as illustrated by the following two Corollaries.

**COROLLARY 3.9.** *Let* $T$ *be a unilateral shift on* $H^2(\mathcal{L})$ *and* $A$ *an operator on* $H^2(\mathcal{L})$ *strictly bounded by* $\gamma$ *satisfying* $T^*A^*AT \le A^*A$. *Then* $Q(z) = (D_A^{-2}\Pi_o^*)(z)$ *is an outer function and* $Q(z)^{-1}$ *is in* $H^\infty(\mathcal{L}, \mathcal{L})$.

**PROOF.** Since $T^*A^*AT \le A^*A$, it follows that there exists a contraction $T'$ on $\mathcal{H}'$ the closed range of $A$ such that $T'A = AT$. By applying the Theorem 3.4, the function $Q(z) = (D_A^{-2}\Pi_o^*)(z)$ is outer and $Q(z)^{-1}$ is in $H^\infty(\mathcal{L}, \mathcal{L})$. This completes the proof.

The previous Corollary leads to a simple proof for an explicit form for the classical spectral factorization theorem for strictly positive Toeplitz operators; see Clancey-Gohberg [10].

**COROLLARY 3.10.** *Let* $T_o$ *be a strictly positive Toeplitz operator on* $H^2(\mathcal{L})$. *Then the function* $Q_o(z) = (T_o^{-1}\Pi_o^*)(z)$ *is an outer function in* $H^\infty(\mathcal{L}, \mathcal{L})$ *and* $Q_o(z)^{-1}$ *is in* $H^\infty(\mathcal{L}, \mathcal{L})$. *Moreover, the function* $F_o(z)$ *in* $H^\infty(\mathcal{L}, \mathcal{L})$ *defined by*

$$F_o(z) = (\Pi_o T_o^{-1}\Pi_o^*)^{1/2} Q_o(z)^{-1} = Q_o(0)^{1/2} Q_o(z)^{-1} \tag{3.27}$$

*is an outer spectral factor for* $T_o$.

**PROOF.** Let $\gamma > 0$ be any positive scalar such that $\|T_o\| < \gamma^2$. As before, let $T$ be the unilateral shift on $H^2(\mathcal{L})$. Let $A$ be the positive square root of $\gamma^2 I - T_o$. Since $T_o$ is strictly positive, $A$ is strictly bounded by $\gamma$. Using the fact that $T_o$ is Toeplitz we have

$$T^*A^*AT = T^*(\gamma^2 I - T_o)T = \gamma^2 I - T_o = A^*A\,. \tag{3.28}$$

According to the previous Corollary the function $(D_A^{-2}\Pi_o^*)(z) = Q(z)$ is outer and $Q(z)^{-1}$ is in $H^\infty(\mathcal{L}, \mathcal{L})$. However, $D_A^2 = T_o$. Thus

$$Q(z) = (D_A^{-2}\Pi_o^*)(z) = (T_o^{-1}\Pi_o^*)(z) = Q_o(z)\,. \tag{3.29}$$

By the previous Corollary $Q_o(z)$ is an outer function and $Q_o(z)^{-1}$ is in $H^\infty(\mathcal{L}, \mathcal{L})$.

To complete the proof it remains to show that the function $F_o(z)$ defined in (3.27) is an outer spectral factor of $T_o$. To see this notice that (3.28) implies that there exists an isometry $T'$ on $H^2(\mathcal{L})$ satisfying $T'A = AT$. (The range of $A$ equals $H^2(\mathcal{L})$ because $\|T_o\| < \gamma^2$.) Since $T'$ is an isometry the central intertwining lifting $B_\gamma$ equals A. Thus $D^2_{B_\gamma} = D^2_A = T_o$. According to Theorem 3.4, and (3.29), the function

$$F_o(z) = (\Pi_o D_A^{-2} \Pi_o^*)^{1/2} Q_o(z)^{-1} = (\Pi_o T_o^{-1} \Pi_o^*)^{1/2} Q_o(z)^{-1}$$

is an outer spectral factor of $T_o$. This readily implies that the function $F_o$ in (3.27) is an outer spectral factor of $T_o$. Since $T_o$ is strictly positive, the operator $F_+$ is invertible. So $F_o(z)^{-1}$ is in $H^\infty(\mathcal{L}, \mathcal{L})$. Therefore, $Q_o(z)$ is in $H^\infty(\mathcal{L}, \mathcal{L})$. The proof is now complete.

## 4. The Four Block Problem

The four block interpolation problem has played a basic role in $H^\infty$ control theory [4,8,12,14,16,17,21,27,38]. In this section we will use Corollary 1.3 to improve upon the Kaftal-Larson-Weiss bounds for the $H^2 - H^\infty$ four block Nehari interpolation problem discussed in [32]. To this end, let $\mathcal{E}$ be finite dimensional, and recall that $L^2(\mathcal{E}, \mathcal{E}')$ is the Hilbert space of all square integrable Lebesgue measurable functions on $[0, 2\pi)$ whose values are operators mapping $\mathcal{E}$ into $\mathcal{E}'$. The $L^2(\mathcal{E}, \mathcal{E}')$ norm of a function $\Psi$ is defined by

$$\|\Psi\|_2^2 = \frac{1}{2\pi}\int_0^{2\pi} \text{trace } (\Psi(t)^*\Psi(t))dt = \|M_\Psi \mid \mathcal{E}\|_2^2 . \tag{4.1}$$

Notice that the $L^2$ norm of $\Psi$ equals the Hilbert-Schmidt norm of the multiplication operator $M_\Psi \mid \mathcal{E}$. To present our $H^2 - H^\infty$ Nehari result, let G be the function defined by

$$G = \begin{bmatrix} G_{11} & G_{12} \\ G_{21} & G_{22} \end{bmatrix} \text{ in } L^\infty(\mathcal{E}_1 \oplus \mathcal{E}_2, \mathcal{E}_1' \oplus \mathcal{E}_2') \tag{4.2}$$

where $\mathcal{E}_1 \oplus \mathcal{E}_2$ is finite dimensional. Now consider the following $H^2$ and $H^\infty$ Nehari optimization problems

$$d_i = \inf \left\{ \left\| \begin{bmatrix} G_{11} + H & G_{12} \\ G_{21} & G_{22} \end{bmatrix} \right\|_i : H \in H^i(\mathcal{E}_1, \mathcal{E}_1') \right\} \qquad (\text{for } i = 2, \infty) \tag{4.3}$$

where $\|\cdot\|_2$ is $L^2$ norm and $\|\cdot\|_\infty$ is the $L^\infty$ norm. Notice that $d_2$ is the distance from G to $H^\infty(\mathcal{E}_1, \mathcal{E}_1') \oplus \{0\}$ in the $L^2$ norm, while $d_\infty$ is the distance from G to $H^\infty(\mathcal{E}_1, \mathcal{E}_1') \oplus \{0\}$ in the $L^\infty$ norm.

By using the projection theorem it is easy to solve the $H^2$ optimization problem in (4.3). To this end, let $P_+$ be the orthogonal projection from $L^2(\cdot)$ onto $H^2(\cdot)$ and let $P_- = I - P_+$. Then for any H in $H^2(\mathcal{E}_1, \mathcal{E}_1')$ it follows that

$$\left\| \begin{bmatrix} G_{11}+H & G_{12} \\ G_{21} & G_{22} \end{bmatrix} \right\|_2^2 = \|G_{11}+H\|_2^2 + \|G_{12}\|_2^2 + \|G_{21}\|_2^2 + \|G_{22}\|_2^2 \geq$$

$$\|G_{11} - P_+G_{11}\|_2^2 + \|G_{12}\|_2^2 + \|G_{21}\|_2^2 + \|G_{22}\|_2^2 \, .$$

Therefore the optimal H in $H^2(\mathcal{E}_1, \mathcal{E}_1')$ solving the $H^2$ optimization problem in (4.3) is given by $H = -P_+G_{11}$. Moreover, the $H^2$ error $d_2$ is given by

$$d_2^2 = \|P_-G_{11}\|_2^2 + \|G_{12}\|_2^2 + \|G_{21}\|_2^2 + \|G_{22}\|_2^2 \, . \tag{4.4}$$

In order to obtain a solution to the $H^\infty$ Nehari optimization problem in (4.3), let A(G) be the operator, with symbol G, mapping $H^2(\mathcal{E}_1) \oplus L^2(\mathcal{E}_2)$ into $K^2(\mathcal{E}_1') \oplus L^2(\mathcal{E}_2')$ defined by $A(G) = (P_- \oplus I)M_G \mid (H^2(\mathcal{E}_1) \oplus L^2(\mathcal{E}_2))$ where $K^2(\mathcal{E}_1') = L^2(\mathcal{E}_1') \ominus H^2(\mathcal{E}_1')$. To be precise,

$$A(G)\begin{bmatrix} x \\ y \end{bmatrix} = \begin{bmatrix} P_-(G_{11}x + G_{12}y) \\ G_{21}x + G_{22}y \end{bmatrix} \qquad (x \oplus y \text{ in } H^2(\mathcal{E}_1) \oplus L^2(\mathcal{E}_2)) \, . \tag{4.5}$$

By using the commutant lifting theorem, it is easy to show that $d_\infty = \|A(G)\|$; see [16], [17] or Section IX.4 in [18]. This fact is also a consequence of the following result.

**THEOREM 4.1.** *Let G in* $L^\infty(\mathcal{E}_1 \oplus \mathcal{E}_2, \mathcal{E}_1' \oplus \mathcal{E}_2')$ *be the function defined in (4.2) where* $n = \dim(\mathcal{E}_1)$. *Then* $d_\infty = \|A(G)\|$. *Moreover, given any* $\delta > 1$, *there exists an* H *in* $H^\infty$ *satisfying the following mixed* $H^2 - H^\infty$ *bounds:*

$$\left\| \begin{bmatrix} G_{11}+H & G_{12} \\ G_{21} & G_{22} \end{bmatrix} \right\|_\infty \leq \delta d_\infty$$

$$\left\| \begin{bmatrix} G_{11}+H & G_{12} \\ G_{21} & G_{22} \end{bmatrix} \right\|_2 \leq \frac{\delta d_2}{\sqrt{\delta^2 - 1 + \dfrac{d_{21}^2}{n d_\infty^2}}} \tag{4.6}$$

*where* $d_{21}^2 = \|P_-G_{11}\|_2^2 + \|G_{21}\|_2^2$. *In fact, by choosing* $\delta^2 = 2 - d_{21}^2/nd_\infty^2$ *equation (4.6) becomes*

$$\left\|\begin{bmatrix} G_{11}+H & G_{12} \\ G_{21} & G_{22} \end{bmatrix}\right\|_\infty \le d_\infty \sqrt{2 - \frac{d_{21}^2}{n d_\infty^2}}$$

$$\left\|\begin{bmatrix} G_{11}+H & G_{12} \\ G_{21} & G_{22} \end{bmatrix}\right\|_2 \le d_2 \sqrt{2 - \frac{d_{21}^2}{n d_\infty^2}} \tag{4.7}$$

**PROOF.** First notice that the operator $A(\Psi) = 0$ for any symbol $\Psi$ in $L^\infty(\mathcal{E}_1 \oplus \mathcal{E}_2, \mathcal{E}_1' \oplus \mathcal{E}_2')$ if and only if

$$\Psi \in \begin{bmatrix} H^\infty(\mathcal{E}_1, \mathcal{E}_1') & 0 \\ 0 & 0 \end{bmatrix}. \tag{4.8}$$

Obviously, if $\Psi$ satisfies (4.8), then $A(\Psi) = 0$. To verify the other half, assume that $\Psi$ is given by

$$\Psi = \begin{bmatrix} \psi_{11} & \psi_{12} \\ \psi_{21} & \psi_{22} \end{bmatrix}$$

and $A(\Psi) = 0$. Clearly this implies that both $\psi_{21}$ and $\psi_{22}$ are zero. Since $A(\Psi)(0 \oplus y) = 0$ for all $y$ on $L^2(\mathcal{E}_2)$, we see that $P\_\psi_{12} e^{-int} \mathcal{E}_2 = 0$ for all $n \ge 0$. Hence $\psi_{12} = 0$. So the only nonzero term in $\Psi$ is $\psi_{11}$. Finally, using $0 = A(\Psi) = P\_\psi_{11} \,|\, H^2(\mathcal{E}_1)$, we see that $\psi_{11}$ is in $H^\infty(\mathcal{E}_1, \mathcal{E}_1')$. Therefore $A(\Psi) = 0$ if and only if $\Psi$ satisfies (4.8).

Now let S be the unilateral shift on $H^2(\mathcal{E}_1)$ and $V_2$ the bilateral shift (multiplication by $e^{it}$) on $L^2(\mathcal{E}_2)$. Let T be the isometry on $\mathcal{H} = H^2(\mathcal{E}_1) \oplus L^2(\mathcal{E}_2)$ defined by $T = S \oplus V_2$. Let $V_1'$ and $V_2'$ be the bilateral shifts on $L^2(\mathcal{E}_1')$ and $L^2(\mathcal{E}_2')$, respectively. Let $T'$ on $\mathcal{H}' = K^2(\mathcal{E}_1') \oplus L^2(\mathcal{E}_2')$ be the co-isometry defined by $T' = P\_V_1' \,|\, K^2(\mathcal{E}_1') \oplus V_2'$. Obviously, $V' = V_1' \oplus V_2'$ is the minimal isometric dilation of $T'$. Recall that all minimal isometric dilations of $T'$ are isomorphic; see [45] or Theorem VI.3.2 in [18]. This implies that there exists a unitary operator W mapping $\mathcal{K}'$ onto $L^2(\mathcal{E}_1') \oplus L^2(\mathcal{E}_2')$ satisfying $V'W = WU'$ and $W \,|\, \mathcal{H}' = I$ where $U'$ on $\mathcal{K}'$ is the Sz.-Nagy-Schäffer minimal isometric dilation of $T'$ in (1.2). It is easy to show that $T'A = AT$, where $A = A(G)$. According to Corollary 1.3, the central intertwining lifting $WB_\gamma$ mapping $\mathcal{H} = H^2(\mathcal{E}_1) \oplus L^2(\mathcal{E}_2)$ into $W\mathcal{K}' = L^2(\mathcal{E}_1') \oplus L^2(\mathcal{E}_2')$ of A satisfies the bounds in (1.25) and (1.26) where $\gamma = \delta\|A\|$. Thus $\|B_\gamma\| \le \delta\|A\|$.

Using $V'WB_\gamma = WB_\gamma T$ it follows that $WB_\gamma = M_{G_\gamma} \,|\, \mathcal{H}$ where $G_\gamma$ is a function in $L^\infty(\mathcal{E}_1 \oplus \mathcal{E}_2, \mathcal{E}_1' \oplus \mathcal{E}_2')$ and $\|B_\gamma\| = \|G_\gamma\|_\infty$. This and $W \,|\, \mathcal{H}' = I$ imply that

$$A(G_\gamma) = P_{\mathcal{H}'} M_{G_\gamma} \mid \mathcal{H} = P_{\mathcal{H}'} W B_\gamma = A(G) .$$

Therefore $A(G_\gamma - G) = 0$. In other words, $\Psi = G_\gamma - G$ satisfies (4.8). So there exists an H in $H^\infty(\mathcal{E}_1, \mathcal{E}_1')$ such that

$$G_\gamma = \begin{bmatrix} G_{11} + H & G_{12} \\ G_{21} & G_{22} \end{bmatrix} . \tag{4.9}$$

Moreover, $\|G_\gamma\|_\infty = \|B_\gamma\| \le \delta\|A\|$. However, if $G'$ is any function given by the right hand side of (4.9) where H is any function in $H^\infty(\mathcal{E}_1, \mathcal{E}_1')$, then $A(G) = A = A(G')$. This implies that $\|A\| \le \|G'\|_\infty$. By taking the infimum of $\|G'\|_\infty$ over all H in $H^\infty(\mathcal{E}_1, \mathcal{E}_1')$ we see that $\|A\| \le d_\infty$. Therefore we have

$$\|A\| \le d_\infty \le \|G_\gamma\|_\infty = \|B_\gamma\| \le \delta\|A\| .$$

For $\delta = 1$, we obtain $d_\infty = \|A\|$.

Notice that $\mathcal{E}_1 = \ker(T^*) = \mathcal{L}$. Moreover,

$$d_{21}^2 = \|P\_G_{11}\|_2^2 + \|G_{21}\|_2^2 = \|A \mid \mathcal{E}_1\|_2^2 . \tag{4.10}$$

So the inequality in equation (1.26) of Corollary 1.3 becomes

$$\|B_\gamma \mid \mathcal{E}_1\|_2^2 \le \frac{\delta^2 \|A \mid \mathcal{E}_1\|_2^2}{\delta^2 - 1 + \dfrac{d_{21}^2}{n d_\infty^2}} . \tag{4.11}$$

According to (4.9) we see that

$$\|B_\gamma \mid \mathcal{E}_2\|_2^2 = \|M_{G_\gamma} \mid \mathcal{E}_2\|_2^2 = \|M_G \mid \mathcal{E}_2\|_2^2 = \|G_{12}\|_2^2 + \|G_{22}\|_2^2 .$$

Using this in equation (4.11) along with (4.10), and the fact that $\delta^2(\delta^2 - 1 + d_{21}^2/nd_\infty^2)^{-1}$ is greater than one (because $d_{21}^2 \le nd_\infty^2$), we have

$$\|G_\gamma\|_2^2 = \|M_{G_\gamma} | (\mathcal{E}_1 \oplus \mathcal{E}_2)\|_2^2 = \|B_\gamma | \mathcal{E}_1\|_2^2 + \|B_\gamma | \mathcal{E}_2\|_2^2 \le$$

$$\frac{\delta^2 \|A | \mathcal{E}_1\|^2}{\delta^2 - 1 + \dfrac{d_{21}^2}{n d_\infty^2}} + \|G_{12}\|_2^2 + \|G_{22}\|_2^2 \le$$

$$\frac{\delta^2 (\|P_- G_{11}\|_2^2 + \|G_{21}\|_2^2 + \|G_{12}\|_2^2 + \|G_{22}\|_2^2)}{\delta^2 - 1 + \dfrac{d_{21}^2}{n d_\infty^2}} = \frac{\delta^2 d_2^2}{\delta^2 - 1 + \dfrac{d_{21}^2}{n d_\infty^2}} .$$

Obviously $\|G_\gamma\|_\infty = \|B_\gamma\| \le \delta d_\infty$. This along with the form of $G_\gamma$ in (4.9) completes the proof.

To complete this section we will show how Corollary 3.7 can be used to give an explicit formula to solve a two block $H^2 - H^\infty$ Nehari problem. To this end, let $G = [G_{11}^*, G_{21}^*]^*$ be a fixed function in $L^\infty(\mathcal{E}_1, \mathcal{E}_1' \oplus \mathcal{E}_2')$. In this case, $\mathcal{E}_2 = \{0\}$, and the $H^2 - H^\infty$ optimization problems in (4.3) reduces to

$$d_i = \inf \left\{ \left\| \begin{bmatrix} G_{11} + H \\ G_{21} \end{bmatrix} \right\|_i : H \in H^\infty(\mathcal{E}_1, \mathcal{E}_1') \right\} \qquad (\text{for } i = 2, \infty) . \qquad (4.12)$$

In this setting, the operator $A = A(G)$ mapping $H^2(\mathcal{E}_1)$ into $K^2(\mathcal{E}_1') \oplus L^2(\mathcal{E}_2')$ is defined by $A = (P_- \oplus I) M_G | H^2(\mathcal{E}_1)$. As before, it is easy to show that $T'A = AT$ where $T = S$ is now the unilateral shift on $H^2(\mathcal{E}_1)$ and $T'$ is the compression of $V_1' \oplus V_2'$ to $\mathcal{H}' = K^2(\mathcal{E}_1') \oplus L^2(\mathcal{E}_2')$. Obviously, $d_\infty = \|A\|$ and

$$d_2^2 = \|P_- G_{11}\|_2^2 + \|G_{21}\|_2^2 = \|A | \mathcal{E}_1\|_2^2 = d_{21}^2 . \qquad (4.13)$$

Finally, notice that in this case $\mathcal{L} = \ker(T^*) = \mathcal{E}_1$ and $\Pi_0$ is the operator from $H^2(\mathcal{E}_1)$ onto $\mathcal{E}_1$ defined by $\Pi_0 = P_{\mathcal{E}_1}$, which picks out the zero degree $\mathcal{E}_1$ component of a function in $H^2(\mathcal{E}_1)$. The following result provides us with a simple formula to compute the central intertwining lifting $B_\gamma$ for the two block problem corresponding to (4.12).

**THEOREM 4.2.** *Let* $G = [G_{11}^*, G_{21}^*]^*$ *be a function in* $L^\infty(\mathcal{E}_1, \mathcal{E}_1' \oplus \mathcal{E}_2')$ *and* $A = A(G)$ *its corresponding linear operator, where* $n = \dim(\mathcal{E}_1)$ *is finite. Let* $\delta > 1$ *and* $\gamma = \delta d_\infty$. *Let* $P(e^{it})$ *and* $Q(z)$ *be the functions defined by*

$$P(e^{it}) = (A D_A^{-2} \Pi_0^*)(e^{it}) \text{ and } Q(z) = (D_A^{-2} \Pi_0^*)(z) \qquad (4.14)$$

*Then the central intertwining lifting* $B_\gamma$ *of* $A$ *is given by* $W B_\gamma = M_{G_\gamma} | H^2(\mathcal{E}_1)$ *where* $W$ *is a unitary operator and* $G_\gamma$ *is the function in* $L^\infty(\mathcal{E}_1, \mathcal{E}_1' \oplus \mathcal{E}_2')$ *defined by*

$$G_\gamma(e^{it}) = P(e^{it})Q(e^{it})^{-1} . \tag{4.15}$$

*In particular, this* $G_\gamma$ *admits a decomposition of the form*

$$G_\gamma = \begin{bmatrix} G_{11} + H \\ G_{21} \end{bmatrix} \tag{4.16}$$

*where* H *is a function in* $H^\infty(\mathcal{E}_1, \mathcal{E}_1')$. *Furthermore, the central intertwining lifting* $G_\gamma$ *satisfies the following* $H^2 - H^\infty$ *bounds*

$$\left\| \begin{bmatrix} G_{11} + H \\ G_{21} \end{bmatrix} \right\|_\infty \leq \delta d_\infty \quad \text{and} \quad \left\| \begin{bmatrix} G_{11} + H \\ G_{21} \end{bmatrix} \right\|_2 \leq \frac{\delta d_2}{\sqrt{\delta^2 - 1 + \dfrac{d_2^2}{n d_\infty^2}}} . \tag{4.17}$$

*Finally, if we choose* $\delta^2 = 2 - d_2^2/nd_\infty^2$ *we obtain*

$$\begin{aligned} \left\| \begin{bmatrix} G_{11} + H \\ G_{21} \end{bmatrix} \right\|_\infty &\leq d_\infty \sqrt{2 - \frac{d_2^2}{n d_\infty^2}} \\ \left\| \begin{bmatrix} G_{11} + H \\ G_{21} \end{bmatrix} \right\|_2 &\leq d_2 \sqrt{2 - \frac{d_2^2}{n d_\infty^2}} . \end{aligned} \tag{4.18}$$

Before proving this result let us first notice that according to Theorem 3.4 the function Q(z) is an outer function and $Q(z)^{-1}$ is in $H^\infty(\mathcal{E}_1, \mathcal{E}_1)$. Moreover, the function $F(z) = Q(0)^{1/2}Q(z)^{-1}$ is an outer spectral factor for the Toeplitz operator $\gamma^2 I - B_\gamma^* B_\gamma$, or equivalently, F(z) is an outer spectral factor for $\gamma^2 I - G_\gamma^* G_\gamma$.

**PROOF OF THEOREM 4.2.** Obviously $V' = V_1' \oplus V_2'$ is a minimal isometric dilation of $T'$. Let $U'$ on $\mathcal{K}' = \mathcal{H}' \oplus H^2(\mathcal{D}')$ be the Sz.-Nagy-Schäffer minimal isometric dilation of $T'$ defined in (1.2). As before, let W be the unitary operator mapping $\mathcal{K}'$ onto $L^2(\mathcal{E}_1') \oplus L^2(\mathcal{E}_2')$ satisfying $W | \mathcal{H}' = I$ and $V'W = WU'$. Since $V'$ is a bilateral shift, $U'$ is a bilateral shift. Using $W | \mathcal{H}' = I$ we have $A = WA$. So according to Corollary 3.7 the central intertwining lifting $WB_\gamma = M_{G_\gamma} | H^2(\mathcal{E}_1)$ where $G_\gamma$ is the function in $L^\infty(\mathcal{E}_1, \mathcal{E}_1' \oplus \mathcal{E}_2')$ defined in (4.15). The fact that $G_\gamma$ admits a decomposition of the form (4.16) and satisfies the $H^2 - H^\infty$ bounds in (4.17) readily follows from (4.13) and the proof of Theorem 4.1. This completes the proof.

**REMARK 4.3.** In many applications it is hard to invert the infinite dimensional operator $D_A^{-2}$. To complete this section we will follow some ideas in [8], [12], [21] to help alleviate this problem. Let $T_o$ be the Toeplitz operator on $H^2(\mathcal{E}_1)$ defined by

$$T_o = P_+(\gamma^2 I - M_{G_{21}}^* M_{G_{21}}) \mid H^2(\mathcal{E}_1)\,.$$

We claim that the Toeplitz operator $T_o$ is strictly positive. This follows because A is strictly bounded by $\gamma$. So for some $\varepsilon > 0$ we have for all h in $\mathcal{H}$

$$\varepsilon\|h\|^2 \le \gamma^2\|h\|^2 - \|A(G)h\|^2 = \gamma^2\|h\|^2 - \|G_{21}h\|^2 - \|P_-G_{11}h\|^2 \le (T_o h\,,\, h)\,.$$

Hence $T_o$ is strictly positive. Let $F_o$ in $H^\infty(\mathcal{E}_1, \mathcal{E}_1)$ be the outer spectral factor for $T_o$. Let $A_{11} = A(G_{11})$ be the Hankel operator from $H^2(\mathcal{E}_1)$ to $K^2(\mathcal{E}_1')$ defined by $A_{11} = P_- M_{G_{11}} \mid H^2(\mathcal{E}_1)$. Then

$$D_A^{-2}\Pi_o^* = (T_o - A_{11}^* A_{11})^{-1}\Pi_o^* = (F_{o+}^* F_{o+} - A_{11}^* A_{11})^{-1}\Pi_o^* =$$

$$F_{o+}^{-1}(I - F_{o+}^{*-1} A_{11}^* A_{11} F_{o+}^{-1})^{-1} F_{o+}^{*-1}\Pi_o^* = F_{o+}^{-1}(I - A_{new}^* A_{new})^{-1}\Pi_o^* F_o(0)^{*-1}$$

where $A_{new}$ is the strictly contractive Hankel operator from $H^2(\mathcal{E}_1)$ to $K^2(\mathcal{E}_1')$ defined by $A_{new} = A_{11}F_{o+}^{-1}$. By ignoring $F_o(0)$, the functions P and Q in (4.14) can be replaced by

$$P_n(e^{it}) = (AF_{o+}^{-1}D_{new}^{-2}\Pi_o^*)(e^{it}) \text{ and } Q_n(z) = F_o(z)^{-1}(D_{new}^{-2}\Pi_o^*)(z)\,, \tag{4.19}$$

where $D_{new}^2 = I - A_{new}^* A_{new}$. As before, the central intertwining lifting $WB_\gamma$ of A is given by $M_{G_\gamma} \mid H^2(\mathcal{E}_1)$ where $G_\gamma = P_n Q_n^{-1}$. If we let $P_{new}$ and $Q_{new}$ be the functions defined by

$$P_{new} = (A_{new}D_{new}^{-2}\Pi_o^*)(e^{it}) \quad\text{and}\quad Q_{new} = (D_{new}^{-2}\Pi_o^*)(z)\,, \tag{4.20}$$

then we see that the central intertwining lifting $G_\gamma$ of A is given by (4.16) where

$$G_{11} + H = P_{new}Q_{new}^{-1}F_o\,. \tag{4.21}$$

The formulas for $P_{new}$ and $Q_{new}$ in (4.20) are easier to use than the formulas for P and Q in (4.14) in the rational case. Because if both $G_{11}$ and $G_{21}$ are rational functions, then $A_{11}$ is a finite rank Hankel operator and the outer function $F_o$ is rational. This implies that the Hankel operator $A_{new}$ is finite rank. Therefore one can use standard state space techniques to compute $F_o$ and $D_{new}^{-2}\Pi_o^*$. (The outer factor $F_o$ is usually computed by solving a Riccati equation, and $d_\infty$ is computed by an iteration on $\gamma$; see [8], [12], [21].) So in the rational setting one can easily compute $P_{new}$ and $Q_{new}$ in (4.20) and thus find the central intertwining lifting $B_\gamma$.

Finally, let us notice that in the Nehari or one block setting, that is, when $G_{21} = 0$ or $\mathcal{E}_2' = \{0\}$, our formula for $G_\gamma = PQ^{-1}$ is precisely the central solution for the Nehari problem obtained in [3], [18], [30].

## 5. Optimal Solutions

In this section we will show that the central intertwining liftings do not necessarily minimize the $H^2$ norm subject to an $H^\infty$ constraint. Throughout this section g is a scalar valued function in $L^\infty$. Now consider the following $H^2$ optimization problem: Given a fixed $\delta \geq 1$ find an h in $H^\infty$ satisfying

$$\mu = \inf \{\|g+h\|_2 : h \in H^\infty\} \quad \text{subject to } \|g+h\|_\infty \leq \delta d_\infty . \tag{5.1}$$

As before, $d_2$ is the distance form g to $H^2$ in the $L^2$ norm, while $d_\infty$ is the distance from g to $H^\infty$ in the $L^\infty$ norm.

Now let us convert the previous $H^2 - H^\infty$ optimization problem to an optimization problem involving the intertwining liftings B of A. As before, let $A = A(g)$ be the Hankel operator mapping $H^2$ into $K^2 = L^2 \ominus H^2$ defined by $A = P\_M_g | H^2$. By the projection theorem, $d_2 = \|P\_g\|_2 = \|A1\|$ where 1 is the constant function 1 in $H^2$. Notice also that 1 is a basis for the kernel of $S^*$ where $S = T$ is the unilateral shift on $H^2$. Now let V be the bilateral shift (multiplication by $e^{it}$) on $L^2$ and $T'$ on $\mathcal{H}' = K^2$ be the co-isometry defined by $T' = P_{\mathcal{H}'} V | \mathcal{H}'$. As before, it is easy to verify that $T'A = AT$. Obviously V is the minimal isometric dilation of $T'$. Recall that B is an intertwining lifting of A if B is an operator mapping $H^2$ into $L^2$ satisfying $P_{\mathcal{H}'}B = A$ and $VB = BS$. Since $VB = BS$ it follows that there exists a b in $L^\infty$ satisfying $B = M_b | H^2$. Moreover, $\|B\| = \|b\|_\infty$. Using the fact that $P_{\mathcal{H}'}B = A = A(g)$, it follows that $A(b) = A(g)$. Thus $A(b-g)1 = 0$. This implies that $b = g + h$ where h is in $H^\infty$. So if B is an intertwining lifting of $A = A(g)$, then $B = M_b | H^2$ where $b = g + h$ and h is in $H^\infty$. Moreover, $\|B\| = \|g+h\|_\infty$. On the other hand, if $b = g + h$ where h is in $H^\infty$, then $B = M_b | H^2$ is an intertwining lifting of A. Therefore, the set of all intertwining lifting B of A are given by $B = M_b | H^2$ where b is in the set $\{g + H^\infty\}$.

As before, let $\delta \geq 1$. Now let $\mathcal{L}_\delta$ be the set of all intertwining lifting B of A satisfying $\|B\| \leq \delta \|A\|$. Recall that $\|A\| = d_\infty$. Notice that B is in $\mathcal{L}_\delta$ if and only if $B = M_b | H^2$ where $b = g + h$ for some h in $H^\infty$ and $\|g+h\|_\infty \leq \delta d_\infty$. Obviously $\|B1\| = \|g+h\|_2$. Therefore the $H^2 - H^\infty$ optimization problem in (5.1) is equivalent to the following optimization problem involving the intertwining liftings B of A in $\mathcal{L}_\delta$

$$\mu = \inf \{\|B1\| : B \in \mathcal{L}_\delta\} . \tag{5.2}$$

We claim that there is a unique $B_*$ in $\mathcal{L}_\delta$ solving the previous optimization problem. To prove this first notice that $\mathcal{L}_\delta$ is convex and compact in the weak operator topology. Let $\mathcal{K}_\delta$ be the convex set defined by $\mathcal{K}_\delta = \{B1 : B \in \mathcal{L}_\delta\}$. It easily follows that $\mathcal{K}_\delta$ is weakly compact.

In particular, $\mathcal{K}_\delta$ is weakly closed. Since $\mathcal{K}_\delta$ is convex, this implies that $\mathcal{K}_\delta$ is strongly closed. By the projection theorem, there exists a unique vector x in $\mathcal{K}_\delta$ satisfying $\|x\| = \inf\{\|B1\| : B \in \mathcal{L}_\delta\}$. Because $\mathcal{K}_\delta$ is closed, $x = B_*1$ for some $B_*$ in $\mathcal{L}_\delta$. Notice that $B_*$ is unique. If $x = B1$ for some other B in $\mathcal{L}_\delta$, then $(B_* - B)1 = 0$. Using the fact that 1 is cyclic for the unilateral shift, $B_* = B$. Therefore $B_*$ is the unique operator in $\mathcal{L}_\delta$ solving the optimization problem in (5.2). Summing up the previous analysis proves part of the following result.

**THEOREM 5.1.** *Let* g *be a function in* $L^\infty$ *and* $A = A(g)$ *be its corresponding Hankel operator. Then given any* $\delta \geq 1$, *there exists a unique intertwining lifting* $B_*$ *of* A *solving the optimization problem in (5.2), or equivalently, there exists a unique* $h_*$ *in* $H^\infty$ *solving the* $H^2 - H^\infty$ *optimization problem in (5.1). Moreover, the optimal intertwining lifting* $B_*$ is $M_{b_*} | H^2$ *where* $b_* = g + h_*$. *Furthermore, for* $\delta > 1$ *the following results hold.*

*(i)* *The function* $|b_*(e^{it})| < \delta d_\infty$ *a.e. on a set of positive Lebesgue measure.*

*(ii)* *If* $d_2 < \mu$, *then* $\|b_*\|_\infty = \delta d_\infty$.

**PROOF.** To prove part (i) we use contradiction and assume that $|b_*(e^{it})| = \gamma$ a.e. where $\gamma = \delta d_\infty$. This implies that $\|B_*1\| = \|b_*1\| = \gamma$. So for any B in $\mathcal{L}_\delta$ we have that $\gamma = \|B_*1\| < \|B1\|$. However, by the commutant lifting theorem (that is, by Theorem 1.1 with the corresponding $\gamma = \|A\|$) there exists an intertwining lifting B in $\mathcal{L}_\delta$ of A satisfying $\|A\| = \|B\|$. Because $\delta > 1$ and $d_\infty = \|A\|$, this implies that $\|B1\| \leq \|A\| = d_\infty < \delta d_\infty = \|B_*1\|$. This contradicts the fact that $\|B_*1\| \leq \|B1\|$. Therefore we must have $|b_*(e^{it})| < \gamma$ a.e. on a set of positive Lebesgue measure.

To prove part (ii) we will also use contradiction. To this end, assume that $d_2 < \mu$ and $\|b_*\|_\infty < \gamma$. Let $h_*$ be the unique optimal function in $H^\infty$ given by $b_* = g + h_*$. Now let $H_r^\infty$ be the closed ball in $H^\infty$ of radius r. Because $\|g + h_*\|_\infty < \gamma$, there exists an $r > 0$ such that $\|g + h_* + h\|_\infty \leq \gamma$ for all h in $H_r^\infty$. In fact, r can be taken to be $r = \gamma - \|b_*\|_\infty$. Obviously $b = g + h_* + h$ defines an intertwining lifting $B = M_b | H^2$ in $\mathcal{L}_\delta$. Therefore

$$\|g + h_*\|_2^2 = \|B_*1\| \leq \|B1\|^2 = \|g + h_* + h\|_2^2 .$$

This readily implies that $g + h_*$ solves the following convex optimization problem

$$\|g + h_*\|_2 = \inf\{\|g + h_* + h\|_2 \quad : \quad h \in H_r^\infty\} .$$

Obviously, $h_* + H_r^\infty$ is convex. So by the projection theorem for the convex sets $0 \leq \mathrm{Re}\,(g + h_*, h_* + h - h_*) = \mathrm{Re}\,(g + h_*, h)$ for all h in $H_r^\infty$. By replacing h by ch where c is the appropriate constant, we see that $0 = (g + h_*, h)$ for all h in $H_r^\infty$. Therefore, $g + h_*$ is orthogonal to $H^2$. In other words, $h_* = -P_+ g$. However, this implies that

$$d_2 = \|P_- g\|_2 = \|g - P_+ g\|_2 = \|g + h_*\|_2 = \mu .$$

This contradicts the assumption that $d_2 < \mu$, and completes the proof.

**REMARK 5.2.** In general the central intertwining lifting $B_\gamma$ of A does not equal $B_*$, the optimal intertwining lifting of A solving the optimization problem in (5.2). This is particularly easy to establish in the rational case. To see this, let g be a rational function in $L^\infty$ and $g_\gamma = g + h$ be the function in $L^\infty$ corresponding to the central intertwining lifting $B_\gamma$ of A, that is, $B_\gamma = M_{g_\gamma} | H^2$. Now assume that $\delta > 1$ and $d_2 < \mu$. Since g is rational, A is a finite rank Hankel operator. From this it is easy to show that $Q(z) = (D_A^{-2} 1)(z)$ is a rational function. According to Theorem 3.4, this function Q(z) is an outer function in $H^2$ and $Q(z)^{-1}$ is in $H^\infty$. Because Q(z) is rational, both Q(z) and $Q(z)^{-1}$ are in $H^\infty$. Moreover, Theorem 3.4 also shows that $F(z) = Q(0)^{1/2} Q(z)^{-1}$ is an outer spectral factor for $\gamma^2 I - B_\gamma^* B_\gamma$. Since Q and $Q^{-1}$ are both in $H^\infty$, the operator $F_+$ on $H^2$ is invertible. This fact along with $F_+^* F_+ = \gamma^2 I - B_\gamma^* B_\gamma$ readily implies that $B_\gamma$ is strictly bounded by $\gamma$. In particular, $\|g_\gamma + h\|_\infty = \|B_\gamma\| < \gamma$. However, by part (ii) of the previous theorem, the optimal solution $b_*$ satisfies $\|b_*\|_\infty = \gamma$. Therefore if g is rational, $\delta > 1$ and $d_2 < \mu$, then the central intertwining lifting $B_\gamma$ does not equal the optimal intertwining lifting $B_*$ of A.

We claim that the condition $d_2 < \mu$ occurs quite often. To begin, first notice that for a rational g in $L^\infty$ if $\|A(g)\| = \|g\|_\infty$, then $|g(e^{it})| = \|g\|_\infty$ a.e. Indeed, this follows from a famous result of Adamjan-Arov-Krein [2], which states that if A is a compact Hankel operator, then there exists a unique g in $L^\infty$ satisfying $A = A(g)$ and $\|g\|_\infty = \|A\| = d_\infty$. Moreover, $|g(e^{it})| = d_\infty$ a.e. Now let g be a rational function in $K^2 = L^2 \ominus H^2$ such that $\bar{g}$ is not a constant times a Blaschke product. Then $|g(e^{it})| \neq \|g\|_\infty$ on a set of positive Lebesgue measure and thus $\|A(g)\| < \|g\|_\infty$. Choose any $\delta > 1$ such that $\delta\|A(g)\| = \delta d_\infty < \|g\|_\infty$. We claim that for these $\delta$'s the $H^2$ error $d_2 < \mu$.

To prove this let us use contradiction and assume that $d_2 = \mu$. This implies that the optimal $h_*$ also solves the following $H^2$ optimization problem

$$\|g + h_*\|_2 = \mu = d_2 = \inf \{ \|g + h\|_2 : h \in H^2 \} .$$

By the uniqueness in the projection theorem $h_* = -P_+ g = 0$. Therefore the optimal $b_* = g$. Thus

$$\|g\|_\infty = \|b_*\|_\infty \le \delta d_\infty = \delta \|A(g)\| < \|g\|_\infty .$$

This is a contradiction. So we conclude that for all rational g in $K^2$ such that $\bar{g}$ is not a constant times a Blaschke product, $d_2 < \mu$ as long as $\delta - 1 > 0$ is small enough. In this case the optimal $B_*$ is different from the central intertwining lifting $B_\gamma$ of A provided by Theorem 1.1.

We conclude this section by mentioning that other results in this direction involving state feedback are given in [42].

## REFERENCES

[1] V. M. Adamjan, D. Z. Arov and M. G. Krein, Infinite Hankel matrices and generalized problems of Carathéodory - Fejér and I. Schur, *Functional Anal. i Prilozen*, **2** (1968), pp. 1-19 (Russian).

[2] V. M. Adamjan, D. Z. Arov and M. G. Krein, Analytic properties of Schmidt pairs for a Hankel operator and the generalized Schur-Takagi problem, *Math USSR Sbornick*, **15** (1971), pp. 31-73.

[3] V. M. Adamjan, D. Z. Arov and M. G. Krein, Infinite Hankel block matrices and related extension problems, *Izv. Akad. Nauk. Armjan SSR, Matematika*, **6** (1971), pp. 87-112, (English Translation *Amer. Math. Soc. Trans.*, **III** (1978), pp. 133-156).

[4] J. A. Ball and N. Cohen, Sensitivity minimization in an $H^\infty$ norm: parameterization of all suboptimal solutions, *Int. J. Control*, **46** (1987), pp. 785-816.

[5] J. A. Ball, I. Gohberg, and L. Rodman, *Interpolation for Rational Matrix Functions*, Birkhauser-Verlag, Basel, 1990.

[6] J. A. Ball and J. W. Helton, A Beurling-Lax theorem for the Lie group U(m, n) which contains most classical interpolation theory, *J. Operator Theory*, **9** (1983), pp. 107-142.

[7] D. S. Bernstein and W. M. Haddad, LQG control with an $H_\infty$ performance bound: a Riccati equation approach, *IEEE Trans. on Automat. Contr.*, **34** (1989), pp. 293-305.

[8] C. C. Chu, J. C. Doyle and E. B. Lee, The general distance problem in $H_\infty$ optimal control theory, *Int. J. Control*, **44** (1986), pp. 565-596.

[9] C.K. Chui and G. Chen, *Signal Processing and Systems Theory*, Springer-Verlag, Berlin, 1992.

[10] K. Clancey and I. Gohberg, *Factorization of Matrix Functions and Singular Integral Operators*, Birkhauser, Basel, Switzerland, 1981.

[11] C. Davis, W. M. Kahan and H. F. Weinberger, Norm-preserving dilations and their applications to optimal error bounds, *SIAM J. Numer. Anal.*, **19** (1982), pp. 445-469.

[12] J. C. Doyle and B. A. Francis, Linear control theory with an $H_\infty$ optimality criterion, *SIAM J. Control and Optimization*, **25** (1987), pp. 815-844.

[13] J. C. Doyle, B. A. Frances, A. Tannenbaum, *Feedback Control Theory*, MacMillan, New York, 1991.

[14] J. C. Doyle, K. Glover, P. P. Khargonekar, and B. A. Francis, State-space solutions to standard $H_2$ and $H_\infty$ and control problems, *IEEE Trans. on Automat. Contr.*, **34**, (1989), pp. 831-847.

[15] H. Dym, I. Gohberg, A maximum entropy principle for contractive interpolants, *J. Functional Analysis*, **65**, pp. 83-125.

[16] A. Fentuch and B. A. Francis, Uniformly optimal control of linear feedback systems, *Automatica*, **21** (1985), pp. 563-574.

[17] A. Fentuch and B. A. Francis, Distance formulas for operator algebras arising in optimal control problems, *Topics in Operator Theory and Interpolation; Operator Theory: Advances and Applications*, **29** (1988), Ed. I Gohberg, pp. 151-170.

[18] C. Foias and A. E. Frazho, *The Commutant Lifting Approach to Interpolation Problems,* Operator Theory Advances and Applications, **44**, Birkhauser-Verlag, Basel, 1990.

[19] C. Foias and A. E. Frazho, Commutant and lifting and simultaneous $H^\infty$ and $L^2$ suboptimization, *SIAM J. Math. Anal.*, **23** (1992), pp. 984-994.

[20] C. Foias, A. E. Frazho and A. Tannenbaum, On combined $H^\infty - H^2$ suboptimal interpolants, submitted for publication.

[21] B. A. Francis, *A Course in* $H^\infty$ *Control Theory,* Lecture Notes in Control and Information Sciences, Springer-Verlag, New York, 1987.

[22] A. E. Frazho and S. M. Kherat, Applications of the Schur algorithm to mixed $H^2 - H^\infty$ Nehari problems,*J. Mathematical Analysis and Applications*, to appear.

[23] A. E. Frazho and S. M. Kherat, On mixed $H^2$–$H^\infty$ interpolation, *Integral Equation and Operator Theory*, to appear.

[24] A. E. Frazho and M. A. Rotea, A remark on mixed $L^2/L^\infty$ bounds, *Integral Equation and Operator Theory*, **15** (1992), pp. 343-348.

[25] P. A. Fuhrmann, *Linear Systems and Operators in Hilbert Space*, McGraw-Hill, New York, 1981.

[26] K. Glover, All optimal Hankel-norm approximations of linear multivariable systems and their $L_\infty$-Error bounds, *Int. J. Cont.*, **39** (1984), pp. 1115-1193.

[27] K. Glover, D. J. N. Limebeer, J. C. Doyle, E. M. Kasenally and M. G. Safonov, A characterization of all solutions to the four block general distance problem, *SIAM J. Control and Optimization*, **29** (1991), pp. 283-324.

[28] K. Glover and D. Mustafa, Derivation of the maximum entropy $H_\infty$-controller and a state-space formula for its entropy, *Int. Jour. Control.*, **50** (1989), pp. 899-916.

[29] I. Gohberg, M. A. Kaashoek and Van Schagen, Rational contractive and unitary interpolants in realized form, *Integral Equations and Operator Theory*, **11** (1988), pp. 105-127.

[30] I. Gohberg. M. A. Kaashoek, H. J. Woerdeman, The band method for positive and strictly contractive extension problems: an alternative version and new applications, *Integral Equations and Operators Theory*, **12** (1989), pp. 343-3829.

[31] G. H. Golub and C. F. Van Loan, *Matrix Computations*, The Johns Hopkins University Press, Baltimore, second edition, 1989.

[32] C. Gu, On the four block problem and simultaneous $H^\infty$ and $L^2$ suboptimization *Mathematical Analysis and Applications*, to appear.

[33] S. Janson, Interpolation of subcouples and quotient couples, preprint.

[34] V. Kaftal, D. Larson and G. Weiss, Quasitriangular subalgebras of semifinite Von Neumann algebras are closed, *J. Functional Analysis*, **107** (1992), pp.387-401.

[35] P. P. Khargonekar and M. A. Rotea, Mixed $H^2/H^\infty$ control: a convex optimization approach, *IEEE Trans. Autom. Control*, **36** (1991), pp. 824-837.

[36] S. V. Kisliakov, Real interpolation of Hardy spaces on the disk and on the bidisc, preprint.

[37] D. Mustafa, Relations between maximum-entropy/$H_\infty$ control and combined $H_\infty$/LQG control, *Systems and Control Letters*, **12** (1989), pp. 193-203.

[38] D. Mustafa and K. Glover, *Minimum Entropy* $H_\infty$ *Control*, Lecture Notes in Control and Information Sciences, Springer-Verlag, New York, 1990.

[39] K. M. Nagpal and P. P. Khargonekar, Filtering and smoothing in an $H^\infty$ setting, *IEEE Trans. Autom. Control*, **36** (1991), pp. 152-166.

[40] G. Pisier, Interpolation between $H^p$ Spaces and non-commutative generalization. I, *Pacific Journ. of Math.*, **155** (1992), pp. 341-368.

[41] M. Rotea and A. E. Frazho, Bounds on solutions to $H^\infty$ algebraic Riccati equations and $H^2$ properties of $H^\infty$ central solution, *Systems and Control Letters*, **19** (1992), pp. 341-352.

[42] M. A. Rotea and P. P. Khargonekar, $H_2$-optimal control with an $H_\infty$-constraint: the state-feedback case, *Automatica*, **27** (1991), pp. 307-316.

[43] D. Sarason, Generalized interpolation in $H^\infty$, *Trans. American Math. Soc.*, **127** (1967), pp. 179-203.

[44] B. Sz.-Nagy and C. Foias, Dilation des commutants d'opérateurs, *C. R. Acad. Sci. Paris, série A*, **266** (1968), pp. 493-495.

[45] B. Sz.-Nagy and C. Foias, *Harmonic Analysis of Operators on Hilbert Space,* North-Holland Publishing Co., Amsterdam, 1970.

[46] N. J. Young, An algorithm for the super-optimal sensitivity-minimising controller, *Proc. Workshop on New Perspectives in Industrial Control System Design Using* $H_\infty$ *Methods*, Oxford University Press, London, 1986.

Department of Mathematics
Indiana University
Bloomington, Indiana

School of Aeronautics and Astronautics
Purdue University
West Lafayette, IN 47907, U.S.A.

Department of Mathematics
Georgia Institute of Technology
Atlanta, Georgia

MSC 1992: Primary, 47A20, 47A57, 93B36.

Operator Theory:
Advances and Applications, Vol. 64




# On Mixed $H^2 - H^\infty$ Tangential Interpolation

A. E. Frazho and S. M. Kherat

The central solution for the commutant lifting theorem is used to solve mixed $H^2$–$H^\infty$ tangential interpolation problems. These include an $H^2$–$H^\infty$ tangential Carathéodory interpolation problem, an $H^2$–$H^\infty$ tangential Nevanlinna-Pick interpolation problem and an $H^2$–$H^\infty$ tangential Hermite-Fejer interpolation problem. Explicit computational formulas in terms of state space realizations are given.

## 1. Introduction

$H^\infty$ tangential interpolation problems have received a great of interest in the past; see [1,2,8,9,10,15,24,30,31,32,33,39,40,41,45] just to mention a few. More precisely, references [8,9,10,39,40,41] have presented some nice existence results, while [1,15,24,30,31,32,33,45] have given some explicit computational formulas. In this paper, our emphasis is slightly different. Here, we will use the central solution for the commutant lifting theorem to solve some mixed $H^2$–$H^\infty$ tangential interpolation problems. Explicit computational formulas in terms of state space realizations will also be given.

Our mixed $H^2$–$H^\infty$ approach is motivated by the nice $H^2$–$H^\infty$ Nehari result obtained by Kaftal, Larson and Weiss [26]. A proof of their result based on the Dym-Gohberg maximum entropy approach [7] is given in [17]. Using the central solution for the commutant lifting theorem, [11] obtained a general solution to a two-sided $H^2$–$H^\infty$ Nehari problem (or standard control problem). For some stronger $H^2$ bounds and exact $H^2$ formulas see [12]. Some state space formulas for a mixed $H^2$–$H^\infty$ Nehari problem are given in [11] and [16]. Using skew-Toeplitz techniques [13] presented an algorithm to solve some $H^2$–$H^\infty$ nonrational interpolation problems. In this paper, we will use the central solution for the commutant lifting theorem to solve a mixed $H^2$–$H^\infty$ tangential interpolation problem.

$H^\infty$ tangential interpolation problems have played an important role in control theory [1,5,14,30,31,32]. Realization theory has also played a basic role in $H^\infty$ interpolation problems [1,5,10,14,19,21,32,37]. Mixed $H^2$–$H^\infty$ problems are receiving some attention in the control literature [3,6,20,29,36,37,38,50]. For this reason, it is believed that some of our results may be useful in control theory.

We will follow the standard notation for Hilbert spaces in [47] and [10]. For example, $P_{\mathcal{H}}$ is the orthogonal projection onto the space $\mathcal{H}$. If C is an operator mapping $\mathcal{E}$ into $\mathcal{H}$, then $\|C\|$ is the usual norm of C, while $\|C\|_2$ is the Hilbert-Schmidt norm of C, that is, $\|C\|_2^2 = \sum_{i=1}^{n} \|Ce_i\|^2$ where $\{e_i : 1 \le i \le n\}$ is an orthonormal basis for $\mathcal{E}$. Throughout, $H^2(\mathcal{E})$ is the Hardy space of all analytic functions in the open unit disc D, with values in $\mathcal{E}$ whose Taylor coefficients are square summable. Moreover, $H^\infty(\mathcal{E}_1, \mathcal{E}_2)$ is the Hardy space of all uniformly bounded analytic functions in D whose values are linear operators mapping $\mathcal{E}_1$ into $\mathcal{E}_2$. If F is in $H^\infty(\mathcal{E}_1, \mathcal{E}_2)$, then $F_+$ denotes the multiplication operator from $H^2(\mathcal{E}_1)$ to $H^2(\mathcal{E}_2)$ defined by $F_+h = Fh$ where h is in $H^2(\mathcal{E}_1)$. It is well known that $\|F_+\| = \|F\|_\infty$ where $\|F\|_\infty$ is the $H^\infty$ norm of F. The $H^2$ norm of F is defined by

$$\|F\|_2^2 = \frac{1}{2\pi} \int_0^{2\pi} \text{trace } F(e^{it})^* F(e^{it}) dt = \|F_+ | \mathcal{E}_1\|_2^2 .$$

In this paper, we always assume that $\mathcal{E}$, $\mathcal{E}_1$ and $\mathcal{E}_2$ are all finite dimensional spaces. Therefore, the previous integral is well defined. Throughout S is the unilateral shift on $H^2(\mathcal{E})$, while $S_1$ and $S_2$ are the unilateral shifts on $H^2(\mathcal{E}_1)$ and $H^2(\mathcal{E}_2)$, respectively. Finally, recall that an operator $\Gamma$ mapping $H^2(\mathcal{E})$ to $H^2(\mathcal{E}_1)$ is Hankel if and only if

$$S_1^* \Gamma = \Gamma S \tag{1.1}$$

In other words, a Hankel operator is an operator interwinning the adjoint of the unilateral shift with the unilateral shift.

Let $\Gamma_1$ be a Hankel operator mapping $H^2(\mathcal{E})$ into $H^2(\mathcal{E}_1)$ and $\Gamma_2$ a Hankel operator mapping $H^2(\mathcal{E})$ into $H^2(\mathcal{E}_2)$. We say that a function F in $H^\infty(\mathcal{E}_1, \mathcal{E}_2)$ *tangentially interpolates the Hankel operator* $\Gamma_2$ *with the Hankel operator* $\Gamma_1$ *if* $F_+^*\Gamma_2 = \Gamma_1$. In Section 4, we will show that this tangential interpolation problem generalizes the classical Hermite-Fejer tangential interpolation problem. Obviously there does not always exist a function tangentially interpolating two Hankel operators $\Gamma_2$ with $\Gamma_1$. For example, if $\Gamma_2 = 0$ and $\Gamma_1$ is nonzero. The following lemma which is a classical application of the commutant lifting theorem gives a necessary and sufficient condition for the existence of an F tangentially interpolating $\Gamma_2$ with $\Gamma_1$.

**LEMMA 1.1** . *Let* $\Gamma_1$ *be a Hankel operator mapping* $H^2(\mathcal{E})$ *into* $H^2(\mathcal{E}_1)$ *and* $\Gamma_2$ *a Hankel operator mapping* $H^2(\mathcal{E})$ *into* $H^2(\mathcal{E}_2)$*. Then, there exists an* F *in* $H^\infty(\mathcal{E}_1, \mathcal{E}_2)$ *tangentially interpolating* $\Gamma_2$ *with* $\Gamma_1$ *if and only if* $\Gamma_1^*\Gamma_1 \le \gamma^2\Gamma_2^*\Gamma_2$ *for some finite scalar* $\gamma$*. Moreover, in this case there exists an* F *tangentially interpolating* $\Gamma_2$ *with* $\Gamma_1$ *and satisfying* $\|F\|_\infty \le \gamma$.

PROOF . Assume that $\Gamma_1 = F_+^*\Gamma_2$ for in F in $H^\infty(\mathcal{E}_1, \mathcal{E}_2)$. This implies that for all h in $H^2(\mathcal{E})$

$$\|\Gamma_1 h\|^2 = \|F_+^*\Gamma_2 h\|^2 \le \|F_+^*\|^2\|\Gamma_2 h\|^2 = \|F\|_\infty^2\|\Gamma_2 h\|^2 . \tag{1.2}$$

Therefore, $\Gamma_1^*\Gamma_1 \le \gamma^2\Gamma_2^*\Gamma_2$ where $\gamma = \|F\|_\infty$. The other half is a simple application of the commutant lifting theorem [10,46,47]. To this end, let $\mathcal{H}_1$ be the closed range of $\Gamma_1$ and $\mathcal{H}_2$ the closed range of $\Gamma_2$. Since $S_i^*\Gamma_i = \Gamma_i S_i$ for i=1,2, it follows that $\mathcal{H}_1$ is an invariant subspace for $S_1^*$ and $\mathcal{H}_2$ is an invariant subspace for $S_2^*$. Now assume that $\Gamma_1^*\Gamma_1 \le \gamma^2\Gamma_2^*\Gamma_2$. This implies that $\|\Gamma_1 h\| \le \gamma\|\Gamma_2 h\|$. So, there exists an operator A mapping $\mathcal{H}_1$ into $\mathcal{H}_2$ satisfying $A^*\Gamma_2 = \Gamma_1$ and $\|A\| \le \gamma$. Since $S_i^*\Gamma_i = \Gamma_i S_i$ for i=1,2, it readily follows that $S_1^*A^* = A^*S_2^* | \mathcal{H}_2$, where the | means restricted to. By the commutant lifting theorem (see Corollary 1.4 on page 156 in [10]), there exists an operator B mapping $H^2(\mathcal{E}_1) \to H^2(\mathcal{E}_2)$, interwining with the unilateral shifts $S_1$ and $S_2$, satisfying $B^* | \mathcal{H}_2 = A^*$ and $\|B^*\| = \|A^*\|$. Since $S_2B = BS_1$, the operator $B = F_+$ where F is in $H^\infty(\mathcal{E}_1, \mathcal{E}_2)$. Thus, $\Gamma_1 = B^*\Gamma_2 = F_+^*\Gamma_2$. Therefore, F tangentially interpolates $\Gamma_2$ with $\Gamma_1$. Finally, $\|F\|_\infty = \|F_+\| = \|B\| = \|A\| \le \gamma$. This completes the proof.

Assume that there exists a function F tangentially interpolating $\Gamma_2$ with $\Gamma_1$, or equivalently $\Gamma_1^*\Gamma_1 \le \gamma^2\Gamma_2^*\Gamma_2$, for some finite $\gamma$. Let $d_\infty$ be the smallest $H^\infty$ norm of all functions F tangentially interpolating $\Gamma_2$ with $\Gamma_1$, that is,

$$d_\infty = \inf\left\{\|F\|_\infty : F_+^*\Gamma_2 = \Gamma_1 \quad \text{and} \quad F \in H^\infty(\mathcal{E}_1, \mathcal{E}_2)\right\}. \tag{1.3}$$

Likewise, let $d_2$ be the smallest $H^2$ norm of all functions F tangentially interpolating $\Gamma_2$ with $\Gamma_1$, that is,

$$d_2 = \inf\left\{\|F\|_2 : F_+^*\Gamma_2 = \Gamma_1 \quad \text{and} \quad F \in H^\infty(\mathcal{E}_1, \mathcal{E}_2)\right\}. \tag{1.4}$$

In this paper, we will use the central solution for the commutant lifting theorem to solve the following $H^2$–$H^\infty$ tangential interpolation problem : Given two Hankel operators $\Gamma_1$ and $\Gamma_2$ and any $\delta > 1$, find an F in $H^\infty(\mathcal{E}_1, \mathcal{E}_2)$ that tangentially interpolates $\Gamma_2$ with $\Gamma_1$, and at the same time satisfies the Kaftal-Larson-Weiss bounds [26]

$$\|F\|_\infty \le \delta\, d_\infty \qquad \text{and} \qquad \|F\|_2 \le \frac{\delta\, d_2}{\sqrt{\delta^2 - 1}}. \tag{1.5}$$

In particular, if we choose $\delta = \sqrt{2}$, then there exists a function F tangentially interpolating $\Gamma_2$ with $\Gamma_1$ and satisfying

$$\|F\|_\infty \le \sqrt{2}\, d_\infty \qquad \text{and} \qquad \|F\|_2 \le \sqrt{2}\, d_2 .$$

The bounds in (1.5) were first introduced by Kaftal-Larson-Weiss [26], to solve a mixed $H^2$–$H^\infty$ Nehari problem. For a maximal entropy proof of their result see [17]. By using the commutant lifting theorem, their result was generalized to the two-sided Nehari setting in [11]. For some stronger bounds see [12].

Let A be an operator mapping $\mathcal{H}$ into $\mathcal{H}'$. We say that an operator B mapping $\mathcal{K}$ into $\mathcal{K}'$ is a lifting of A if $\mathcal{H}$ is a subspace of $\mathcal{K}$ and $\mathcal{H}'$ is a subspace of $\mathcal{K}'$ and $B^* | \mathcal{H}' = A^*$, or equivalently, $P_{\mathcal{H}'}B = AP_{\mathcal{H}}$. Notice that B is a lifting of A if and only if B admits a matrix representation of the form

$$B = \begin{bmatrix} A & 0 \\ * & * \end{bmatrix} : \begin{bmatrix} \mathcal{H} \\ \mathcal{K} \ominus \mathcal{H} \end{bmatrix} \to \begin{bmatrix} \mathcal{H}' \\ \mathcal{K}' \ominus \mathcal{H}' \end{bmatrix}.$$

In many applications $\mathcal{H} = \mathcal{K}$. So in this case B is a lifting of A if and only if $P_{\mathcal{H}'}B = A$, or equivalently, $B^* | \mathcal{H}' = A^*$. Obviously $U'$ on $\mathcal{K}'$ is an isometric lifting of a contraction $T'$ on $\mathcal{H}'$ if $U'$ is an isometry on $\mathcal{K}'(\supseteq \mathcal{H}')$ and $U'^* | \mathcal{H}' = T'^*$, or equivalently, $P_{\mathcal{H}'}U' = T'P_{\mathcal{H}'}$. It is well known that any contraction admits an isometric lifting; see for example [10,47]. Finally, let T on $\mathcal{H}$ and $T'$ on $\mathcal{H}'$ be contractions. Then $I(T, T')$ denotes the set of all operators interwining T with $T'$, that is, $I(T, T')$ is the set of all operators A mapping $\mathcal{H}$ into $\mathcal{H}'$ satisfying $AT = T'A$. The following $H^2$–$H^\infty$ version of the commutant lifting theorem will be crucial in our solution to the $H^2$–$H^\infty$ tangential interpolation problem.

**THEOREM 1.2** *([11]). Let* $A_0$ *be an operator in* $I(T, T')$ *where* T *on* $\mathcal{H}$ *is an isometry and* T′ *on* $\mathcal{H}'$ *is a contraction. Let* U′ *on* $\mathcal{K}'$ *be an isometric lifting of* T′ *and* δ *a scalar such that* $\delta \geq 1$. *Then, there exists an operator* B *in* $I(T, U')$ *lifting* $A_0$ *and satisfying the following bounds*

$$\left.\begin{array}{ll}\text{(i)} & \|B\| \leq \delta \|A_0\| \\ \text{(ii)} & \|Be\| \leq \dfrac{\delta \|A_0 e\|}{\sqrt{\delta^2 - 1}} \quad \text{for all} \quad e \in \ker(T^*)\end{array}\right\} \tag{1.6}$$

For some stronger bounds on B see [12]. An explicit formula for B will be given in the following section. For now, let us establish an existence result for our tangential interpolation problem.

**THEOREM 1.3.** *Let* $\Gamma_1$ *mapping* $H^2(\mathcal{E})$ *into* $H^2(\mathcal{E}_1)$ *and* $\Gamma_2$ *mapping* $H^2(\mathcal{E})$ *into* $H^2(\mathcal{E}_2)$ *be two specified Hankel operators. Then, there exists a function* F *in* $H^\infty(\mathcal{E}_1, \mathcal{E}_2)$ *tangentially interpolating* $\Gamma_2$ *with* $\Gamma_1$ *and satisfying* $\|F\|_\infty \leq \gamma$ *if and only if* $\Gamma_1^*\Gamma_1 \leq \gamma^2\Gamma_2^*\Gamma_2$. *Moreover, in this case given any scalar* $\delta \geq 1$, *there exists an* F *tangentially interpolating* $\Gamma_2$ *with* $\Gamma_1$ *and satisfying the following* $H^2$–$H^\infty$ *constraints*

$$\|F\|_\infty \leq \delta\, d_\infty \quad \text{and} \quad \|F\|_2 \leq \frac{\delta\, d_2}{\sqrt{\delta^2 - 1}}. \tag{1.7}$$

*Furthermore,* $d_\infty$ *and* $d_2$ *are given by*

$$d_\infty = \|A\| \quad \text{and} \quad d_2 = \|AP_1 | \mathcal{E}\|_2 \tag{1.8}$$

*where* A *is the operator mapping* $\mathcal{H}_1$ *into* $\mathcal{H}_2$, *uniquely defined by* $\Gamma_1 = A^*\Gamma_2$, *and* $\mathcal{H}_1$ *(respectively* $\mathcal{H}_2$*) is the closed range of* $\Gamma_1$ *(respectively* $\Gamma_2$*) and* $P_1 = P_{\mathcal{H}_1}$ *is the orthogonal projection onto* $\mathcal{H}_1$.

PROOF. The necessary part of the theorem is clear. To prove the sufficient part, assume that $\Gamma_1^*\Gamma_1 \leq \gamma^2\Gamma_2^*\Gamma_2$. As before, there exists an operator A mapping $\mathcal{H}_1$ into $\mathcal{H}_2$, such that $\Gamma_1 = A^*\Gamma_2$ and $\|A\| \leq \gamma$. Obviously, $\mathcal{H}_1$ and $\mathcal{H}_2$ are invariant subspaces for $S_1^*$ and $S_2^*$

respectively. Let $T_1$ on $\mathcal{H}_1$ and $T_2$ on $\mathcal{H}_2$ be the contractions defined by $T_1^* = S_1^* | \mathcal{H}_1$ and $T_2^* = S_2^* | \mathcal{H}_2$. Clearly $S_1$ and $S_2$ are isometric liftings of $T_1$ and $T_2$, respectively. Moreover, using $A^*\Gamma_2 = \Gamma_1$ and $S_i^*\Gamma_i = \Gamma_i S_i$ for i=1,2, it follows that $T_1^*A^* = A^*T_2^*$, or equivalently, $T_2A = AT_1$. Now let $A_0$ be the operator mapping $H^2(\mathcal{E}_1)$ into $\mathcal{H}_2$ defined by $A_0 = AP_1$. Notice that $A_0$ is a lifting of A and $A_0^* = A^*$. In particular, $A_0^*\Gamma_2 = \Gamma_1$. Furthermore, $A_0$ is in $I(S, T_2)$. This follows because $T_2A_0 = T_2AP_1 = AT_1P_1 = AP_1S_1 = A_0S_1$. Because $S_2$ is an isometric lifting of $T_2$, Theorem 1.2 shows that there exists an operator B in $I(S_1, S_2)$ lifting of $A_0$ and satisfying the bounds in (1.6). Since $S_2B = BS_1$, there exists an F in $H^\infty(\mathcal{E}_1, \mathcal{E}_2)$ satisfying $B = F_+$. This implies that $F_+^*\Gamma_2 = B^*\Gamma_2 = A_0^*\Gamma_2 = \Gamma_1$. The first bound (i) in (1.6) along with $||A|| = ||A_0||$ gives

$$||F||_\infty = ||F_+|| = ||B|| \leq \delta \, ||A|| . \tag{1.9}$$

Using the fact that the kernel of $S_1^*$ is $\mathcal{E}_1$, the second bound (ii) in (1.6) gives

$$||Fe|| = ||F_+e|| = ||Be|| \leq \frac{\delta \, ||AP_1e||}{\sqrt{\delta^2 - 1}} . \tag{1.10}$$

This readily implies that

$$||F||_2 = ||F_+ | \mathcal{E}_1||_2 \leq \frac{\delta \, ||AP_1 | \mathcal{E}_1||_2}{\sqrt{\delta^2 - 1}} . \tag{1.11}$$

So it remains to prove that $d_\infty = ||A||$ and $d_2 = ||AP_1 | \mathcal{E}_1||_2$.

To this end, assume that F is any function in $H^\infty$ tangentially interpolating $\Gamma_2$ with $\Gamma_1$. Then $F_+^*\Gamma_2 = \Gamma_1 = A^*\Gamma_2$ Thus, $A^* = F_+^* | \mathcal{H}_2$. In particular, $||A|| = ||F_+^* | \mathcal{H}_2|| \leq ||F||_\infty$. So $||A|| \leq d_\infty$. On the other hand, by choosing $\delta = 1$ in (1.9) we see that there exists at least one F in $H^\infty$ tangentially interpolating $\Gamma_2$ with $\Gamma_1$ and satisfying $d_\infty \leq ||F||_\infty \leq ||A|| \leq d_\infty$. Therefore, $||A|| = d_\infty$.

To prove that $d_2 = ||AP_1 | \mathcal{E}_1||_2$, let F be any function in $H^\infty$ tangentially interpolating $\Gamma_2$ with $\Gamma_1$. Recall that $F_+^* | \mathcal{H}_2 = A^*$, or equivalently, $AP_1 = P_2F_+$, where $P_2 = P_{\mathcal{H}_2}$ is the orthogonal projection onto $\mathcal{H}_2$. Thus, $AP_1e = P_2F_+e$ for all e in $\mathcal{E}_1$. Moreover,

$$||Fe||^2 = ||F_+e||^2 \geq ||P_2F_+e||^2 = ||AP_1e||^2 . \tag{1.12}$$

Thus, $d_2^2 \geq ||AP_1 | \mathcal{E}_1||_2^2$. On the other hand, by (1.11) there exists a tangentially interpolating function F satisfying

$$d_2 \le \|F\|_2 \le \frac{\delta \, \|AP_1 \mid \mathcal{E}_1\|_2}{\sqrt{\delta^2 - 1}} . \tag{1.13}$$

So letting $\delta$ approach infinity, we see that $d_2 \le \|AP_1 \mid \mathcal{E}_1\|_2$. Combining this with the fact that $\|AP_1 \mid \mathcal{E}_1\|_2 \le d_2$, it is clear that $\|AP_1 \mid \mathcal{E}_1\|_2 = d_2$. This completes the proof.

**2. Formulas for the Central Solution.**

In this section, we will give an explicit formula for the central solution to the commutant lifting theorem that will be useful in computing a function F in $H^\infty$ that tangentially interpolates $\Gamma_2$ with $\Gamma_1$, and, at the same time, satisfies the Kaftal-Larson-Weiss $H^2$–$H^\infty$ bounds in (1.7). To begin, let T′ be a contraction on $\mathcal{H}'$. Recall that U′ on $\mathcal{K}'(\supseteq\mathcal{H}')$ is a *minimal isometric dilation* of T′ if U′ is an isometric lifting of T′ (that is, $U'^* \mid \mathcal{H}' = T'^*$) and $\mathcal{H}'$ is cyclic for U′

$$\mathcal{K}' = \bigvee_0^\infty U'^n \mathcal{H}' . \tag{2.1}$$

It is well known that all minimal isometric dilations of a contraction are unitarily equivalent; see Theorem 3.2 page 133 in [10] or [47]. Any contraction T′ on $\mathcal{H}'$ admits a minimal isometric dilation. The Sz.-Nagy-Schaffer minimal isometric dilation of T′ is the isometry U′ on $\mathcal{K}' = \mathcal{H}' \oplus H^2(\mathcal{D}_{T'})$ defined by

$$U' = \begin{bmatrix} T' & 0 \\ D_{T'} & S' \end{bmatrix} \quad \text{on} \quad \begin{bmatrix} \mathcal{H}' \\ H^2(\mathcal{D}_{T'}) \end{bmatrix} \tag{2.2}$$

where $D_{T'}$ is the positive square root of $I - T'^*T'$ and $\mathcal{D}_{T'}$ is the closed range of $D_{T'}$ and S′ is the unilateral shift on $H^2(\mathcal{D}_{T'})$. If $h \oplus g$ is in $\mathcal{H}' \oplus H^2(\mathcal{D}_{T'})$, then

$$U' \begin{bmatrix} h \\ g \end{bmatrix} = \begin{bmatrix} T'h \\ D_{T'}h + zg \end{bmatrix} . \tag{2.3}$$

The operator $D_{T'}$ in (2.2) embeds $D_{T'}h$ into the subspace of constant functions in $H^2(\mathcal{D}_{T'})$. It is easy to verify that U′ in (2.2) is indeed a minimal isometric dilation of T′.

Let T be an isometry on $\mathcal{H}$ and T′ a contraction on $\mathcal{H}'$. Let $A_0$ be an operator in $I(T, T')$ bounded by $\gamma$, that is $\|A_0\| \le \gamma$. Let $D_{A_0}$ be the positive square root of $\gamma^2 I - A_0^*A_0$ and $\mathcal{D}_{A_0}$

be the closed range of $D_{A_0}$. Let $\mathcal{F}$ be the closed range of $D_{A_0}T\mathcal{H}$ and $\omega$ the isometry from $\mathcal{F}$ into $\mathcal{D}_{T'} \oplus \mathcal{D}_{A_0}$ defined by

$$\omega D_{A_0} T = \begin{bmatrix} D_{T'} A_0 \\ D_{A_0} \end{bmatrix} . \tag{2.4}$$

Throughout, $\omega$ is the operator from $\mathcal{D}_{A_0}$ to $\mathcal{D}_{T'} \oplus \mathcal{D}_{A_0}$ defined by $\omega = \omega P_{\mathcal{F}}$ where $P_{\mathcal{F}}$ is the orthogonal projection onto the subspace $\mathcal{F}$. By using $T'A_0 = A_0 T$, it is easy to verify that $\omega | \mathcal{F}$ is indeed an isometry (see Section 1 of Chapter V in [10]). In fact, this follows from

$$\begin{aligned} &\|D_{A_0} Th\|^2 = \gamma^2 \|Th\|^2 - \|A_0 Th\|^2 = \gamma^2 \|h\|^2 - \|T'A_0 h\|^2 = \\ &\|A_0 h\|^2 - \|T'A_0 h\|^2 + \gamma^2 \|h\|^2 - \|A_0 h\|^2 = \|D_{T'} A_0 h\|^2 + \|D_{A_0} h\|^2 = \|\omega D_{A_0} Th\|^2 . \end{aligned}$$

Let $\Pi_1$ mapping $\mathcal{D}_{T'} \oplus \mathcal{D}_{A_0}$ into $\mathcal{D}_{T'}$ and $\Pi_2$ mapping $\mathcal{D}_{T'} \oplus \mathcal{D}_{A_0}$ into $\mathcal{D}_{A_0}$ be the contractions defined by

$$\Pi_1 \begin{bmatrix} d' \\ h \end{bmatrix} = d' \qquad \text{and} \qquad \Pi_2 \begin{bmatrix} d' \\ h \end{bmatrix} = h . \tag{2.5}$$

The operator $\Pi_1$ picks out the first component of $\mathcal{D}_{T'} \oplus \mathcal{D}_{A_0}$ while $\Pi_2$ picks out the second. Let $B_\gamma$ be the operator from $\mathcal{H}$ to $\mathcal{K}' = \mathcal{H}' \oplus H^2(\mathcal{D}_{T'})$ defined by

$$B_\gamma = \begin{bmatrix} A_0 \\ \Pi_1 \omega (I - z\Pi_2 \omega)^{-1} D_{A_0} \end{bmatrix} . \tag{2.6}$$

Notice that for any h in $\mathcal{H}$ the operator $B_\gamma$ becomes

$$B_\gamma h = \begin{bmatrix} A_0 h \\ \sum_{n=0}^{\infty} z^n \Pi_1 \omega (\Pi_2 \omega)^n D_{A_0} h \end{bmatrix} . \tag{2.7}$$

The operator $B_\gamma$ is the central solution to the commutant lifting theorem with respect to the bound $\gamma$; see also equation (4.10) on page 454 in [10]. The following theorem (in a slightly different form) is taken from [11]; see also [12] for some stronger bounds.

**THEOREM 2.1.** *Let T on $\mathcal{H}$ be an isometry, T′ on $\mathcal{H}'$ a contraction and U′ on $\mathcal{H}' \oplus H^2(\mathcal{D}_{T'})$ the Sz.-Nagy-Schaffer minimal isometric dilation of T′. Let $A_0$ be an operator in $I(T, T')$ and set $\gamma = \delta\|A\|$ where $\delta$ is a scalar such that $\delta \geq 1$. Then the central solution $B_\gamma$ is an operator in $I(T, U')$ lifting of $A_0$ and satisfying the following bounds*

$$\left.\begin{aligned} &\text{(i)}\ \ \|B_\gamma\| \leq \delta\,\|A_0\| \\ &\text{(ii)}\ \ \|B_\gamma e\| \leq \frac{\delta\,\|A_0 e\|}{\sqrt{\delta^2 - 1}} \qquad (\text{for all} \quad e \in \ker(T^*)) \end{aligned}\right\} \tag{2.8}$$

Now let A be an operator mapping $\mathcal{H}_1$ into $\mathcal{H}'$ satisfying $T'A = AT_1$ where T′ on $\mathcal{H}'$ and $T_1$ on $\mathcal{H}_1$ are both contractions. Let T on $\mathcal{H}(\supseteq \mathcal{H}_1)$ be an isometric lifting of $T_1$, that is $T^* \mid \mathcal{H}_1 = T_1^*$, or equivalently, $P_1 T = T_1 P_1$, where $P_1 = P_{\mathcal{H}_1}$ is the orthogonal projection onto $\mathcal{H}_1$. Finally, let $A_0 = AP_1$. Notice that $T'A_0 = T'AP_1 = AT_1P_1 = A_0T$. In other words, $A_0$ is an operator in $I(T, T')$. According to Theorem 2.1, the central solution $B_\gamma$ is an operator in $I(T, U')$ lifting $A_0$ and satisfying the bounds (2.8). Notice that $B_\gamma$ is a lifting of A because $B_\gamma$ is a lifting of $A_0$ and $A_0$ is a lifting of A. This $B_\gamma$ is precisely what we need to compute a solution to our tangential interpolation problem. The following result gives us an explicit formula in terms of the operator A to compute the central solution $B_\gamma$.

**COROLLARY 2.2.** *Let T on $\mathcal{H}$ be an isometric lifting of $T_1$ on $\mathcal{H}_1$ and U′ on $\mathcal{H}' \oplus H^2(\mathcal{D}_{T'})$ the Sz.-Nagy-Schaffer minimal isometric dilation of a contraction T′ on $\mathcal{H}'$. Let A be an operator in $I(T_1, T')$. Let $\delta > 1$ and set $\gamma = \delta\,\|A\|$. Finally, let $B_\gamma$ be the operator from $\mathcal{H}$ to $\mathcal{H}' \oplus H^2(\mathcal{D}_{T'})$ defined by*

$$B_\gamma = \begin{bmatrix} AP_1 \\ D_{T'}AT_A^*(I - zT_A^*)^{-1}P_1 + \gamma^2 D_{T'}A(I - zT_A^*)^{-1}D_{AT_1}^{-2}P_1T^*P_{\mathcal{M}}(I - zT^*)^{-1} \end{bmatrix} \tag{2.9}$$

*where $\mathcal{M}$ is the orthogonal complement of $\mathcal{H}_1$ and $T_A^*$ is the operator on $\mathcal{H}_1$ defined by*

$$T_A^* = (\gamma^2 I - T_1^*A^*AT_1)^{-1}T_1^*(\gamma^2 I - A^*A)\,. \tag{2.10}$$

*Then $B_\gamma$ is an operator in $I(T, U')$ lifting of A and satisfying the bounds*

$$\left.\begin{array}{ll}\text{(i)}\ \|B_\gamma\| \le \delta\,\|A\| & \\ \text{(ii)}\ \|B_\gamma e\| \le \dfrac{\delta\,\|AP_1 e\|}{\sqrt{\delta^2 - 1}} & (\text{for all} \quad e \in \ker(T^*))\end{array}\right\} \tag{2.11}$$

*Moreover, the operator* $T_A^*$ *is similar to a contraction. Furthermore, if* T *has no eigenvalues on the unit circle, then* $T_A^*$ *has no eigenvalues on the unit circle. In particular, if* $\mathcal{H}_1$ *is finite dimensional and* T *is a unilateral shift, then all the eigenvalues of* $T_A^*$ *are in the open unit disc.*

PROOF. Let $A_0 = AP_1$. Because T is a lifting of $T_1$, it follows that $A_0$ is in $I(T, T')$. Obviously $\|A\| = \|A_0\|$. To prove this Corollary we will use $A_0 = AP_1$ to show that the $B_\gamma$ in (2.6) can be converted to the $B_\gamma$ in (2.9). Since $\delta > 1$, it follows that $\|A_0\| < \gamma = \delta\|A\|$. This implies that $D_{A_0}$ is onto $\mathcal{H}$. Therefore, the subspace $\mathcal{F}$ equals the range of $X = D_{A_0}T$. So the orthogonal projection $P_{\mathcal{F}}$ is given by

$$\begin{aligned} P_{\mathcal{F}} &= X(X^*X)^{-1}X^* = D_{A_0}T(T^*D_{A_0}^2T)^{-1}T^*D_{A_0} \\ &= D_{A_0}T(\gamma^2 I - T^*A_0^*A_0T)^{-1}T^*D_{A_0}\,. \end{aligned} \tag{2.12}$$

Using this along with the definitions of $\Pi_2$ and $\omega$ we have

$$\Pi_2\omega = \Pi_2\omega P_{\mathcal{F}} = D_{A_0}(\gamma^2 I - T^*A_0^*A_0T)^{-1}T^*D_{A_0}\,. \tag{2.13}$$

Moreover, $\Pi_1\omega$ becomes

$$\Pi_1\omega = \Pi_1\omega P_{\mathcal{F}} = D_{T'}A_0(\gamma^2 I - T^*A_0^*A_0T)^{-1}T^*D_{A_0}\,. \tag{2.14}$$

Now let $D_{A_0T}^2$ equal $(\gamma^2 I - T^*A_0^*A_0T)$. Then, using (2.13) and (2.14) we have

$$\begin{aligned} \Pi_1\omega(I - z\Pi_2\omega)^{-1}D_{A_0} &= D_{T'}A_0D_{A_0T}^{-2}T^*D_{A_0}(I - zD_{A_0}D_{A_0T}^{-2}T^*D_{A_0})^{-1}D_{A_0} \\ &= D_{T'}A_0D_{A_0T}^{-2}T^*D_{A_0}^2(I - zD_{A_0T}^{-2}T^*D_{A_0}^2)^{-1}\,. \end{aligned} \tag{2.15}$$

If we set $T_0^* = D_{A_0T}^{-2}T^*D_{A_0}^2$, then, the previous equation shows that the bottom term for $B_\gamma$ in (2.6) becomes

$$\Pi_1\omega(I - z\Pi_2\omega)^{-1}D_{A_0} = D_{T'}A_0T_0^*(I - zT_0^*)^{-1}\,. \tag{2.16}$$

In order to find the inverses in (2.16), we will convert to matrices. Recall that $\mathcal{M}$ is the orthogonal complement of $\mathcal{H}_1$ in $\mathcal{H}$. Using $T'A_0 = A_0 T$ and $T'A = AT_1$ along with $A_0 = AP_1$ we have

$$\begin{aligned} D_{A_0T}^2 &= (\gamma^2 I - A_0^* T'^* T' A_0) = \gamma^2 I - P_1 A^* T'^* T' A P_1 \\ &= \gamma^2 P_{\mathcal{M}} + \gamma^2 P_1 - T_1^* A^* A T_1 P_1 \, . \end{aligned} \tag{2.17}$$

Therefore a matrix representation of $D_{A_0T}^2$ with respect to $\mathcal{H}_1 \oplus \mathcal{M}$ is given by

$$D_{A_0T}^2 = \begin{bmatrix} \gamma^2 - T_1^* A^* A T_1 & 0 \\ 0 & \gamma^2 \end{bmatrix} \quad \text{on} \quad \begin{bmatrix} \mathcal{H}_1 \\ \mathcal{M} \end{bmatrix} . \tag{2.18}$$

Since $\mathcal{H}_1$ is invariant for $T^*$ is follows that $P_{\mathcal{M}} T^* \mid \mathcal{H}_1 = 0$. So a matrix representation for $T^*$ is given by

$$T^* = \begin{bmatrix} T^* & P_1 T^* \mid \mathcal{M} \\ 0 & P_{\mathcal{M}} T^* \mid \mathcal{M} \end{bmatrix} \quad \text{on} \quad \begin{bmatrix} \mathcal{H}_1 \\ \mathcal{M} \end{bmatrix} . \tag{2.19}$$

Finally, using $A_0 = AP_1$, a matrix representation for $D_{A_0}^2$ is given by

$$D_{A_0}^2 = \begin{bmatrix} D_A^2 & 0 \\ 0 & \gamma^2 \end{bmatrix} \quad \text{on} \quad \begin{bmatrix} \mathcal{H}_1 \\ \mathcal{M} \end{bmatrix} \tag{2.20}$$

where $D_A$ is the positive square root of $\gamma^2 I - A^* A$. Using (2.18), (2.19), (2.20) and $T_0^* = D_{A_0T}^{-2} T^* D_{A_0}^2$, it follows that a matrix representation for $T_0^*$ is given by

$$T_0^* = \begin{bmatrix} T_A^* & \gamma^2 D_{AT_1}^{-2} P_1 T^* \\ 0 & P_{\mathcal{M}} T^* \end{bmatrix} \quad \text{on} \quad \begin{bmatrix} \mathcal{H}_1 \\ \mathcal{M} \end{bmatrix} \tag{2.21}$$

where $D_{AT_1}^2 = \gamma^2 I - T_1^* A^* A T_1^*$ and $T_A$ is the operator on $\mathcal{H}_1$ defined by

$$T_A^* = (\gamma^2 I - T_1^* A^* A T_1)^{-1} T_1^* (\gamma^2 I - A^* A) \, . \tag{2.22}$$

Here we defined $T_A^*$ by (2.22) to be consistent with the notation in Chapter XIV of [10]. Obviously the matrix representation for $D_{T'} A_0 = D_{T'} A P_1$ is given by

$$D_{T'}A_0 = \begin{bmatrix} D_{T'}A \ , & 0 \end{bmatrix} \quad : \quad \mathcal{H}_1 \oplus \mathcal{M} \to \mathcal{D}_{T'} \, . \tag{2.23}$$

Therefore according to (2.21) and (2.23)

$$D_{T'}A_0T_0^* = \begin{bmatrix} D_{T'}AT_A^* \ , & \gamma^2 D_{T'}AD_{AT_1}^{-2}P_1T^* \end{bmatrix} . \tag{2.24}$$

Notice that because $\mathcal{H}_1$ is invariant for $T^*$, we have $P_{\mathcal{M}}T^*P_{\mathcal{M}} = P_{\mathcal{M}}T^*$. In particular, for z in the open unit disc D we have

$$(I - zP_{\mathcal{M}}T^*P_{\mathcal{M}})^{-1}P_{\mathcal{M}} = P_{\mathcal{M}}(I - zT^*)^{-1} \, . \tag{2.25}$$

So using this in (2.21) we have

$$(I - zT_0^*)^{-1} =$$

$$\begin{bmatrix} (I - zT_A^*)^{-1} & z\gamma^2(I - zT_A^*)^{-1}D_{AT_1}^{-2}P_1T^*P_{\mathcal{M}}(I - zT^*)^{-1} \\ 0 & P_{\mathcal{M}}(I - zT^*)^{-1} \end{bmatrix} \quad \text{on} \quad \begin{bmatrix} \mathcal{H}_1 \\ \mathcal{M} \end{bmatrix} \tag{2.26}$$

Finally, using (2.16), (2.24) and (2.26), we have

$$\Pi_1\omega(I - z\Pi_2\omega)^{-1}D_{A_0} = \begin{bmatrix} \Psi_1 \ , & \Psi_2 \end{bmatrix} \tag{2.27}$$

where $\Psi_1$ is the operator from $\mathcal{H}_1$ to $H^2(\mathcal{D}_{T'})$ given by

$$\Psi_1 = D_{T'}AT_A^*(I - zT_A^*)^{-1} \tag{2.28}$$

and $\Psi_2$ is the operator from $\mathcal{M}$ to $H^2(\mathcal{D}_{T'})$ given by

$$\begin{aligned} \Psi_2 &= \gamma^2 D_{T'}AT_A^*z(I - zT_A^*)^{-1}D_{AT_1}^{-2}P_1T^*P_{\mathcal{M}}(I - zT^*)^{-1} \\ &\quad + \gamma^2 D_{T'}AD_{AT_1}^{-2}P_1T^*P_{\mathcal{M}}(I - zT^*)^{-1} \\ &= \gamma^2 D_{T'}A(I - zT_A^*)^{-1}D_{AT_1}^{-2}P_1T^*P_{\mathcal{M}}(I - zT^*)^{-1} \, . \end{aligned} \tag{2.29}$$

Finally, using $P_{\mathcal{M}}T^*P_{\mathcal{M}} = P_{\mathcal{M}}T^*$, (this follows because $\mathcal{H}_1$ is invariant for $T^*$) we have

$$\Pi_1\omega(I - z\Pi_2\omega)^{-1}D_{A_0} = \Psi_1P_1 + \Psi_2P_{\mathcal{M}} =$$

$$D_{T'}AT_A^*(I - zT_A^*)^{-1}P_1 + \gamma^2 D_{T'}A(I - zT_A^*)^{-1}D_{AT_1}^{-2}P_1T^*P_{\mathcal{M}}(I - zT^*)^{-1} \, .$$

Therefore the last term of the central solution $B_\gamma$ for the commutant lifting theorem in (2.6) is given by the previous equation. This along with Theorem 2.1 readily shows that $B_\gamma$ in (2.9) is a lifting of A in $I(T, U')$ satisfying the bounds in (2.11).

Now let us show that $T_A^*$ is similar to a contraction. For another proof of this fact see page 473 of [10]. To this end, notice that by (2.13) and the definition of $T_0^*$ we have

$$\Pi_2\omega = D_{A_0}(\gamma^2 I - T^* A_0^* A_0 T)^{-1} T^* D_{A_0}^2 D_{A_0}^{-1} = D_{A_0} T_0^* D_{A_0}^{-1} . \tag{2.30}$$

Therefore, $T_0^*$ is similar to the contraction $\Pi_2\omega$. Moreover, because $\mathcal{H}_1$ is a reducing subspace for $D_{A_0}$ we have $P_1 D_{A_0} = D_A P_1$. Since $\mathcal{H}_1$ is invariant for $T_0^*$ and $T_0^* | \mathcal{H}_1 = T_A^*$, we have

$$P_1 \Pi_2 \omega | \mathcal{H}_1 = P_1 D_{A_0} T_0^* D_{A_0}^{-1} | \mathcal{H}_1 = D_A P_1 T_0^* P_1 D_A^{-1} = D_A T_A^* D_A^{-1} . \tag{2.31}$$

Hence $T_A^*$ is similar to the contraction $P_1 \Pi_2 \omega | \mathcal{H}_1$ .

To complete the proof, assume that T has no eigenvalues on the unit circle. Since $T_A^*$ is similar to $P_1 \Pi_2 \omega | \mathcal{H}_1$, it is sufficient to show that $P_1 \Pi_2 \omega | \mathcal{H}_1$ has no eigenvalues on the unit circle. To this end, assume that there exists a nonzero f in $\mathcal{H}_1$ satisfying

$$P_1 \Pi_2 \omega f = \lambda f \qquad \text{where} \qquad |\lambda| = 1 . \tag{2.32}$$

Then, clearly $\|f\| = \|P_1 \Pi_2 \omega f\| \le \|\Pi_2 \omega f\| \le \|\omega f\| = \|P_{\mathcal{F}} f\| \le \|f\|$, and thus f is in $\mathcal{F}$, or equivalently, $f = D_{A_0} T h$ for some h in $\mathcal{H}$. Moreover, $P_1 \Pi_2 \omega f = \Pi_2 \omega f$. By (2.32) we see that

$$\lambda D_{A_0} T h = P_1 \Pi_2 \omega D_{A_0} T h = \Pi_2 \begin{bmatrix} D_{T'} A h \\ D_{A_0} h \end{bmatrix} = D_{A_0} h . \tag{2.33}$$

This implies that $\lambda D_{A_0} T h = D_{A_0} h$. Recall that $D_{A_0}$ is invertible because $\|\gamma\| > \|A\|$. Thus, $Th = \bar{\lambda} h$, that is $\bar{\lambda}$ is an eigenvalue of T, which is not possible. This completes the proof.

In many applications, T is a unilateral shift. In this case, the wandering subspace $\mathcal{E}_1 = \mathcal{H} \ominus T\mathcal{H}$ is cyclic for T. So, if T is a unilateral shift, one can completely determine the central solution $B_\gamma$ in (2.9), by computing $B_\gamma | \mathcal{E}_1$. This follows because $\mathcal{E}_1$ is cyclic for T and $B_\gamma T^n e = U'^n B e$ for all $n \ge 0$. The following result provides us with an explicit formula for computing $B_\gamma | \mathcal{E}_1$.

**COROLLARY 2.3.** *Let* T *on* $\mathcal{H}$ *be an isometric lifting of* $T_1$ *on* $\mathcal{H}_1$. *Let* T′ *on* $\mathcal{H}'$ *be a contraction and* A *an operator in* $I(T_1, T')$. *Finally, let* $\delta > 1$ *and set* $\gamma = \delta\|A\|$. *Then for* e *in the kernel of* $T^*$, *the central interwining lifting* $B_\gamma$ *in (2.9) becomes*

$$B_\gamma e = \begin{bmatrix} AP_1 e \\ -D_{T'}AT_A^*(I - zT_A^*)^{-1}(\gamma^2 I - A^*A)^{-1}A^*AP_1 e \end{bmatrix}. \tag{2.34}$$

PROOF. Using $T^*e = 0$, the definition of $T_A^*$ in (2.10), along with $P_{\mathcal{M}} = I - P_1$ in (2.9) we see that the bottom term in $B_\gamma$ becomes

$$\begin{aligned}
&\Pi_1\omega(I - z\Pi_2\omega)^{-1}D_{A_0}e = \\
&D_{T'}AT_A^*(I - zT_A^*)^{-1}P_1e + \gamma^2 D_{T'}A(I - zT_A^*)^{-1}D_{AT_1}^{-2}P_1T^*(I - P_1)(I - zT^*)^{-1}e = \\
&D_{T'}AT_A^*(I - zT_A^*)^{-1}P_1e - \gamma^2 D_{T'}A(I - zT_A^*)^{-1}D_{AT_1}^{-2}T^*P_1e = \\
&D_{T'}A(I - zT_A^*)^{-1}D_{AT_1}^{-2}T^*(D_A^2 - \gamma^2 I)P_1e = \\
&-D_{T'}A(I - zT_A^*)^{-1}D_{AT_1}^{-2}T^*A^*AP_1e = -D_{T'}A(I - zT_A^*)^{-1}T_A^*D_A^{-2}A^*AP_1e.
\end{aligned}$$

This completes the proof.

The following corollary will be helpful in solving our $H^2$–$H^\infty$ tangential interpolation problem. In this Corollary, we will not use the characteristic function, because it is not needed in our approach. For some similar results involving the characteristic function see [10] and [15].

**COROLLARY 2.4.** *Let the unilateral shifts* $S_1$ *on* $H^2(\mathcal{E}_1)$ *and* $S_2$ *on* $H^2(\mathcal{E}_2)$ *be isometric liftings of the contractions* $T_1$ *on* $\mathcal{H}_1$ *and* $T_2$ *on* $\mathcal{H}_2$ *respectively. Let* $A$ *be an operator in* $I(T_1, T_2)$ *and* $\gamma = \delta\|A\|$ *where* $\delta$ *is any scalar such that* $\delta > 1$. *Let* $F$ *be the function defined by*

$$F(z) = AP_1 \mid \mathcal{E}_1 - D_2AT_A^*(I - zT_A^*)^{-1}(\gamma^2 I - A^*A)^{-1}A^*AP_1 \mid \mathcal{E}_1 \tag{2.35}$$

*where* $D_2$ *is the operator from* $\mathcal{H}_2$ *to* $H^2(\mathcal{E}_2)$ *defined by*

$$D_2h = (S_2 - T_2)h \qquad (h \in \mathcal{H}_2). \tag{2.36}$$

*Then* $F$ *is in* $H^\infty(\mathcal{E}_1, \mathcal{E}_2)$ *and the operator* $F_+$ *is a lifting of* $A$ *satisfying the following* $H^2$–$H^\infty$ *bounds*

$$\|F\|_\infty \le \delta\,\|A\| \quad \text{and} \quad \|F\|_2 \le \frac{\delta\,\|AP_1 \mid \mathcal{E}_1\|_2}{\sqrt{\delta^2 - 1}}. \tag{2.37}$$

PROOF. Let $T' = T_2$ and $S_1 = T$ in Corollary 2.3. Let $\mathcal{K}_2$ be the subspace of $H^2(\mathcal{E}_2)$ defined by

$$\mathcal{K}_2 = \bigvee_0^\infty S_2^n \mathcal{H}_2 .$$

Obviously $\mathcal{K}_2$ is invariant for $S_2$. Moreover, the operator $U_2$ on $\mathcal{K}_2$ defined by $U_2 = S_2 | \mathcal{K}_2$ is the minimal isometric dilation of $T_2$. Recall that all minimal isometric dilations are unitarily equivalent. In fact, according to Theorem 3.2 on page 133 in [10], or Theorem 4.1 on page 11 of [47], there exists a unitary operator $\psi$ mapping $\mathcal{H}_2 \oplus H^2(\mathcal{D}_{T_2})$ onto $\mathcal{K}_2$ satisfying $\psi | \mathcal{H}_2 = I$ and $\psi U' = U_2 \psi = S_2 \psi$ where $U'$ is the Sz.-Nagy-Schaffer dilation of $T_2$; see (2.2). According to Corollary 2.2 and $\psi | \mathcal{H}_2 = I$ we see that $\psi B_\gamma$ is a lifting of A satisfying the bounds in (2.8). Furthermore, $S_2 \psi B_\gamma = \psi U' B_\gamma = \psi B_\gamma S_1$. This implies that $\psi B_\gamma = F_+$ for some F in $H^\infty(\mathcal{E}_1 , \mathcal{E}_2)$. Moreover, because $F = F_+ | \mathcal{E}_1 = \psi B_\gamma | \mathcal{E}_1$, it follows that F satisfies the $H^2$–$H^\infty$ bounds in (2.37).

To complete the proof, it remains to show that $F = \psi B_\gamma | \mathcal{E}_1$ is given by (2.35). To this end, notice that for h in $\mathcal{H}_2$ we have

$$\psi \begin{bmatrix} 0 \\ D_{T_2} h \end{bmatrix} = \psi(U' - T_2)h = (S_2 - T_2)h = D_2 h . \tag{2.38}$$

So by computing $\psi B_\gamma e$ for e in $\mathcal{E}_1$ from equation (2.34) along with

$$\psi \begin{bmatrix} 0 \\ S'g \end{bmatrix} = \psi U' \begin{bmatrix} 0 \\ g \end{bmatrix} = S_2 \psi \begin{bmatrix} 0 \\ g \end{bmatrix} \quad \text{for g in } H^2(\mathcal{D}_{T_2}) ,$$

we have

$$\begin{aligned} \psi B_\gamma e &= \psi \begin{bmatrix} AP_1 e \\ -\sum_{n=0}^{\infty} S'^n D_{T_2} A T_A^{*n+1} (\gamma^2 I - A^* A)^{-1} A^* AP_1 e \end{bmatrix} \\ &= AP_1 e - \sum_{n=0}^{\infty} S_2^n \psi \begin{bmatrix} 0 \\ D_{T_2} A T_A^{*n+1} (\gamma^2 I - A^* A)^{-1} A^* AP_1 e \end{bmatrix} \\ &= AP_1 e - D_2 A T_A^* (I - z T_A^*)^{-1} (\gamma^2 I - A^* A)^{-1} A^* AP_1 e . \end{aligned}$$

Since $\psi B_\gamma e = Fe$, this completes the proof.

### 3. A State Space Approach

In this section we will use realization theory to compute $d_\infty$ and $d_2$, and then give some state space formulas to compute a solution for our $H^2$–$H^\infty$ tangential interpolation problem. Let G be a function in $H^\infty(\mathcal{E}, \mathcal{E}')$. Then $\Gamma(G)$ is the Hankel operator mapping $H^2(\mathcal{E})$ into $H^2(\mathcal{E}')$ with symbol G defined by

$$\Gamma(G)h = P_+ GRh \qquad (\text{for } h \in H^2(\mathcal{E})) \tag{3.1}$$

where R is the unitary operator on $L^2(\mathcal{E})$ defined by $(Rf)(e^{it}) = f(e^{-it})$. It is easy to show that $S'^*\Gamma(G) = \Gamma(G)S$ where S is the unilateral shift on $H^2(\mathcal{E})$ and $S'$ is the unilateral shift on $H^2(\mathcal{E}')$. Throughout this section $G_1$ and $G_2$ are two rational functions in $H^\infty(\mathcal{E}, \mathcal{E}_1)$ and $H^\infty(\mathcal{E}, \mathcal{E}_2)$ respectively. Moreover, $\Gamma_1 = \Gamma(G_1)$ is the Hankel operator from $H^2(\mathcal{E})$ to $H^2(\mathcal{E}_1)$ with symbol $G_1$ while $\Gamma_2 = \Gamma(G_2)$ is the Hankel operator from $H^2(\mathcal{E})$ to $H^2(\mathcal{E}_2)$ with symbol $G_2$.

In this section, we will use some elementary facts from realization theory. For further results on realization theory see [1,10,18,28,44]. In this paper, we use the notation 'tr' to transpose block row matrices into block column matrices. For example,

$$[G_1 , G_2]^{tr} = \begin{bmatrix} G_1 \\ G_2 \end{bmatrix} \quad \text{and} \quad [C_1 , C_2]^{tr} = \begin{bmatrix} C_1 \\ C_2 \end{bmatrix}.$$

Throughout most of this section $\{M, B, [C_1 , C_2]^{tr}\}$ is a minimal realization of $[G_1 , G_2]^{tr}$, that is,

$$\begin{bmatrix} G_1 \\ G_2 \end{bmatrix} = \begin{bmatrix} C_1 \\ C_2 \end{bmatrix}(I - zM)^{-1}B \tag{3.2}$$

where $C_1$ maps $\mathbb{C}^n$ into $\mathcal{E}_1$, $C_2$ maps $\mathbb{C}^n$ into $\mathcal{E}_2$, B maps $\mathcal{E}$ into $\mathbb{C}^n$ and M is a matrix on $\mathbb{C}^n$. Moreover, the pair $\{M, B\}$ is controllable, while the pair $\{[C_1 , C_2]^{tr} , M\}$ is observable. Since $G_1$ and $G_2$ are rational function in $H^\infty$, the operator M is stable. In other words, all the eigenvalues of M are in the open unit disc D. Obviously $\{M, B, C_1\}$ is a realization of $G_1$, while $\{M, B, C_2\}$ is a realization of $G_2$. Now let $W_i$ for i=1,2 be the observability operators from $\mathbb{C}^n$ into $H^2(\mathcal{E}_i)$ defined by $W_i = C_i(I - zM)^{-1}$. Now define the controllability operator $W_c$ from $\mathbb{C}^n$ into $H^2(\mathcal{E})$ by $W_c = B^*(I - zM^*)^{-1}$. Because $\{M, B\}$ is controllable, the controllability operator $W_c$ is one-to-one, or equivalently, $W_c^*$ is onto $\mathbb{C}^n$. The observability grammians $Q_1$ and $Q_2$ are defined by

$$Q_1 \triangleq W_1^* W_1 = \sum_{k=0}^{\infty} M^{*k} C_1^* C_1 M^k \quad \text{and} \quad Q_2 \triangleq W_2^* W_2 = \sum_{k=0}^{\infty} M^{*k} C_2^* C_2 M^k . \tag{3.3}$$

Since M is stable, the observability grammians $Q_1$ and $Q_2$ are well defined and uniquely satisfy the following Lyapunov equations

$$Q_1 = M^* Q_1 M + C_1^* C_1 \quad \text{and} \quad Q_2 = M^* Q_2 M + C_2^* C_2 . \tag{3.4}$$

Finally, it is noted that we do not need the controllability grammian in our tangential interpolation problem.

It is well known that the Hankel operators $\Gamma_1$ and $\Gamma_2$ admit a factorization of the form (see [1],[10],[19],[28])

$$\Gamma_1 = W_1 W_c^* \quad \text{and} \quad \Gamma_2 = W_2 W_c^* . \tag{3.5}$$

To verify this fact in our setting, notice that $S_i^* W_i = W_i M$ and $S^* W_c = W_c M^*$, or equivalently, $MW_c^* = W_c S$. Therefore $S_i^* W_i W_c^* = W_i W_c^* S$, for i=1,2. So $W_i W_c^*$ is a Hankel operator. However, for e in $\mathcal{E}$ we have $W_i W_c^* e = W_i Be = G_i Be = \Gamma_i e$. Hence, for $n \geq 0$ we see that

$$\Gamma_i S^n e = S_i^{*n} \Gamma_i e = S_i^{*n} W_i W_c^* e = W_i W_c^* S^n e .$$

Since $\mathcal{E}$ is cyclic for S we obtain the desired factorization in (3.5).

The following result uses the observability grammian $Q_2$ to determine the existence of an F tangentially interpolating $\Gamma_2$ with $\Gamma_1$.

**THEOREM 3.1.** *Let* $\{M, B, [C_1, C_2]^{tr}\}$ *be a minimal realization of* $[G_1, G_2]^{tr}$, *with the corresponding observability grammians* $Q_1$ *and* $Q_2$. *Then there exists a function* F *in* $H^\infty(\mathcal{E}_1, \mathcal{E}_2)$ *tangentially interpolating* $\Gamma(G_2)$ *with* $\Gamma(G_1)$ *if and only if* $Q_2$ *is strictly positive. In fact, there exists a function* F *tangentially interpolating* $\Gamma(G_2)$ *with* $\Gamma(G_1)$ *and satisfying* $\|F\|_\infty \leq \gamma$ *if and only if* $Q_1 \leq \gamma^2 Q_2$. *Moreover, in this case, given any* $\delta > 1$, *there exists a function* F *in* $H^\infty(\mathcal{E}_1, \mathcal{E}_2)$ *tangentially interpolating* $\Gamma(G_2)$ *with* $\Gamma(G_1)$ *and satisfying the following* $H^2$–$H^\infty$ *bounds*

$$\|F\|_\infty \le \delta\, d_\infty \quad \text{and} \quad \|F\|_2 \le \frac{\delta\, d_2}{\sqrt{\delta^2 - 1}} \,. \tag{3.6}$$

PROOF. By Lemma 1.1 there exists an F in $H^\infty(\mathcal{E}_1, \mathcal{E}_2)$ tangentially interpolating $\Gamma_2$ with $\Gamma_1$ if and only if $\Gamma_1^*\Gamma_1 \le \gamma^2 \Gamma_2^*\Gamma_2$ for some finite $\gamma$. Using (3.3) and (3.5) we see that this is equivalent to

$$(Q_1 W_c^* f, W_c^* f) \le \gamma^2 (Q_2 W_c^* f, W_c^* f) \qquad (f \in H^2(\mathcal{E}))\,. \tag{3.7}$$

However, $\{M, B\}$ is controllable, or equivalently, the controllability operator $W_c^*$ is onto. Therefore $\Gamma_1^*\Gamma_1 \le \gamma^2\Gamma_2^*\Gamma_2$ if and only if $Q_1 \le \gamma^2 Q_2$. According to Theorem 1.3 to complete the proof it remains to show that $Q_1 \le \gamma^2 Q_2$ if and only if $Q_2$ is strictly positive. Clearly, if the kernel of $Q_2$ is zero, then there exists a finite $\gamma$ such that $Q_1 \le \gamma^2 Q_2$. In fact, any $\gamma^2 > \|\sqrt{Q_2^{-1}}\, Q_1 \sqrt{Q_2^{-1}}\,\|$ will do. On the other hand, if $Q_1 \le \gamma^2 Q_2$ and x is in the kernel of $Q_2$, then x is also in the kernel of $Q_1$. This along with (3.3) show that

$$\left\| \begin{bmatrix} C_1 \\ C_2 \end{bmatrix} (I - zM)^{-1} x \right\|^2 = \|W_1 x\|^2 + \|W_2 x\|^2 = (Q_1 x, x) + (Q_2 x, x) = 0\,. \tag{3.8}$$

Because $\{M, B, [C_1, C_2]^{tr}\}$ is a minimal realization, $\{[C_1, C_2]^{tr}, M\}$ is observable. Therefore x=0. So the kernel of $Q_2$ is zero and $Q_2$ is strictly positive. This completes the proof.

The following theorem allows us to compute $d_2$ and $d_\infty$ in terms of $C_1$, $Q_1$ and $Q_2$.

**THEOREM 3.2.** *Let* $\{M, B, [C_1, C_2]^{tr}\}$ *be a minimal realization of* $[G_1, G_2]^{tr}$. *Assume that there exists an* F *in* $H^\infty(\mathcal{E}_1, \mathcal{E}_2)$ *interpolating two finite rank Hankel operators* $\Gamma(G_2)$ *with* $\Gamma(G_1)$, *or equivalently,* $Q_2 > 0$. *Then*

$$d_\infty^2 = \lambda_{max}(Q_2^{-1}Q_1) \qquad \text{and} \qquad d_2^2 = \operatorname{trace}(C_1 Q_2^{-1} C_1^*) \tag{3.9}$$

*where* $\lambda_{max}$ *denotes the largest eigenvalue. Moreover, given any* $\delta > 1$, *the operator* $\delta^2 d_\infty^2 Q_2 - Q_1$ *is strictly positive.*

PROOF. Recall that from Theorem 1.3 we have $d_\infty = \|A\|$ where A is the operator from $\mathcal{H}_1$ to $\mathcal{H}_2$ defined by $A^*\Gamma_2 = \Gamma_1$. However, using $\Gamma_2 = W_2 W_c^*$ and $\Gamma_1 = W_1 W_c^*$ along with the fact that $W_c^*$ is onto we have

$$A^* W_2 = W_1 . \tag{3.10}$$

This implies that

$$\begin{aligned} d_\infty^2 = \|A\|^2 &= \sup \{\|A^* W_2 x\|^2 \ : \ \|W_2 x\| = 1\} \\ &= \sup \{\|W_1 x\|^2 \ : \ \|W_2 x\| = 1\} = \sup \{(Q_1 x , x) \ : \ (Q_2 x , x) = 1\} \\ &= \sup \{(Q_1 x , x) \ : \ \|\sqrt{Q_2}\, x\| = 1\} \\ &= \sup \{(\sqrt{Q_2^{-1}}\, Q_1 \sqrt{Q_2^{-1}}\, y , y) \ : \ \|y\| = 1\} . \end{aligned}$$

Therefore $d_\infty^2$ is the largest eigenvalue of $\sqrt{Q_2^{-1}}\, Q_1 \sqrt{Q_2^{-1}}$ . However, notice that

$$\sqrt{Q_2^{-1}} \left[\sqrt{Q_2^{-1}}\, Q_1 \sqrt{Q_2^{-1}}\right] \sqrt{Q_2} = Q_2^{-1} Q_1$$

So $\sqrt{Q_2^{-1}}\, Q_1 \sqrt{Q_2^{-1}}$ is similar to $Q_2^{-1} Q_1$ . Hence $d_\infty^2$ is the largest eigenvalue of $Q_2^{-1} Q_1$ (Formulas of the form $\lambda_{max}(Q_2^{-1} Q_1)$ also appear in some $H^\infty$ half plane interpolation problems; see [24].) Finally, notice that for any $\delta > 1$ we have that $\delta^2 d_\infty^2 I - \sqrt{Q_2^{-1}}\, Q_1 \sqrt{Q_2^{-1}}$ is strictly positive. Therefore $\delta^2 d_\infty^2 Q_2 - Q_1$ is strictly positive.

In order to compute $d_2$ we need an expression for A. Notice that for x and y in $\mathbb{C}^n$

$$(W_2 x , A W_1 y) = (A^* W_2 x , W_1 y) = (W_1 x , W_1 y) = (x , Q_1 y) = (W_2 x , W_2 Q_2^{-1} Q_1 y) .$$

This implies that

$$A W_1 = W_2 Q_2^{-1} Q_1 . \tag{3.11}$$

According to Theorem 1.3 we have $d_2 = \|A P_1 \mid \mathcal{E}_1\|_2$. So now we have to compute $P_1 e$ for e in $\mathcal{E}_1$. Since $\mathcal{H}_1$ equals the range of $W_1$, there exists an x in $\mathbb{C}^n$ such that $P_1 e = W_1 x$. In other words, by the Projection Theorem $e - W_1 x$ is orthogonal to $W_1 \mathbb{C}^n$, or equivalently, $W_1^* e - W_1^* W_1 x$ is orthogonal to $\mathbb{C}^n$. Since $W_1^* e = C_1^* e$ and $W_1^* W_1 = Q_1$, it follows that $C_1^* e - Q_1 x$ is orthogonal to $\mathbb{C}^n$, or equivalently, $C_1^* e = Q_1 x$. (This also shows that the range of $C_1^*$ is contained in the range of $Q_1$.) Therefore, $x = Q_1^{-r} C_1^* e$ where $Q_1^{-r}$ is the restricted inverse of $Q_1$, that is, $Q_1^{-r} y = 0$ if y is in the kernel of $Q_1$, and if y is in the range of $Q_1$, then $Q_1^{-r} y = x$ where $y = Q_1 x$ and x is in the range of $Q_1$. By using $P_1 e = W_1 Q_1^{-r} C_1^* e$ in (3.11) we have

$$A P_1 e = W_2 Q_2^{-1} C_1^* e \qquad (e \in \mathcal{E}_1) . \tag{3.12}$$

Finally, by Theorem 1.3 we have

$$d_2^2 = \sum_{i=1}^{k} \|AP_1 e_i\|^2 = \sum_{i=1}^{k} (W_2 Q_2^{-1} C_1^* e_i \,,\, W_2 Q_2^{-1} C_1^* e_i)$$
$$= \sum_{i=1}^{k} (C_1 Q_2^{-1} C_1^* e_i \,,\, e_i) = \operatorname{trace}\,(C_1 Q_2^{-1} C_1^*) \,,$$

where $\{e_i\}$ is an orthonormal basis for $\mathcal{E}_1$. This completes the proof.

**REMARK 3.3 .** Assume that there exists a function F tangentially interpolating $\Gamma(G_2)$ with $\Gamma(G_1)$ where $G_1$ and $G_2$ are rational. Then the $H^2$ optimal function $F_2$ tangentially interpolating $\Gamma_2$ with $\Gamma_1$ is given by

$$F_2 \;=\; AP_1 \,|\, \mathcal{E}_1 \;=\; W_2 Q_2^{-1} C_1^* \,. \tag{3.13}$$

In other words, this $F_2$ is the only function in $H^\infty(\mathcal{E}_1 \,,\, \mathcal{E}_2)$ tangentially interpolating $\Gamma_2$ with $\Gamma_1$ and satisfying $d_2 = \|F_2\|_2$ . Notice that $F_2$ is in $H^\infty(\mathcal{E}_1 \,,\, \mathcal{E}_2)$ because $G_1$ and $G_2$ are rational functions in $H^\infty$. (If $G_1$ or $G_2$ were not rational, then the optimal $H^2$ solution $F_2$ may not be in $H^\infty$.)

By Theorem 1.3 and (3.12) it follows that $d_2 = \|F_2\|_2$ . So it remains to show that $F_{2+}^* \Gamma_2 = \Gamma_1$, or equivalently, $F_{2+}^* \,|\, \mathcal{H}_2 = A^*$. To this end, let $h_2$ be in $\mathcal{H}_2$ and e in $\mathcal{E}_1$. For all $n \geq 0$ we have

$$(F_{2+}^* h_2 \,,\, S_1^n e) = (h_2 \,,\, S_2^n F_{2+} e) = (S_2^{*n} h_2 \,,\, AP_1 e) =$$
$$(A^* T_2^{*n} h_2 \,,\, e) = (T_1^{*n} A^* h_2 \,,\, e) = (A^* h_2 \,,\, S_1^n e) \,.$$

Because $\mathcal{E}_1$ is cyclic for $S_1$ we have $F_{2+}^* \,|\, \mathcal{H}_2 = A^*$. Therefore, $F_2$ is an $H^2$ optimal function tangentially interpolating $\Gamma_2$ with $\Gamma_1$. To prove uniqueness, let F be any other function tangentially interpolating $\Gamma_2$ with $\Gamma_1$. Then $A^* = F_+^* \,|\, \mathcal{H}_2$. In other words, $AP_1 = P_2 F_+$, where $P_2$ is the orthogonal projection onto $\mathcal{H}_2$. Thus,

$$d_2 = \|AP_1 \,|\, \mathcal{E}_1\|_2 = \|P_2 F_+ \,|\, \mathcal{E}_1\|_2 \leq \|F\|_2 \,.$$

Moreover, we have equality if and only if $\|P_2 F_+ \,|\, \mathcal{E}_1\|_2 = \|F_+ \,|\, \mathcal{E}_1\|_2$ , or equivalently, $F = AP_1 \,|\, \mathcal{E}_1 = F_2$ . So $F_2$ is the only function in $H^\infty$ tangentially interpolating $\Gamma_2$ with $\Gamma_1$ and satisfying $d_2 = \|F_2\|$.

The following theorem gives an $H^2$–$H^\infty$ solution F to our tangential interpolation problem in terms of a minimal realization of $[G_1, G_2]^{tr}$.

**THEOREM 3.4.** *Let* $\{M, B, [C_1, C_2]^{tr}\}$ *be a minimal realization for* $[G_1, G_2]^{tr}$ *with the corresponding observability grammians* $Q_1$ *and* $Q_2$. *Given any* $\delta > 1$, *set* $\gamma = \delta d_\infty$. *If* $Q_2 > 0$, *then a function* F *in* $H^\infty(\mathcal{E}_1, \mathcal{E}_2)$ *tangentially interpolating* $\Gamma(G_2)$ *with* $\Gamma(G_1)$ *and satisfying the* $H^2$–$H^\infty$ *bounds in (3.6) is given by the realization*

$$F(z) = \hat{C}(I - z\hat{M})^{-1}\hat{B}, \tag{3.14}$$

*where* $\hat{B}$ *mapping* $\mathcal{E}_1$ *into* $\mathbb{C}^n$, $\hat{M}$ *on* $\mathbb{C}^n$ *and* $\hat{C}$ *mapping* $\mathbb{C}^n$ *into* $\mathcal{E}_2$ *are the operators defined by*

$$\hat{C} = \gamma^2 C_2(\gamma^2 Q_2 - Q_1 + C_1^* C_1)^{-1} \qquad \text{and} \qquad \hat{B} = C_1^*$$
$$\hat{M} = (\gamma^2 Q_2 - Q_1)M(\gamma^2 Q_2 - Q_1 + C_1^* C_1)^{-1}. \tag{3.15}$$

*Moreover,* $\hat{M}$ *has all of its eigenvalues inside the unit disc. Furthermore, if* $Q_2$ *is singular, then there exists no* F *tangentially interpolating* $\Gamma_2$ *with* $\Gamma_1$.

Notice that the term $\gamma^2 Q_2 - Q_1 + C_1^* C_1$ in (3.15) is also given by

$$\gamma^2 Q_2 - M^* Q_1 M = \gamma^2 Q_2 - Q_1 + C_1^* C_1, \tag{3.16}$$

where we used the Lyapunov equation $Q_1 = M^* Q_1 M + C_1^* C_1$.

PROOF. To derive the state space expression for F, the proof uses equation (2.35) of Corollary 2.4 and operates from the right side to the left to compute the individual terms of that expression in terms of known quatities $M$, $C_1$, $C_2$, $Q_1$ and $Q_2$ of the minimal realization. To this end, notice that by using equation (3.12), and $A^* W_2 = W_1$ we have

$$A^* A P_1 | \mathcal{E}_1 = W_1 Q_2^{-1} C_1^*. \tag{3.17}$$

The next expression we need to derive in (2.35) is $(\gamma^2 I - A^* A)^{-1} W_1$. By (3.11)

$$(\gamma^2 I - A^* A)W_1 = \gamma^2 W_1 - A^* W_2 Q_2^{-1} Q_1 = W_1(\gamma^2 I - Q_2^{-1} Q_1). \tag{3.18}$$

This readily implies that

$$(\gamma^2 I - A^*A)^{-1}W_1 = W_1(\gamma^2 I - Q_2^{-1}Q_1)^{-1} . \tag{3.19}$$

Equation (3.9) in Theorem 3.2 guarantees that $\gamma^2 I - Q_2^{-1}Q_1$ is invertible.

We claim that

$$T_2W_2 = W_2Q_2^{-1}M^*Q_2 . \tag{3.20}$$

This is a consequence of the following computation where x and y are in $\mathbb{C}^n$

$$(T_2W_2x , W_2y) = (W_2x , T_2^*W_2y) = (W_2x , W_2My) = (x , Q_2My) =$$
$$(M^*Q_2x , y) = (Q_2Q_2^{-1}M^*Q_2x , y) = (W_2Q_2^{-1}M^*Q_2x , W_2y) .$$

We still need to perform some further calculations, that is, by (3.11)

$$\begin{aligned}(\gamma^2 I - A^*T_2^*T_2A)W_1 &= \gamma^2W_1 - A^*T_2^*T_2W_2Q_2^{-1}Q_1\\ &= \gamma^2W_1 - A^*T_2^*W_2Q_2^{-1}M^*Q_1\\ &= \gamma^2W_1 - A^*W_2MQ_2^{-1}M^*Q_1\\ &= W_1(\gamma^2 I - MQ_2^{-1}M^*Q_1) . \end{aligned}\tag{3.21}$$

We claim that $\gamma^2 I - MQ_2^{-1}M^*Q_1$ is invertible. If $(\gamma^2 I - MQ_2^{-1}M^*Q_1)x = 0$ for some x in $\mathbb{C}^n$, then equation (3.21) along with $\|A\| < \gamma$ and the fact that $(\gamma^2 I - A^*T_2^*T_2A)$ is invertible imply that $W_1x = 0$. Therefore, $Q_1x = W_1^*W_1x = 0$. Hence $\gamma^2 Ix = 0$, or equivalently, $x = 0$. This proves our invertibility claim. (Here we have assumed that $\gamma \neq 0$, or equivalently, $d_\infty \neq 0$. If $d_\infty = 0$, then the trivial function $F = 0$ tangentially interpolates $\Gamma_2$ with $\Gamma_1$. So in this case $\Gamma_1 = 0$ and our $H^2$–$H^\infty$ tangential interpolation problem is trivial.) Finally, by taking the inverse in (3.21), we obtain one of the equations that we have been looking for

$$(\gamma^2 I - A^*T_2^*T_2A)^{-1}W_1 = W_1(\gamma^2 I - MQ_2^{-1}M^*Q_1)^{-1} . \tag{3.22}$$

Now we are ready to compute an expression for $T_A^*$ which plays a key role in (2.35). Using (3.18) and $T_1^*W_1 = S_1^*W_1 = W_1M$ along with (3.22) we have

$$\begin{aligned}T_A^*W_1 &= (\gamma^2 I - T_1^*A^*AT_1)^{-1}T_1^*(\gamma^2 I - A^*A)W_1\\ &= (\gamma^2 I - A^*T_2^*T_2A)^{-1}T_1^*W_1(\gamma^2 I - Q_2^{-1}Q_1)\\ &= (\gamma^2 I - A^*T_2^*T_2A)^{-1}W_1M(\gamma^2 I - Q_2^{-1}Q_1)\\ &= W_1(\gamma^2 I - MQ_2^{-1}M^*Q_1)^{-1}M(\gamma^2 I - Q_2^{-1}Q_1) = W_1N \end{aligned}\tag{3.23}$$

where $N = (\gamma^2 I - MQ_2^{-1}M^*Q_1)^{-1}M(\gamma^2 I - Q_2^{-1}Q_1)$. We claim that N is stable, that is, all the eigenvalues of N are in the open unit disc. If $Nx = \lambda x$ for some nonzero x in $\mathbb{C}^n$, then equation (3.23) implies that $T_A^* W_1 x = \lambda W_1 x$. Notice that the operator $W_1$ may not be one-to-one. If $W_1 x$ is nonzero, then $\lambda$ is an eigenvalue for $T_A^*$. In this case Corollary 2.2 implies that $|\lambda| < 1$. On the other hand, if $W_1 x = 0$, then $Q_1 x = W_1^* W_1 x = 0$. This implies that $Nx = (\gamma^2 I - MQ_2^{-1}M^*Q_1)^{-1}M\gamma^2 x = \lambda x$. By taking the inverse along with $Q_1 x = 0$ we obtain $M\gamma^2 x = \lambda\gamma^2 x$. (As before, we assume that $\gamma \neq 0$, or equivalently, $d_\infty \neq 0$.) Therefore, $\lambda$ is an eigenvalue for M. Thus $|\lambda| < 1$. So we conclude that N is stable. (It is an interesting exercise to prove that N is stable without using $T_A^*$.) By taking the appropriate inverses in (3.23) for z in the open unit disc, we obtain

$$(I - zT_A^*)^{-1}W_1 = W_1(I - zN)^{-1}. \tag{3.24}$$

Finally, we need an expression for $D_2AW_1$ to compute (2.35). By using $AW_1 = W_2Q_2^{-1}Q_1$ and the definition of $D_2$ in (2.36) we have

$$\begin{aligned} D_2AW_1 = D_2W_2Q_2^{-1}Q_1 &= (S_2 - T_2)W_2Q_2^{-1}Q_1 \\ &= (zW_2 - W_2Q_2^{-1}M^*Q_2)Q_2^{-1}Q_1 \\ &= W_2(zI - Q_2^{-1}M^*Q_2)Q_2^{-1}Q_1 \\ &= C_2(I - zM)^{-1}(zI - Q_2^{-1}M^*Q_2)Q_2^{-1}Q_1 \end{aligned} \tag{3.25}$$

where the third equality follows from the formula for $T_2W_2$ in (3.20).

Substituting (3.12), (3.17), (3.19), (3.23), (3.24) and (3.25) into equation (2.35) of Corollary 2.4 we obtain

$$\begin{aligned} F(z) = W_2Q_2^{-1}C_1^* - W_2(zI - Q_2^{-1}M^*Q_2)Q_2^{-1}Q_1(\gamma^2 I - MQ_2^{-1}M^*Q_1)^{-1} \times \\ \times M(\gamma^2 I - Q_2^{-1}Q_1)(I - zN)^{-1}(\gamma^2Q_2 - Q_1)^{-1}C_1^*. \end{aligned} \tag{3.26}$$

Now let $\hat{M}$ be the matrix on $\mathbb{C}^n$ defined by

$$\begin{aligned} \hat{M} &= (\gamma^2Q_2 - Q_1)N(\gamma^2Q_2 - Q_1)^{-1} = (\gamma^2Q_2 - Q_1)(\gamma^2 I - MQ_2^{-1}M^*Q_1)^{-1}MQ_2^{-1} \\ &= (\gamma^2Q_2 - Q_1)M(\gamma^2Q_2 - M^*Q_1M)^{-1}. \end{aligned} \tag{3.27}$$

According to (3.16), this is precisely the operator $\hat{M}$ in (3.15). Moreover, because $\hat{M}$ is similar to N, the operator $\hat{M}$ is stable.

Finally, using the $\hat{M}$ in (3.27) in the previous expression (3.26) for F(z) along with $W_2 = C_2(I - zM)^{-1}$, we obtain

$$F(z) = C_2(I - zM)^{-1}\Psi(z)(I - z\hat{M})^{-1}C_1^* \tag{3.28}$$

where $\Psi(z)$ is the function given by

$$\begin{aligned}\Psi(z) &= Q_2^{-1}(I - z\hat{M}) - Q_2^{-1}(zI - M^*)Q_1(\gamma^2 I - MQ_2^{-1}M^*Q_1)^{-1}MQ_2^{-1} \\ &= Q_2^{-1}(I - z\hat{M}) - Q_2^{-1}(zI - M^*)Q_1M(\gamma^2 Q_2 - M^*Q_1M)^{-1} \\ &= Q_2^{-1}\left[(I - z\hat{M})(\gamma^2 Q_2 - M^*Q_1M) - (zI - M^*)Q_1M\right](\gamma^2 Q_2 - M^*Q_1M)^{-1} \\ &= Q_2^{-1}\left[\gamma^2 Q_2 - M^*Q_1M - z(\gamma^2 Q_2 - Q_1)M - zQ_1M + M^*Q_1M\right](\gamma^2 Q_2 - M^*Q_1M)^{-1} \\ &= \gamma^2 (I - zM)(\gamma^2 Q_2 - M^*Q_1M)^{-1} .\end{aligned} \tag{3.29}$$

Substituting $\Psi(z)$ into our expression for F in (3.28) we have

$$F(z) = \gamma^2 C_2(\gamma^2 Q_2 - M^*Q_1M)^{-1}(I - z\hat{M})^{-1} C_1^* . \tag{3.30}$$

According to (3.16), this is precisely the F(z) given by (3.14). Therefore the F(z) in (2.35) of Corollary 2.4 can be computed by the realization in (3.30). So according to Corollary 2.4 and (1.8), the F(z) in (3.30) tangentially interpolates $\Gamma_2$ with $\Gamma_1$ and satisfies the $H^2$–$H^\infty$ bounds in (3.6). The proof is now complete.

As a matter of consistency, it is interesting to point out that in equation (3.14) as $\delta$ approaches infinity, the expression for F becomes

$$F(z) = C_2(I - zM)^{-1}Q_2^{-1}C_1^* = W_2Q_2^{-1}C_1^* = F_2(z)$$

which is precisely the $H^2$ optimal solution given in Remark 3.3.

The following corollary gives an alternate method to compute the function F in (3.14) tangentially interpolating $\Gamma(G_2)$ with $\Gamma(G_1)$. Recall that D is the open unit disc in the complex plane.

**COROLLARY 3.5.** *Let* $\{M , B , [C_1 , C_2]^{tr}\}$ *be a minimal realization for* $[G_1 , G_2]^{tr}$ *with the corresponding observability grammians* $Q_1$ *and* $Q_2$. *Assume that there exists a function tangentially interpolating* $\Gamma(G_2)$ *with* $\Gamma(G_1)$, *or equivalently,* $Q_2 > 0$; *and set* $\gamma = \delta\, d_\infty$ . *Then the function* F *in (3.14) tangentially interpolating* $\Gamma(G_2)$ *with* $\Gamma(G_1)$, *and at the same time, satisfying the* $H^2$–$H^\infty$ *bounds in (3.6) can also be computed by*

$$F(z) = P(z)\,Q(z)^{-1} \tag{3.31}$$

*where $P(z)$ and $Q(z)$ are the analytic functions in $\bar{D}$ defined by*

$$P(z) = \gamma^2 C_2 (I - zM)^{-1} (\gamma^2 Q_2 - Q_1)^{-1} C_1^* \qquad \text{and}$$
$$Q(z) = C_1 (I - zM)^{-1} (\gamma^2 Q_2 - Q_1)^{-1} C_1^* + I\,. \tag{3.32}$$

*Moreover,* $Q(z)^{-1}$ *is also analytic in* $\bar{D}$. *In particular,* $Q(z)$ *is an outer function.*

PROOF. According to (3.14) and (3.15) we have

$$\begin{aligned} F(z) &= \gamma^2 C_2 (\gamma^2 Q_2 - Q_1 + C_1^* C_1)^{-1} \left[ I - z(\gamma^2 Q_2 - Q_1) M (\gamma^2 Q_2 - Q_1 + C_1^* C_1)^{-1} \right]^{-1} C_1^* \\ &= \gamma^2 C_2 \left[ \gamma^2 Q_2 - Q_1 + C_1^* C_1 - z(\gamma^2 Q_2 - Q_1) M \right]^{-1} C_1^* \\ &= \gamma^2 C_2 \left[ (\gamma^2 Q_2 - Q_1)(I - zM) + C_1^* C_1 \right]^{-1} C_1^*\,. \end{aligned} \tag{3.33}$$

The following matrix inversion lemma [28] will be used to simplify equation (3.33)

$$(X + YZ)^{-1} = X^{-1} - X^{-1} Y (Z X^{-1} Y + I)^{-1} Z X^{-1}\,. \tag{3.34}$$

Using $X = (\gamma^2 Q_2 - Q_1)(I - zM)$, $Y = C_1^*$ and $Z = C_1$ in the bracketed expression of equation (3.34) we have

$$Z X^{-1} Y + I = C_1 (I - zM)^{-1} (\gamma^2 Q_2 - Q_1)^{-1} C_1^* + I = Q(z)\,. \tag{3.35}$$

Using the matrix inversion lemma (3.34) along with (3.35) in (3.33) we have

$$\begin{aligned} F(z) &= \gamma^2 C_2 (I - zM)^{-1} (\gamma^2 Q_2 - Q_1)^{-1} C_1^* \\ &\quad - \gamma^2 C_2 (I - zM)^{-1} (\gamma^2 Q_2 - Q_1)^{-1} C_1^* Q(z)^{-1} C_1 (I - zM)^{-1} (\gamma^2 Q_2 - Q_1)^{-1} C_1^* \\ &= \gamma^2 C_2 (I - zM)^{-1} (\gamma^2 Q_2 - Q_1)^{-1} C_1^* \left[ Q(z) - C_1 (I - zM)^{-1} (\gamma^2 Q_2 - Q_1)^{-1} C_1^* \right] Q(z)^{-1} \\ &= \gamma^2 C_2 (I - zM)^{-1} (\gamma^2 Q_2 - Q_1)^{-1} C_1^* Q(z)^{-1} = P(z) Q(z)^{-1}\,. \end{aligned} \tag{3.36}$$

Because $M$ is stable, equation (3.35) implies that $P(z)$ and $Q(z)$ are analytic functions in $\bar{D}$. To complete the proof, it remains to show that $Q(z)^{-1}$ is analytic in $\bar{D}$. To this end, notice that

$$I - Z(X + YZ)^{-1} Y = (I + Z X^{-1} Y)^{-1} = Q(z)^{-1}\,. \tag{3.37}$$

However, by following the computations in (3.33) backwards we now have

$$Z(X + YZ)^{-1}Y = C_1 \left[(\gamma^2 Q_2 - Q_1)(I - zM) + C_1^* C_1\right]^{-1} C_1^*$$

$$= C_1(\gamma^2 Q_2 - Q_1 + C_1^* C_1)^{-1} \left[I - z(\gamma^2 Q_2 - Q_1)M(\gamma^2 Q_2 - Q_1 + C_1^* C_1)^{-1}\right]^{-1} C_1^* . \quad (3.38)$$

Substituting (3.38) into (3.37) along with the definition of $\hat{M}$, we have

$$I - C_1(\gamma^2 Q_2 - Q_1 + C_1^* C_1)^{-1}(I - z\hat{M})^{-1} C_1^* = Q(z)^{-1} . \quad (3.39)$$

Because $\hat{M}$ is stable, it readily follows that $Q(z)^{-1}$ is analytic in $\overline{D}$. This completes the proof.

In some applications, the realization $\{M, B, [C_1 , C_2]^{tr}\}$ is controllable, stable and not necessarily observable. (By stable, we mean that all the eigenvalues of M are in D.) The following theorem shows how to compute an F tangentially interpolating $\Gamma_2$ with $\Gamma_1$ and satisfying the $H^2$–$H^\infty$ bounds from a controllable, stable realization. The closed range of an operator C is denoted by ran(C).

**COROLLARY 3.6.** *Let* $\{M \text{ on } \mathbb{C}^n, B, [C_1 , C_2]^{tr}\}$ *be a controllable, stable realization for* $[G_1 , G_2]^{tr}$ *and let* $Q_1$ *and* $Q_2$ *be the corresponding observability grammians. Then, there exists an* F *in* $H^\infty(\mathcal{E}_1 , \mathcal{E}_2)$ *tangentially interpolating* $\Gamma(G_2)$ *with* $\Gamma(G_1)$ *and satisfying* $\|F\|_\infty \le \gamma$, *if and only if* $Q_1 \le \gamma^2 Q_2$, *for some* $\gamma > 0$. *The condition* $Q_1 \le \gamma^2 Q_2$ *for some finite* $\gamma$ *is equivalent to* $\operatorname{ran}(Q_1) \subseteq \operatorname{ran}(Q_2)$. *In this case,*

$$d_\infty^2 = \lambda_{max} \left[(V^* Q_2 V)^{-1} V^* Q_1 V\right] \quad \text{and} \quad d_2^2 = \operatorname{trace} \left[C_1 V(V^* Q_2 V)^{-1} V^* C_1^*\right] \quad (3.40)$$

*where* V *is any isometry mapping* $\mathbb{C}^k$ *into* $\mathbb{C}^n$ *whose range equals the range of* $Q_2$. *Moreover, given any* $\delta > 1$ *with* $\gamma = \delta\, d_\infty$, *then a function* F *in* $H^\infty(\mathcal{E}_1 , \mathcal{E}_2)$ *tangentially interpolating* $\Gamma(G_2)$ *with* $\Gamma(G_1)$ *and satisfying the* $H^2$–$H^\infty$ *bounds in (3.6) is given by the realization*

$$F(z) = \hat{C}_o (I - z\hat{M}_o)^{-1} \hat{B}_o , \quad (3.41)$$

*where* $\hat{B}_o$ *mapping* $\mathcal{E}_1$ *into* $\mathbb{C}^k$, $\hat{M}_o$ *on* $\mathbb{C}^k$ *and* $\hat{C}_o$ *mapping* $\mathbb{C}^k$ *into* $\mathcal{E}_2$ *are the operators defined by*

$$\hat{C}_o = \gamma^2 C_2 V \left[V^*(\gamma^2 Q_2 - Q_1 + C_1^* C_1)V\right]^{-1} \quad \text{and} \quad \hat{B}_o = V^* C_1^*$$
$$\hat{M}_o = V^*(\gamma^2 Q_2 - Q_1)MV \left[V^*(\gamma^2 Q_2 - Q_1 + C_1^* C_1)V\right]^{-1} . \quad (3.42)$$

*Furthermore, the function* F(z) *in (3.42) can also be computed by*

$$F(z) = P(z)\,Q(z)^{-1} \tag{3.43}$$

*where P(z) and Q(z) are the analytic functions in* $\overline{D}$ *defined by*

$$P(z) = \gamma^2 C_2 (I - zM)^{-1} V(\gamma^2 V^* Q_2 V - V^* Q_1 V)^{-1} V^* C_1^* \quad \text{and}$$
$$Q(z) = C_1 (I - zM)^{-1} V(\gamma^2 V^* Q_2 V - V^* Q_1 V)^{-1} V^* C_1^* + I\,. \tag{3.44}$$

*Finally, the function* $Q(z)^{-1}$ *is analytic in* $\overline{D}$.

PROOF. According to Theorem 1.3, there exists an F tangentially interpolating $\Gamma_2$ with $\Gamma_1$ and satisfying $\|F\|_\infty \le \gamma$, if and only if $\Gamma_1^*\Gamma_1 \le \gamma^2 \Gamma_2^* \Gamma_2$. However, using $\Gamma_i = W_i W_c^*$ for i=1,2 and the fact that $W_c^*$ is onto (because of the controllability assumption), we see that $\Gamma_1^*\Gamma_1 \le \gamma^2 \Gamma_2^*\Gamma_2$ if and only if $Q_1 \le \gamma^2 Q_2$. So it remains to show that $Q_1 \le \gamma^2 Q_2$ if and only if $\text{ran}(Q_1) \subseteq \text{ran}(Q_2)$. To prove this, assume that $Q_1 \le \gamma^2 Q_2$. If x is in the kernel of $Q_2$, then obviously x is in the kernel of $Q_1$. Thus, $\ker(Q_2) \subseteq \ker(Q_1)$, or equivalently, $\text{ran}(Q_1) \subseteq \text{ran}(Q_2)$. On the other hand, if $\text{ran}(Q_1) \subseteq \text{ran}(Q_2)$, then obviously $Q_1 \le \gamma^2 Q_2$ for some $\gamma$. If fact any $\gamma^2 \ge \|\sqrt{Q_2^{-r}}\, Q_1 \sqrt{Q_2^{-r}}\|$ will work.

From now on, we assume that there exists a function F tangentially interpolating $\Gamma_2$ with $\Gamma_1$. To complete the proof, we will extract a minimal realization from $\{M, B, [C_1, C_2]^{tr}\}$ and substitute this minimal realization into (3.14) of the previous theorem to obtain (3.41). To this end, let $\mathcal{X}_o$ be the observability space defined by

$$\mathcal{X}_o = \bigvee_{n=0}^{\infty} M^{*n} [C_1^*, C_2^*] \begin{bmatrix} \mathcal{E}_1 \\ \mathcal{E}_2 \end{bmatrix}. \tag{3.45}$$

We claim that $\mathcal{X}_o$ equals th range of $Q_2$. To see this, notice that the orthogonal complement $\mathcal{X}_o^\perp$ of $\mathcal{X}_o$ equals the kernel of

$$W = \begin{bmatrix} C_1 \\ C_2 \end{bmatrix} (I - zM)^{-1} = \begin{bmatrix} W_1 \\ W_2 \end{bmatrix} \tag{3.46}$$

where W is viewed as an operator from $\mathbb{C}^n$ to $H^2(\mathcal{E}_1 \oplus \mathcal{E}_2)$. Thus $\mathcal{X}_o^\perp = \ker(W_1) \cap \ker(W_2)$. Recall that $W_1^* W_1 = Q_1$ and $W_2^* W_2 = Q_2$ and $\ker(Q_2) \subseteq \ker(Q_1)$. Therefore $\mathcal{X}_o^\perp = \ker(Q_2)$. Hence, $\mathcal{X}_o$ equals the range of $Q_2$, which proves our claim.

Obviously, $\mathcal{X}_o$ is an invariant subspace for $M^*$ and thus, $P_oM = P_oMP_o$, where $P_o$ is the orthogonal projection onto $\mathcal{X}_o$. Because the set $\{M, B, [C_1, C_2]^{tr}\}$ is controllable and $\mathcal{X}_o$ is its observability subspace, it follows that the set

$$\Sigma_o = \{P_oM_o \mid \mathcal{X}_o \text{ on } \mathcal{X}_o, P_oB, [C_1 \mid \mathcal{X}_o, C_2 \mid \mathcal{X}_o]^{tr}\} \tag{3.47}$$

is a controllable and observable realization of $[G_1, G_2]^{tr}$. Moreover, using the fact that $Q_i = P_oQ_iP_o$ along with $P_oM = P_oMP_o$ and $\operatorname{ran}(C_i^*) \subseteq \mathcal{X}_o$ in the Lyapunov equation (3.4) for $Q_i$ we have

$$P_oQ_iP_o = P_oM^*P_oQ_iP_oMP_o + P_oC_i^*C_iP_o \qquad (\text{for } i{=}1,2)\,. \tag{3.48}$$

Because $\operatorname{ran}(Q_1) \subseteq \operatorname{ran}(Q_2) = \mathcal{X}_o$, it follows that $P_oQ_1 \mid \mathcal{X}_o = Q_1 \mid \mathcal{X}_o$ and $P_oQ_2 \mid \mathcal{X}_o = Q_2 \mid \mathcal{X}_o$. Therefore, the positive operators $Q_1 \mid \mathcal{X}_o$ and $Q_2 \mid \mathcal{X}_o$ on $\mathcal{X}_o$ are the corresponding observability grammians for the minimal realization $\Sigma_o$ in (3.47) of $[G_1, G_2]^{tr}$.

Let V be any isometry mapping $\mathbb{C}^k$ into $\mathbb{C}^n$ whose range is $\mathcal{X}_o$. Define the operators $M_o$ on $\mathbb{C}^k$, $B_o$ mapping $\mathcal{E}$ into $\mathbb{C}^k$ and $[C_{o1}, C_{o2}]^{tr}$ mapping $\mathbb{C}^k$ into $\mathcal{E}_1 \oplus \mathcal{E}_2$ by

$$M_o = V^*MV \;,\; B_o = V^*B \;,\; C_{o1} = C_1V \text{ and } C_{o2} = C_2V\,. \tag{3.49}$$

Obviously, the set $\{M_o, B_o, [C_{o1}, C_{o2}]^{tr}\}$ is a minimal realization of $[G_1, G_2]^{tr}$. Moreover, $V^*Q_1V$ and $V^*Q_2V$ are the corresponding observability grammians for $\{M_o, B_o, [C_{o1}, C_{o2}]^{tr}\}$; see equation (3.48). By Theorem 3.2 we obtain (3.40).

According to Theorem 3.4, an F tangentially interpolating $\Gamma_2$ with $\Gamma_1$ and satisfying the $H^2$–$H^\infty$ bounds in (3.6) is given by the realization

$$F(z) = \hat{C}_o\,(I - z\hat{M}_o)^{-1}\,\hat{B}_o\,, \tag{3.50}$$

where

$$\begin{aligned} &\hat{C}_o = \gamma^2C_{o2}(\gamma^2 Q_{o2} - Q_{o1} + C_{o1}^*C_{o1})^{-1} \qquad \text{and} \qquad \hat{B}_o = C_{o1}^* \\ &\hat{M}_o = (\gamma^2Q_{o2} - Q_{o1})M_o(\gamma^2 Q_{o2} - Q_{o1} + C_{o1}^*C_{o1})^{-1}\,. \end{aligned} \tag{3.51}$$

Obviously, we have

$$\left[\gamma^2Q_{o2} - Q_{o1} + C_{o1}^*C_{o1}\right]^{-1} = \left[V^*(\gamma^2Q_2 - Q_1 + C_1^*C_1)V\right]^{-1}. \tag{3.52}$$

Therefore, the operator $\hat{C}_o$ in (3.51) becomes

$$\hat{C}_o = \gamma^2 C_2 V\left[V^*(\gamma^2 Q_2 - Q_1 + C_1^* C_1)V\right]^{-1}. \tag{3.53}$$

Using (3.52) in the expression for $\hat{M}_o$ in (3.51) along with $Q_i P_o = Q_i$ for i=1,2 we have

$$\begin{aligned}\hat{M}_o &= V^*(\gamma^2 Q_2 - Q_1)VV^*MV\left[V^*(\gamma^2 Q_2 - Q_1 + C_1^* C_1)V\right]^{-1}\\ &= V^*(\gamma^2 Q_2 - Q_1)MV\left[V^*(\gamma^2 Q_2 - Q_1 + C_1^* C_1)V\right]^{-1}.\end{aligned} \tag{3.54}$$

These are precisely the equations in (3.43). By Theorem 3.4 this F in (3.50) tangentially interpolates $\Gamma_2$ with $\Gamma_1$ and satisfies the $H^2$–$H^\infty$ bounds in (3.6). To complete the proof, it remains to establish (3.43) and (3.44).

According to Corollary 3.5, we know that $F(z) = P(z)\,Q(z)^{-1}$ where

$$\begin{aligned}P(z) &= \gamma^2 C_2 V(I - zV^*MV)^{-1}(\gamma^2 V^* Q_2 V - V^* Q_1 V)^{-1} V^* C_1^* \qquad \text{and}\\ Q(z) &= C_1 V(I - zV^*MV)^{-1}(\gamma^2 V^* Q_2 V - V^* Q_1 V)^{-1} V^* C_1^* + I.\end{aligned} \tag{3.55}$$

Recall that $P_o = VV^*$ and that $C_i = C_i P_o$ for i=1,2. Moreover, $P_o M = P_o M P_o$ and thus $(P_o M)^n = P_o M^n$ for $n \geq 0$. Using these observations, we have for i=1,2

$$\begin{aligned}&C_i V(I - zV^*MV)^{-1} = C_i P_o(I - zVV^*M)^{-1}V = C_i P_o(I - zP_o M)^{-1}V =\\ &\sum_{n=0}^{\infty} C_i P_o(zP_o M)^n V = C_i \sum_{n=0}^{\infty} P_o(zM)^n V = C_i(I - zM)^{-1}V.\end{aligned} \tag{3.56}$$

Substituting (3.56) into (3.55) we obtain P(z) and Q(z) defined in (3.44). Corollary 3.5 guarantees that $Q(z)^{-1}$ is analytic in $\bar{D}$. This completes the proof.

It is emphasized that we have chosen to represent P and Q in (3.44) in terms of $C_i(I - zM)^{-1}$ rather than expressing them as in (3.55). This is because in the next section, we can easily compute $C_i(I - zM)^{-1}$ in terms of the data whereas computing P(z) and Q(z) by (3.55) would be more difficult to compute even though $(I - zV^*MV)^{-1}$ is of lower dimension.

**REMARK 3.7.** We can use the results in [12] to show that our function F tangentially interpolating $\Gamma_2$ with $\Gamma_1$ satisfies an even stronger $H^2$ bound than the one given in (3.6). This follows because according to [12], the central contractive intertwining lifting $B_\gamma$ of $A_0 = AP_1$ in (2.6) satisfies an even stronger bound than the one given in (2.11). To be precise, $\|B_\gamma\| \leq \delta\|A\|$ and

$$\|B_\gamma e\|^2 \leq \frac{\|AP_1 e\|^2(\delta^2 \|A\|^2 - \|AP_1 e\|^2)}{\|A\|^2(\delta^2 - 1)} \qquad (e \in \ker(T^*) \text{ and } \|e\| = 1). \tag{3.57}$$

Now let

$$\lambda_m^2 = \max \min \|AP_1 e_i\|^2 \tag{3.58}$$

where the minimum is taken over an orthonormal basis $\{e_i\}_{i=1}^k$ for $\mathcal{E}_1$ and the maximum is taken over all orthonormal bases for $\mathcal{E}_1$. It is an interesting exercise to show that for any operator L mapping a k-dimensional space $\mathcal{E}_1$ into $\mathcal{H}$, then

$$\frac{1}{k}\|L\|_2^2 = \frac{1}{k}\operatorname{tr}(L^*L) = \max \min \|Le_i\|^2 . \tag{3.59}$$

(One can prove this result by using the fact that there exists a unitary transformation U such that $U^*L^*LU$ has all of its diagonal elements equal. This type of transformation plays an important role in finite wordlength and round off error problems; see [25] and [49] for further details.) In particular, (3.59) and $d_2 = \|AP_1 \mid \mathcal{E}_1\|_2$ imply that $\lambda_m^2 = d_2^2/k$. So by choosing an orthonormal basis $\{e_i\}_{i=1}^k$ which solves the max-min problem in (3.58), we see that $\|AP_1 e_i\|^2 \geq \lambda_m^2 = d_2^2/k$ for all i. Thus $-\|AP_1 e_i\|^2 \leq -d_2^2/k$ for all i. Using this along with the definition of the Hilbert-Schmidt norm $\|B_\gamma \mid \mathcal{E}_1\|_2^2 = \sum\|B_\gamma e_i\|^2$ in (3.57) we have

$$\|B_\gamma \mid \mathcal{E}_1\|_2^2 \leq \frac{\|AP_1 \mid \mathcal{E}_1\|_2^2(\delta^2 \|A\|^2 - d_2^2/k)}{\|A\|^2(\delta^2 - 1)} . \tag{3.60}$$

Therefore, our function F tangentially interpolating $\Gamma_2$ with $\Gamma_1$ satisfies the following $H^2$ bound

$$\|F\|_2^2 \leq \frac{d_2^2(\delta^2 d_\infty^2 - d_2^2/k)}{d_\infty^2\,(\delta^2 - 1)} . \tag{3.61}$$

Notice that when k approaches infinity, we get the $H^2$ bound in (3.6). However, if $\mathcal{E}_1$ is one-dimensional, then equation (3.61) simplifies to

$$\|F\|_2^2 \leq \frac{d_2^2(\delta^2 d_\infty^2 - d_2^2)}{d_\infty^2\,(\delta^2 - 1)} . \tag{3.62}$$

For futher results concerning these bounds and central commutant lifting see [12].

**4. Applications of The $H^2$–$H^\infty$ Tangential Interpolation Problem.**

In this section, we will use our theory to solve an $H^2$–$H^\infty$ tangential Hermite-Fejer interpolation problem. To begin, let $\mathcal{U} = \{U_{i,j} \ : \ 1 \le i \le n \ \text{ and } \ 0 \le j < d_i\}$ be a set of operators mapping $\mathcal{E}$ into $\mathcal{E}_2$ and $\mathcal{Y} = \{Y_{i,j} \ : \ 1 \le i \le n \ \text{ and } \ 0 \le j < d_i\}$ be a set of operators mapping $\mathcal{E}$ into $\mathcal{E}_1$. Let $\alpha_1, \alpha_2, \ldots, \alpha_n$ be n distinct complex numbers in the open unit disc D. We say that a function G in $H^\infty(\mathcal{E}_2, \mathcal{E}_1)$ tangentially interpolates $\mathcal{U}$ with $\mathcal{Y}$ if

$$\begin{bmatrix} Y_{i,0} \\ Y_{i,1} \\ Y_{i,2} \\ \vdots \\ Y_{i,d_i-1} \end{bmatrix} = \begin{bmatrix} G(\alpha_i) & 0 & \cdots & 0 \\ \frac{G^{(1)}(\alpha_i)}{1!} & G(\alpha_i) & \cdots & 0 \\ \vdots & \vdots & \vdots & \vdots \\ \frac{G^{(d_i-1)}(\alpha_i)}{(d_i-1)!} & \frac{G^{(d_i-2)}(\alpha_i)}{(d_i-2)!} & \cdots & G(\alpha_i) \end{bmatrix} \begin{bmatrix} U_{i,0} \\ U_{i,1} \\ U_{i,2} \\ \vdots \\ U_{i,d_i-1} \end{bmatrix} \tag{4.1}$$

for all $1 \le i \le n$ where $G^{(k)}(\alpha)$ is the k-th derivative of G evaluated at $\alpha$. Notice that there may not exist a function G tangentially interpolating $\mathcal{U}$ with $\mathcal{Y}$. For example, if $U_{i,j} = 0$ for all i and j and $\mathcal{Y}$ is nonzero. Now assume that there exists a function G tangentially interpolating $\mathcal{U}$ with $\mathcal{Y}$. Then $\tilde{d}_\infty$ is the smallest $H^\infty$-norm of all functions tangentially interpolating $\mathcal{U}$ with $\mathcal{Y}$, that is,

$$\tilde{d}_\infty = \inf \{ \|G\|_\infty \ : \ G \in H^\infty(\mathcal{E}_2, \mathcal{E}_1) \text{ and G tangentially interpolates } \mathcal{U} \text{ with } \mathcal{Y} \}. \tag{4.2}$$

Similarly, $\tilde{d}_2$ is the smallest $H^2$-norm, of all functions tangentially interpolating $\mathcal{U}$ with $\mathcal{Y}$, that is,

$$\tilde{d}_2 = \inf \{ \|G\|_2 \ : \ G \in H^\infty(\mathcal{E}_2, \mathcal{E}_1) \text{ and G tangentially interpolates } \mathcal{U} \text{ with } \mathcal{Y} \}. \tag{4.3}$$

In this section, we will use our previous theory to construct a G (whenever one exists) tangentially interpolating $\mathcal{U}$ with $\mathcal{Y}$ and satisfying the following Kaftal-Larson-Weiss $H^2$–$H^\infty$ bounds

$$\|G\|_\infty \le \delta \tilde{d}_\infty \quad \text{and} \quad \|G\|_2 \le \frac{\delta \tilde{d}_2}{\sqrt{\delta^2 - 1}}. \tag{4.4}$$

where, as before, $\delta > 1$.

Notice that the previous problem includes many $H^2$–$H^\infty$ classical tangential interpolation problems. The following list contains only a few.

- Setting n=1 and $\alpha_1 = 0$ leads to an $H^2$–$H^\infty$ tangential Carathéodory interpolation problem : Given a set of operators $\{U_j = U_{1,j} \ : \ 0 \le j < d\}$ mapping $\mathcal{E}$ into $\mathcal{E}_2$ and $\{Y_j = Y_{1,j} \ : \ 0 \le j < d\}$ mapping $\mathcal{E}$ into $\mathcal{E}_1$, find (if possible) an analytic function $G = \sum_0^\infty G_n z^n$ in $H^\infty(\mathcal{E}_2, \mathcal{E}_1)$ such that

$$\begin{bmatrix} Y_0 \\ Y_1 \\ \vdots \\ Y_{d-1} \end{bmatrix} = \begin{bmatrix} G_0 & 0 & \dots & 0 \\ G_1 & G_0 & \dots & 0 \\ \vdots & \vdots & \vdots & \vdots \\ G_{d-1} & G_{d-2} & \dots & G_0 \end{bmatrix} \begin{bmatrix} U_0 \\ U_1 \\ \vdots \\ U_{d-1} \end{bmatrix} \tag{4.5}$$

and at the same time G satisfies the $H^2$–$H^\infty$ bounds in (4.4).

- Setting $d_1 = d_2 = \dots = d_n = 1$ leads to a $H^2$–$H^\infty$ tangential Nevanlinna-Pick interpolation problem : Given n distinct complex numbers $\alpha_1, \alpha_2, \dots, \alpha_n$ in the open unit disc, a set of operators $\{U_i = U_{i,0} \ : \ 1 \le i \le n\}$ mapping $\mathcal{E}$ into $\mathcal{E}_2$ and $\{Y_i = Y_{i,0} \ : \ 1 \le i \le n\}$ mapping $\mathcal{E}$ into $\mathcal{E}_1$, find (if possible) an analytic function G(z) in $H^\infty(\mathcal{E}_2, \mathcal{E}_1)$ such that

$$Y_i = G(\alpha_i)\, U_i \qquad (\text{for } 1 \le i \le n). \tag{4.6}$$

and at the same time G satisfies the $H^2$–$H^\infty$ bounds in (4.4).

Throughout this paper, the functions x(i,j) in $H^\infty$ defined by

$$x(i,j) = \frac{z^j}{(1 - \alpha_i z)^{j+1}} \qquad (\text{for } 1 \le i \le n \quad \text{and} \quad 0 \le j < d_i) \tag{4.7}$$

will play a basic role in our approach. It is well known that these functions form a generalized eigenvector chain for the backward shift $S^*$ on $H^2$, that is,

$$\begin{aligned} (S^* - \alpha_i)x(i,j) &= x(i,j-1) \qquad \text{if} \quad 0 < j < d_i \\ &= 0 \qquad \text{if} \quad j = 0 \end{aligned} \tag{4.8}$$

(see Section 1 of Chapter X in [10].) Using the Cauchy integral formula it is easy to show that these functions x(i,j) also have a derivative reproducing property in $H^\infty$, that is, given any f in $H^2$, we have

$$\frac{f^{(n)}(\alpha)}{n!} = (f\,,\, z^n(1-\bar{\alpha}z)^{-(n+1)}) \qquad (\text{for} \quad \alpha \in D \text{ and } n \geq 0) \tag{4.9}$$

In particular,

$$\frac{f^{(j)}(\bar{\alpha}_i)}{j!} = (f\,,\, x(i,j)) \qquad (\text{for } 1 \leq i \leq n \quad \text{and} \quad 0 \leq j < d_i)\,. \tag{4.10}$$

Let $\ell_k^2(\mathcal{E}) = \bigoplus_1^k \mathcal{E}$ be the Hilbert space formed by the set of all k-tuples with elements in $\mathcal{E}$. Let $d = d_1 + d_2 + \cdots + d_n$. Throughout this section $\Phi_1$ is the operator from $\ell_d^2(\mathcal{E}_1)$ into $H^2(\mathcal{E}_1)$ and $\Phi_2$ is the operator from $\ell_d^2(\mathcal{E}_2)$ into $H^2(\mathcal{E}_2)$ defined by the following block row matrices

$$\begin{aligned} \Phi_1 &= \left[x(1,d_1-1)I_1\,,\, x(1,d_1-2)I_1\,,\, \ldots\,,\, x(1,0)I_1\,,\, \ldots\,,\, x(n,d_n-1)I_1\,,\, \ldots\,,\, x(n,0)I_1\right] \\ \Phi_2 &= \left[x(1,d_1-1)I_2\,,\, x(1,d_1-2)I_2\,,\, \ldots\,,\, x(1,0)I_2\,,\, \ldots\,,\, x(n,d_n-1)I_2\,,\, \ldots\,,\, x(n,0)I_2\right]. \end{aligned} \tag{4.11}$$

where $I_1$ is the identity on $\mathcal{E}_1$ and $I_2$ is the identity on $\mathcal{E}_2$. Notice that the operators $\Phi_1$ and $\Phi_2$ are formed by the x(i,j)'s where the j's move in decreasing order. Let $G_1$ in $H^\infty(\mathcal{E}\,,\,\mathcal{E}_1)$, and $G_2$ in $H^\infty(\mathcal{E}\,,\,\mathcal{E}_2)$ be the functions defined by

$$\begin{aligned} G_1 &= \Phi_1 \left[Y_{1,0}^*\,,\, Y_{1,1}^*\,,\, \ldots\,,\, Y_{1,d_1-1}^*\,,\, Y_{2,0}^*\,,\, \ldots\,,\, Y_{2,d_2-1}^*\,,\, \ldots\,,\, Y_{n,0}^*\,,\, \ldots\,,\, Y_{n,d_n-1}^*\right]^* \\ G_2 &= \Phi_2 \left[U_{1,0}^*\,,\, U_{1,1}^*\,,\, \ldots\,,\, U_{1,d_1-1}^*\,,\, U_{2,0}^*\,,\, \ldots\,,\, U_{2,d_2-1}^*\,,\, \ldots\,,\, U_{n,0}^*\,,\, \ldots\,,\, U_{n,d_n-1}^*\right]^*. \end{aligned} \tag{4.12}$$

If F is a function in $H^\infty(\mathcal{E}_1\,,\,\mathcal{E}_2)$, then $\tilde{F}$ is the function in $H^\infty(\mathcal{E}_2\,,\,\mathcal{E}_1)$, defined by $\tilde{F}(z) = F(\bar{z})^*$. In this section, we will show that G tangentially interpolates $\mathcal{U}$ with $\mathcal{Y}$ if and only if $\tilde{G} = F$ tangentially interpolates $\Gamma(G_2)$ with $\Gamma(G_1)$. Thus, $\tilde{d}_\infty = d_\infty$ and $\tilde{d}_2 = d_2$. Once we establish this result, we can use our previous theory to compute a $G = \tilde{F}$ which tangentially interpolates $\mathcal{U}$ with $\mathcal{Y}$ and satisfies the $H^2$–$H^\infty$ bounds in (4.4).

Let $J_{i\mathcal{E}}$ be the block Jordan operator on $\ell_{d_i}^2(\mathcal{E})$ defined by

$$J_{i\mathcal{E}} = \begin{bmatrix} \alpha_i I & 0 & 0 & \dots & 0 \\ I & \alpha_i I & 0 & \dots & 0 \\ 0 & I & \alpha_i I & \dots & 0 \\ \vdots & \vdots & \vdots & \vdots & \vdots \\ 0 & 0 & 0 & \dots & \alpha_i I \end{bmatrix}. \tag{4.13}$$

Recall that if T is any stable matrix on $\mathbb{C}^m$ and $f(\lambda) = \sum_0^\infty f_n \lambda^n$ is in $H^\infty$, then f(T) is the matrix defined by $f(T) = \sum_0^\infty f_n T^n$. It is well known (see [23]) that if f is any function in $H^\infty$, then

$$f(J_{i\mathcal{E}}) = \begin{bmatrix} f(\alpha_i)I & 0 & \dots & 0 \\ \frac{f^{(1)}(\alpha_i)}{1!}I & f(\alpha_i)I & \dots & 0 \\ \vdots & \vdots & \vdots & \vdots \\ \frac{f^{(d_i-1)}(\alpha_i)}{(d_i-1)!}I & \frac{f^{(d_i-2)}(\alpha_i)}{(d_i-2)!}I & \dots & f(\alpha_i)I \end{bmatrix}. \tag{4.14}$$

Notice that $(I - zJ_{i\mathcal{E}})^{-1} = f(J_{i\mathcal{E}})$ where $f(\lambda)$ is the function given by $f(\lambda) = (I - z\lambda)^{-1}$. So using (4.14) we readily obtain for z in D

$$(I - zJ_{i\mathcal{E}})^{-1} = \begin{bmatrix} x(i,0)I & 0 & \dots & 0 \\ x(i,1)I & x(i,0)I & \dots & 0 \\ \vdots & \vdots & \vdots & \vdots \\ x(i,d_i-1)I & x(i,d_i-2)I & \dots & x(i,0)I \end{bmatrix}. \tag{4.15}$$

Finally, let $J_{\mathcal{E}}$ be the block Jordan matrix on $\ell_d^2(\mathcal{E})$ defined by

$$J_{\mathcal{E}} = \operatorname{diag}\left[J_{1\mathcal{E}}, J_{2\mathcal{E}}, \dots, J_{n\mathcal{E}}\right]. \tag{4.16}$$

Let $C_1$ be the block row matrix mapping $\ell_d^2(\mathcal{E})$ into $\mathcal{E}_1$ and $C_2$ be the block row matrix mapping $\ell_d^2(\mathcal{E})$ into $\mathcal{E}_2$ defined by

$$\begin{aligned} C_1 &= \left[Y_{1,d_1-1}, Y_{1,d_1-2}, \dots, Y_{1,0}, Y_{2,d_2-1}, \dots, Y_{n,d_n-1}, \dots, Y_{n,0}\right] \\ C_2 &= \left[U_{1,d_1-1}, U_{1,d_1-2}, \dots, U_{1,0}, U_{2,d_2-1}, \dots, U_{n,d_n-1}, \dots, U_{n,0}\right]. \end{aligned} \tag{4.17}$$

Notice that the block row matrices $C_1$ and $C_2$ are formed by $Y_{i,j}$ and $U_{i,j}$ where the j moves in decreasing order. Finally, let B be the block column matrix mapping $\mathcal{E}$ into $\ell_d(\mathcal{E})$ defined by

$$B = \begin{bmatrix} I, 0, \dots, 0, I, 0, \dots, 0, I, 0, \dots, 0 \end{bmatrix}^* \tag{4.18}$$

where the identity appears in the positions $1$, $d_1+1$, $d_1+d_2+1$, ..., $d_1+d_2+\dots+d_{n-1}+1$ and zeros everywhere else. Notice that the pair $(J_{\mathcal{E}}, B)$ is controllable. However, the pair $([C_1, C_2]^{tr}, J_{\mathcal{E}})$ may not be observable. Furthermore, by consulting (4.15) along with the definitions of $C_1$, $C_2$, B and $G_1$, $G_2$, we see that

$$\begin{bmatrix} G_1 \\ G_2 \end{bmatrix} = \begin{bmatrix} C_1 \\ C_2 \end{bmatrix} (I - zJ_{\mathcal{E}})^{-1} B . \tag{4.19}$$

In other words, $\{J_{\mathcal{E}}, B, [C_1, C_2]^{tr}\}$ is a stable, controllable realization of $[G_1, G_2]^{tr}$. This realization is not necessarily observable.

Let $\{a_i\}$ be a set of scalars and $\{B_i\}$ a set of operators mapping $\mathcal{K}_1$ into $\mathcal{K}_2$. Then a straight forward computation shows that

$$\begin{bmatrix} B_1, B_2, \dots, B_m \end{bmatrix} \begin{bmatrix} a_1 I & 0 & \dots & 0 \\ a_2 I & a_1 I & \dots & 0 \\ \vdots & \vdots & \vdots & \vdots \\ a_m I & a_{m-1} I & \dots & a_1 I \end{bmatrix} = \begin{bmatrix} a_m I, a_{m-1} I, \dots, a_1 I \end{bmatrix} \begin{bmatrix} B_m & 0 & \dots & 0 \\ B_{m-1} & B_m & \dots & 0 \\ \vdots & \vdots & \vdots & \vdots \\ B_1 & B_2 & \dots & B_m \end{bmatrix} . \tag{4.20}$$

Now let $L_{iY}$ from $\ell^2_{d_i}(\mathcal{E})$ to $\ell^2_{d_i}(\mathcal{E}_1)$ and $L_{iU}$ from $\ell^2_{d_i}(\mathcal{E})$ to $\ell^2_{d_i}(\mathcal{E}_2)$ be the lower triangular block Toeplitz matrices defined by

$$L_{iY} = \begin{bmatrix} Y_{i,0} & 0 & \dots & 0 \\ Y_{i,1} & Y_{i,0} & \dots & 0 \\ \vdots & \vdots & \vdots & \vdots \\ Y_{i,d_i-1} & Y_{i,d_i-2} & \dots & Y_{i,0} \end{bmatrix} \text{ and } L_{iU} = \begin{bmatrix} U_{i,0} & 0 & \dots & 0 \\ U_{i,1} & U_{i,0} & \dots & 0 \\ \vdots & \vdots & \vdots & \vdots \\ U_{i,d_i-1} & U_{i,d_i-2} & \dots & U_{i,0} \end{bmatrix} . \tag{4.21}$$

Let $L_Y$ from $\ell^2_d(\mathcal{E})$ into $\ell^2_d(\mathcal{E}_1)$ and $L_U$ from $\ell^2_d(\mathcal{E})$ into $\ell^2_d(\mathcal{E}_2)$ be the block lower triangular matrices defined by

$$L_Y = \operatorname{diag}\begin{bmatrix} L_{1Y}, L_{2Y}, \dots, L_{nY} \end{bmatrix} \text{ and } L_U = \operatorname{diag}\begin{bmatrix} L_{1U}, L_{2U}, \dots, L_{nU} \end{bmatrix} . \tag{4.22}$$

By applying (4.20) and using the definitions of $C_1$, $C_2$, $\Phi_1$ and $\Phi_2$ along with equation (4.15) we see that the observability operators $W_1$ from $\ell_d^2(\mathcal{E})$ into $H^2(\mathcal{E}_1)$ and $W_2$ from $\ell_d^2(\mathcal{E})$ into $H^2(\mathcal{E}_2)$ are given by

$$W_1 = C_1(I - zJ_{\mathcal{E}})^{-1} = \Phi_1 L_Y \quad \text{and} \quad W_2 = C_2(I - zJ_{\mathcal{E}})^{-1} = \Phi_2 L_U \,. \tag{4.23}$$

Let $G = \sum_0^\infty G_n z^n$ be in $H^\infty(\mathcal{E}_2, \mathcal{E}_1)$. and T a block stable matrix on $\ell_d^2(\mathcal{E}_2)$. The entries of T are operators on $\mathcal{E}_2$. By G(T) we mean the block matrix mapping $\ell_d^2(\mathcal{E}_2)$ into $\ell_d^2(\mathcal{E}_1)$ defined by

$$G(T) = \sum_{n=0}^\infty \operatorname{diag}\,[G_n, G_n, \dots, G_n]\, T^n$$

where $\operatorname{diag}\,[G_n, G_n, \dots, G_n]$ is the diagonal block matrix mapping $\ell_d^2(\mathcal{E}_2)$ into $\ell_d^2(\mathcal{E}_1)$. By using (4.14) we see that $G(J_{\mathcal{E}_2})$ is the lower triangular block matrix mapping $\ell_d^2(\mathcal{E}_2)$ into $\ell_d^2(\mathcal{E}_1)$ given by

$$G(J_{\mathcal{E}_2}) = \operatorname{diag}\left[G(J_{1\mathcal{E}_2}), G(J_{2\mathcal{E}_2}), \dots, G(J_{n\mathcal{E}_2})\right] \tag{4.24}$$

where $G(J_{i\mathcal{E}_2})$ is the lower triangular Toeplitz matrix mapping $\ell_{d_i}^2(\mathcal{E}_2)$ into $\ell_{d_i}(\mathcal{E}_1)$ defined by the $d_i \times d_i$ block matrix in (4.1). Because the functions x(i,j) are a generalized eigenvector chain for the backward shift (see equation (4.8)), it follows that $S_2^*\Phi_2 = \Phi_2 J_{\mathcal{E}_2}$.

Let F be the function in $H^\infty(\mathcal{E}_1, \mathcal{E}_2)$ defined by $F(z) = \tilde{G}(z)$. In other words, if F admits a power series expansion of the form $F(z) = \sum_{n=0}^\infty F_n z^n$ and $G(z) = \sum_{n=0}^\infty G_n z^n$, then $F_n^* = G_n$ for all n. Using the fact $\tilde{F}(z) = \sum_{n=0}^\infty F_n^* z^n$ along with $S_2^*\Phi_2 = \Phi_2 J_{\mathcal{E}_2}$ and (4.24) we have

$$F_+^*\Phi_2 = \sum_{n=0}^\infty F_n^* S_2^{*n}\Phi_2 = \Phi_2 \sum_{n=0}^\infty \operatorname{diag}\left[G_n, \dots, G_n\right] J_{\mathcal{E}_2}^n = \Phi_2 G(J_{\mathcal{E}_2})\,. \tag{4.25}$$

Notice that $F_n^*\Phi_2 = \Phi_2 \operatorname{diag}\,[G_n, \dots, G_n]$ because $\Phi_2$ is simply a block row matrix whose entries are scalars $x(i,j)\, I_2$. In particular, (4.25) along with the observability operator $W_2 = \Phi_2 L_U$ in (4.23) imply that

$$F_+^* W_2 = \Phi_2 G(J_{\mathcal{E}_2}) L_U . \tag{4.26}$$

Now recall that from (3.5) the Hankel operators $\Gamma(G_1) = W_1 W_c^*$ and $\Gamma(G_2) = W_2 W_c^*$. Because $\{J_{\mathcal{E}}, B, [C_1, C_2]^{tr}\}$ is a controllable realization of $[G_1, G_2]^{tr}$, the operator $W_c^*$ is onto. Thus, $F_+^* \Gamma_2 = \Gamma_1$ if and only if $F_+^* W_2 = W_1$. By (4.26) we see that F tangentially interpolates $\Gamma_2$ with $\Gamma_1$ if and only if

$$\Phi_2 G(J_{\mathcal{E}_2}) L_U = F_+^* W_2 = W_1 = \Phi_1 L_Y . \tag{4.27}$$

However, both $\Phi_1$ and $\Phi_2$ are one-to-one because they are block matrices consisting of generalized eigenvectors for the backward shift. Therefore, F tangentially interpolates $\Gamma_2$ with $\Gamma_1$ if and only if $G(J_{\mathcal{E}_2}) L_U = L_Y$, or equivalently, $G = \tilde{F}$ tangentially interpolates $\mathcal{U}$ with $\mathcal{Y}$. Summing up we obtain the following result.

**THEOREM 4.1.** *Let* $\mathcal{U} = \{U_{i,j} \ : \ 1 \le i \le n \ \text{ and } \ 0 \le j < d_i\}$ *be the set of operators mapping* $\mathcal{E}$ *into* $\mathcal{E}_2$ *and* $\mathcal{Y} = \{Y_{i,j} \ : \ 1 \le i \le n \ \text{ and } \ 0 \le j < d_i\}$ *be the set of operators mapping* $\mathcal{E}$ *into* $\mathcal{E}_1$. *Let* $G_1$ *and* $G_2$ *be the rational functions defined according to (4.12). Then a function* G *in* $H^\infty(\mathcal{E}_2, \mathcal{E}_1)$ *tangentially interpolates* $\mathcal{U}$ *with* $\mathcal{Y}$ *if and only if* $F = \tilde{G}$ *tangentially interpolates* $\Gamma(G_2)$ *with* $\Gamma(G_1)$. *Moreover, if a solution to this tangential interpolation problem exists, then* $\tilde{d}_\infty = d_\infty$ and $\tilde{d}_2 = d_2$.

Now let us give an explicit formula for the observability operators $Q_1$ and $Q_2$. According to (4.23) we have $W_1 = \Phi_1 L_Y$ and $W_2 = \Phi_2 L_U$. Thus

$$Q_1 = L_Y^* \Phi_1^* \Phi_1 L_Y \quad \text{and} \quad Q_2 = L_U^* \Phi_2^* \Phi_2 L_U . \tag{4.28}$$

Notice that we can easily compute $\Phi_1^* \Phi_1$ and $\Phi_2^* \Phi_2$ by using the derivative reproducing property of x(i,j) in (4.10); see also Section XI.4 in [10]. To see this let $x(r,q)I$ be the i-th column of $\Phi_1$ and $\Phi_2$ and $x(s,t)I$ be the j-th column of $\Phi_1$ and $\Phi_2$. Then by the derivative reproducing property in (4.10) the (i,j)-th entry of the gram matrices $\Phi_1^* \Phi_1$ and $\Phi_2^* \Phi_2$ are given by

$$\left[\Phi_1^*\Phi_1\right]_{i,j} = (\, x(s,t)\,,\, x(r,q)\,)\, I_1 = \frac{x^{(q)}(s,t)}{q!}(\bar{\alpha})\, I_1 \quad \text{and}$$
$$\left[\Phi_2^*\Phi_2\right]_{i,j} = \frac{x^{(q)}(s,t)}{q!}(\bar{\alpha})\, I_2\,. \tag{4.29}$$

where $1/\alpha$ is the pole of $x(r,q)$. So we can use (4.28) along with (4.29) to compute $Q_1$ and $Q_2$ without solving the corresponding Lyapunov equations

$$Q_1 = J_{\mathcal{E}}^* Q_1 J_{\mathcal{E}} + C_1^* C_1 \quad \text{and} \quad Q_2 = J_{\mathcal{E}}^* Q_2 J_{\mathcal{E}} + C_2^* C_2\,. \tag{4.30}$$

Finally, by consulting Corollary 3.6 along with the fact that G tangentially interpolates $\mathcal{U}$ with $\mathcal{Y}$ if and only if $\tilde{G}$ tangentially interpolates $\Gamma(G_2)$ with $\Gamma(G_1)$, we readily obtain the main result of the section.

**THEOREM 4.2.** *Let* $\mathcal{U} = \{U_{i,j} \;:\; 1 \le i \le n \;\text{ and }\; 0 \le j < d_i\}$ *be the set of operators mapping* $\mathcal{E}$ *into* $\mathcal{E}_2$ *and* $\mathcal{Y} = \{Y_{i,j} \;:\; 1 \le i \le n \;\text{ and }\; 0 \le j < d_i\}$ *be the set of operators mapping* $\mathcal{E}$ *into* $\mathcal{E}_1$. *Let* $Q_1$ *and* $Q_2$ *be the corresponding observability grammians computed according to (4.28) and (4.29). Then there exists a function* G *in* $H^\infty(\mathcal{E}_2, \mathcal{E}_1)$ *tangentially interpolating* $\mathcal{U}$ *with* $\mathcal{Y}$ *and satisfying* $\|G\|_\infty \le \gamma$ *if and only if* $Q_1 \le \gamma^2 Q_2$ *for some* $\gamma > 0$. *In this case,*

$$\tilde{d}_\infty^2 = \lambda_{max}\left[(V^*Q_2V)^{-1}V^*Q_1V\right] \quad \text{and} \quad \tilde{d}_2^2 = \text{trace}\left[C_1V(V^*Q_2V)^{-1}V^*C_1\right] \tag{4.31}$$

*where V is any isometry mapping* $\mathbb{C}^k$ *into* $\ell_d^2(\mathcal{E})$ *whose range equals the range of* $Q_2$ *(set* V=I *if* $Q_2$ *is invertible). Moreover, given any* $\delta > 1$, *then a function* G *in* $H^\infty(\mathcal{E}_2, \mathcal{E}_1)$ *tangentially interpolating* $\mathcal{U}$ *with* $\mathcal{Y}$ *and satisfying the* $H^2$–$H^\infty$ *bounds in (4.4) is given by the state space realization*

$$G(z) = \tilde{C}\,(I - z\tilde{M})^{-1}\,\tilde{B}\,, \tag{4.32}$$

*where* $\tilde{B}$ *mapping* $\mathcal{E}_2$ *into* $\mathbb{C}^k$, $\tilde{M}$ *on* $\mathbb{C}^k$ *and* $\tilde{C}$ *mapping* $\mathbb{C}^k$ *into* $\mathcal{E}_1$ *are the matrices defined by*

$$\tilde{M} = \left[V^*(\gamma^2Q_2 - Q_1 + C_1^*C_1)V\right]^{-1} V^*J_{\mathcal{E}}^*(\gamma^2Q_2 - Q_1)V$$
$$\tilde{B} = \left[V^*(\gamma^2Q_2 - Q_1 + C_1^*C_1)V\right]^{-1} V^*C_2^* \quad \text{and} \quad \tilde{C} = \gamma^2 C_1 V \tag{4.33}$$

*where* $\gamma = \delta\, d_\infty$, $C_1$ *mapping* $\ell_d^2(\mathcal{E})$ *into* $\mathcal{E}_1$ *and* $C_2$ *mapping* $\ell_d^2(\mathcal{E})$ *into* $\mathcal{E}_2$ *are the block row matrices are defined by (4.17). The function* G *in (4.32) can also be computed by*

$$G(z) = \tilde{Q}(z)^{-1}\,\tilde{P}(z) \tag{4.34}$$

*where* $\tilde{Q}(z)$ *and* $\tilde{P}(z)$ *are the analytic functions in* $\overline{D}$ *defined by*

$$\begin{aligned}\tilde{Q}(z) &= I + C_1 V(\gamma^2 V^* Q_2 V - V^* Q_1 V)^{-1} V^* L_Y^* \tilde{\Phi}_1(z) \qquad \text{and}\\ \tilde{P}(z) &= \gamma^2 C_1 V(\gamma^2 V^* Q_2 V - V^* Q_1 V)^{-1} V^* L_U^* \tilde{\Phi}_2(z)\,.\end{aligned} \tag{4.35}$$

*Finally, the function* $\tilde{Q}(z)^{-1}$ *is also analytic in* $\overline{D}$.

PROOF. To finish the proof, it remains to establish equation (4.35). According to Corollary 3.6 and $G(z) = \tilde{F}(z)$, we see that $G(z) = \tilde{Q}(z)^{-1}\tilde{P}(z)$ where $\tilde{P}(z)$ and $\tilde{Q}(z)$ are now given given by

$$\begin{aligned}\tilde{P}(z) &= \gamma^2 C_1 V(\gamma^2 V^* Q_2 V - V^* Q_1 V)^{-1} V^* (I - zJ_{\mathcal{E}}^*)^{-1} C_2^* \qquad \text{and}\\ \tilde{Q}(z) &= I + C_1 V(\gamma^2 V^* Q_2 V - V^* Q_1 V)^{-1} V^* (I - zJ_{\mathcal{E}}^*)^{-1} C_1^*\end{aligned} \tag{4.36}$$

However, by (4.23) we see that

$$L_Y^* \tilde{\Phi}_1(z) = (I - zJ_{\mathcal{E}}^*)^{-1} C_1^* \qquad \text{and} \qquad L_U^* \tilde{\Phi}_2(z) = (I - zJ_{\mathcal{E}}^*)^{-1} C_2^* \tag{4.37}$$

Substituting (4.37) into (4.36) readily yields (4.35). Obviously, $\tilde{Q}(z)^{-1}$ is analytic in $\overline{D}$ by Corollary 3.6. This completes the proof.

**REMARK 4.3.** According to Remark 3.7, our function G satisfies an even stronger $H^2$ bound than the one given in (4.4). To be precise, the G in (4.32) or (4.34) tangentially interpolating $\mathcal{U}$ with $\mathcal{Y}$, also satisfies $\|G\|_\infty \le \delta\, d_\infty$ and

$$\|G\|_2^2 \le \frac{\tilde{d}_2^2(\delta^2 \tilde{d}_\infty^2 - \tilde{d}_2^2/k)}{\tilde{d}_\infty^2\,(\delta^2 - 1)}$$

where k is the dimension of $\mathcal{E}_1$.

Let us emphasize how our procedure can be applied to the special $H^2$–$H^\infty$ tangential interpolation problems described earlier in this section.

• For the tangential Carathéodory setting ($\alpha_1 = 0$, $d = d_1$ and $n = 1$), the previous theorem is particularly simple to use. In this case, $\mathcal{U} = \{U_j = U_{1,j} \ : \ 0 \le j < d\}$ and $\mathcal{Y} = \{Y_j = Y_{1,j} \ : \ 0 \le j < d\}$. The row operators $C_1$ and $C_2$ now simplify to

$$C_1 = \begin{bmatrix} Y_{d-1} , Y_{d-2} , \dots , Y_0 \end{bmatrix} \quad \text{and} \quad C_2 = \begin{bmatrix} U_{d-1} , U_{d-2} , \dots , U_0 \end{bmatrix} . \tag{4.38}$$

Moreover, the operator $J_{\mathcal{E}}$ is now the lower shift block matrix on $\ell_d^2(\mathcal{E})$, that is

$$J_{\mathcal{E}} = \begin{bmatrix} 0 & 0 & \dots & 0 & 0 \\ I & 0 & \dots & 0 & 0 \\ 0 & I & \dots & 0 & 0 \\ \vdots & \vdots & \vdots & \vdots & \vdots \\ 0 & 0 & \dots & I & 0 \end{bmatrix} . \tag{4.39}$$

The functions $\Phi_1$ and $\Phi_2$ become

$$\Phi_1 = \begin{bmatrix} z^{d-1} I_1 , z^{d-2} I_1 , \dots , I_1 \end{bmatrix} \quad \text{and} \quad \Phi_2 = \begin{bmatrix} z^{d-1} I_2 , z^{d-2} I_2 , \dots , I_2 \end{bmatrix} . \tag{4.40}$$

The observability operators $W_1 = \Phi_1 L_Y$ and $W_2 = \Phi_2 L_U$ are given by

$$W_1 = \begin{bmatrix} z^{d-1} I_1 , z^{d-2} I_1 , \dots , I_1 \end{bmatrix} L_Y \quad \text{and} \quad W_2 = \begin{bmatrix} z^{d-1} I_2 , z^{d-2} I_2 , \dots , I_2 \end{bmatrix} L_U \tag{4.41}$$

where $L_U = L_{1U}$ and $L_Y = L_{1Y}$ are the lower triangular Toeplitz matrices generated by $\{U_j = U_{1,j} \ : \ 0 \le j < d\}$ and $\{Y_j = Y_{1,j} \ : \ 0 \le j < d\}$ respectively; see equation (4.21). In this case the operators $\Phi_1$ and $\Phi_2$ are isometries. Therefore, for the tangential Carathédory interpolation problem the observability grammians $Q_1$ and $Q_2$ are given by

$$Q_1 = W_1^* W_1 = L_Y^* L_Y \qquad \text{and} \qquad Q_2 = W_2^* W_2 = L_U^* L_U . \tag{4.42}$$

So in the tangential Carathéodory setting, one simply applies Theorem 4.2 to the operators $C_1$, $C_2$, $J_{\mathcal{E}}$, $\Phi_1$, $\Phi_2$, $Q_1$ and $Q_2$ defined in (4.38), (4.39), (4.40) and (4.42).

• In the tangential Nevanlinna-Pick setting, $d_i = 1$ for all $i$. In this case $\mathcal{U} = \{U_i = U_{i,1} \ : \ 1 \le i \le n\}$ and $\mathcal{Y} = \{Y_i = Y_{i,1} \ : \ 1 \le i \le n\}$ and the block row operators $C_1$ and $C_2$ simplify to

$$C_1 = \begin{bmatrix} Y_1 , Y_2 , \dots , Y_n \end{bmatrix} \quad \text{and} \quad C_2 = \begin{bmatrix} U_1 , U_2 , \dots , U_n \end{bmatrix}. \tag{4.43}$$

The operators $W_1 = \Phi_1 L_Y$ and $W_2 = \Phi_2 L_U$ become

$$\begin{aligned} W_1 &= \begin{bmatrix} (1-\alpha_1 z)^{-1} Y_1 , (1-\alpha_2 z)^{-1} Y_2 , \dots , (1-\alpha_n z)^{-1} Y_n \end{bmatrix} \quad \text{and} \\ W_2 &= \begin{bmatrix} (1-\alpha_1 z)^{-1} U_1 , (1-\alpha_2 z)^{-1} U_2 , \dots , (1-\alpha_n z)^{-1} U_n \end{bmatrix} \end{aligned} \tag{4.44}$$

Using the fact that

$$( x(j,0) , x(i,0) ) = ( (1-\alpha_j z)^{-1} , (1-\alpha_i z)^{-1} ) = (1-\bar{\alpha}_i \alpha_j)^{-1} \tag{4.45}$$

it readily follows from (4.44) that the observability grammians $Q_1$ and $Q_2$ are given by

$$Q_1 = \begin{bmatrix} \frac{Y_1^* Y_1}{1-\bar{\alpha}_1 \alpha_1} & \cdots & \frac{Y_1^* Y_n}{1-\bar{\alpha}_1 \alpha_n} \\ \vdots & \vdots & \vdots \\ \frac{Y_n^* Y_1}{1-\bar{\alpha}_n \alpha_1} & \cdots & \frac{Y_n^* Y_n}{1-\bar{\alpha}_n \alpha_n} \end{bmatrix} \quad \text{and} \quad Q_2 = \begin{bmatrix} \frac{U_1^* U_1}{1-\bar{\alpha}_1 \alpha_1} & \cdots & \frac{U_1^* U_n}{1-\bar{\alpha}_1 \alpha_n} \\ \vdots & \vdots & \vdots \\ \frac{U_n^* U_1}{1-\bar{\alpha}_n \alpha_1} & \cdots & \frac{U_n^* U_n}{1-\bar{\alpha}_n \alpha_n} \end{bmatrix}. \tag{4.46}$$

Finally, to use formula (4.34) notice that

$$\begin{aligned} L_U^* \tilde{\Phi}_2(z) &= \begin{bmatrix} (1-\alpha_1 \bar{z})^{-1} U_1 , \dots , (1-\alpha_n \bar{z})^{-1} U_n \end{bmatrix}^* \quad \text{and} \\ L_Y^* \tilde{\Phi}_1(z) &= \begin{bmatrix} (1-\alpha_1 \bar{z})^{-1} Y_1 , \dots , (1-\alpha_n \bar{z})^{-1} Y_n \end{bmatrix}^* \end{aligned} \tag{4.47}$$

So in the tangential Nevanlinna-Pick interpolation problem, one simply applies Theorem 4.2 to $J_{\mathcal{E}} = \mathrm{diag}\,(\alpha_1 I , \alpha_2 I , \dots , \alpha_n I)$ along with the operators $C_1$, $C_2$, $Q_1$ and $Q_2$ in (4.43) and (4.46).

## REFERENCES

[1] J. A. Ball, I. Gohberg and Rodman, *Interpolation for Rational Matrix Functions*, Birkhauser-Verlag, Basel, 1990.

[2] J. A. Ball and J. W. Helton, A Beurling-Lax theorem for the Lie group U(m, n) which contains most classical interpolation theory, *J. Operator Theory,* **9** (1983), pp. 107-142.

[3] D. S. Bernstein and W. M. Haddad, LQG control with an $H_\infty$ performance bound: A Riccati equation approach, *IEEE Trans on Automat. Contr.*, **34**, No. 3, (1989), pp. 293-305.

[4] Ph. Delsarte, Y. Genin and Y. Kamp, The Nevanlinna-Pick problem for matrix-valued functions, *SIAM J. Appl. Math.*, **36**, (1979), pp. 47-61.

[5] J. C. Doyle, B. A. Frances, A. Tannenbaum, *Feedback Control Theory,* MacMillan, New York, 1991.

[6] J. C. Doyle, K. Glover, P. P. Khargonekar, and B. A. Francis, State-space solutions to standard $H_2$ and $H_\infty$ and control problems, *IEEE Trans. on Automat. Contr.*, **34**, No. 8, (1989), pp. 831-847.

[7] H. Dym, I. Gohberg, A Maximum Entropy Principle for Contractive Interpolants, *J. Functional Analysis* **65**, (1986), pp. 83-125.

[8] I. P. Fedcina, A Criterion for the solvability of the Nevanlinna-Pick tangent problem, *Mat. Issled.*, **7** (1972), pp. 213-227.

[9] I. P. Fedcina, The tangential Nevanlinna-Pick problem with multiple points, *Akad. Nauk Armjan. S. S. Dokl.*, **61** (1975), pp. 214-218.

[10] C. Foias and A. E. Frazho, *The Commutant Lifting Approach to Interpolation Problems,* Operator Theory Advances and Applications, **44**, Birkhauser-Verlag, Basel, 1990.

[11] C. Foias and A. E. Frazho, Commutant Lifting and Simultaneous $H^\infty$ and $L^2$ Suboptimization, *SIAM J. Math. Anal.*, **23**, No. 4, (1992), pp. 984-994.

[12] C. Foias, A. E. Frazho, and W. S. Li, The Exact $H^2$ Estimate for the Central $H^\infty$ Interpolant, in preparation.

[13] C. Foias, A. E. Frazho and A. Tannenbaum, On Combined $H^\infty$–$H^2$ Suboptimal Interpolants, Submitted for publication.

[14] B. A. Francis, *A Course in $H^\infty$ Control Theory,* Lecture Notes in Control and Information Sciences, Springer-Verlag, New York, 1987.

[15] A. E. Frazho, A Commutant Lifting Theorem Approach to Tangential Interpolation, *J. Mathematical Analysis and Applications,* **162** (1991), pp.159-176.

[16] A. E. Frazho and S. M. Kherat, Applications of the Schur Algorithm to Mixed $H^2$–$H^\infty$ Nehari Problems, *J. Mathematical Analysis and Applications,* To appear.

[17] A. E. Frazho and M. A. Rotea, A Remark on Mixed $L^2 / L^\infty$ Bounds, *Integr. Equat. Oper. Theory,* **15**, (1992), pp. 343-348.

[18] P. A. Fuhrmann, *Linear Systems and Operators in Hilbert Space,* McGraw-Hill, New York, 1981.

[19] K. Glover, All optimal Hankel-norm approximations of linear multivariable systems and their $L_\infty$-error bounds, *Int. J. Control,* **39** (1984), pp. 1115-1193.

[20] K. Glover and D. Mustafa, Derivation of the maximum entropy $H_\infty$-controller and a state-space formula for its entropy, *Int. J. Control,* **50**, No. 3, (1989), pp. 899-916.

[21] I. Gohberg, M. A. Kaashoek and Van Schagen, Rational Contractive and Unitary Interpolants in Realized Form, *Integral Equations and Operator Theory* **11**, (1988), pp. 105-127.

[22] I. Gohberg, M. A. Kaashoek, H. J. Woerdeman, The Band Method for Positive and Strictly Contractive Extension Problems: An Alternative Version and New Applications, *Integral Equations and Operator Theory* **12**, (1989), pp. 343-382.

[23] G. H. Golub and C. F. Van Loan, *Matrix Computations,* The Johns Hopkins University Press, Baltimore, second edition, 1989.

[24] Y. S. Hung, $H^\infty$ interpolation of rational matrices, *Internat. J. Control,* **48**, (1988), pp. 1659-1713.

[25] S. Y. Hwang, Minimum Uncorrelated Unit Noise in State Space Digital Filtering, *IEEE Trans. Acoustics, Speech and signal proc.,* **ASSP-25**, No. 4, (1977), pp. 273-281.

[26] V. Kaftal, D. Larson, and G. Weiss, Quasitriangular subalgebras of semifinite Von Neumann algebras are closed, *J. Functional Analysis,* **107** No. 2 (1992), pp. 387-401.

[27] T. Kailath, A Theorem of I. Schur and its impact on modern signal processing, I. Schur methods in Operator theory and Signal Processing, pp. 9-30 in *Operator Theory: Advances and Applications,* Ed. I. Gohberg, 1986.

[28] T. Kailath, *Linear Systems,* Prentice-Hall, New Jersey, 1980.

[29] P. P. Khargonekar and M. A. Rotea, Mixed $H^2/H^\infty$ Control: A Convex Optimization Approach, *IEEE trans. Autom. Control,* **36** ,No 7, (1991), pp. 824-837.

[30] H. Kimura, Conjugation, interpolation and model matching in $H^\infty$. *Int. J. Control,* **49** (1989), pp. 269-307.

[31] H. Kimura, Directional interpolation approach to $H^\infty$-optimization and robust stablization, *IEEE Trans. Automatic Control,* **32** (1987), pp. 1085-1093.

[32] H. Kimura, Directional interpolation in the state space, *Systems and Control Letters,* **10** (1988), pp. 317-324.

[33] D. J. N. Limebeer and B. D. O. Anderson, An interpolation theory approach to $H^\infty$ controller degree bounds, *Linear Algebra and its Applications,* **98** (1988), pp. 347-386.

[34] D. J. N. Limebeer and G. D. Halikias, A Controller Degree Bound for $H^\infty$-Optimal Control Problems of the Second Kind, *SIAM J. Control and Optimization,* **26,** No. 3, (1988), pp. 646-677.

[35] D. J. N. Limebeer and Y. S. Hung, An Analysis of the pole-zero cancellations in $H^\infty$-optimal control problems of the first kind, *SIAM J. Control and Optimization,* **25** (1987), pp. 1457-1493.

[36] D. Mustafa, Relations between maximum-entropy/$H_\infty$ control and combined $H_\infty$/LQG control, *Syst. Contr. Lett.,* **12**, No. 3, (1989), pp. 193-203.

[37] D. Mustafa and K. Glover, *Minimum Entropy* $H_\infty$ *Control, Lecture notes in Control and Information Sciences,* Springer-Verlag, New York, 1990.

[38] K. M. Nagpal and P. P. Khargonekar, Filtering and Smoothing in an $H^\infty$ Setting, *IEEE Trans., Autom., Control* **36** (1991), pp. 152-166.

[39] M. Rosenblum and J. Rovnyak, An operator-theoretic approach to theorems of the Pick-Nevanlinna and Loewner types I, *Integral Equations and Operator Theory,* **3** (1980), pp. 408-436.

[40] M. Rosenblum and J. Rovnyak, An operator-theoretic approach to theorems of the Pick-Nevanlinna and Loewner types II, *Integral Equations and Operator Theory,* **5** (1982), pp. 870-887.

[41] M. Rosenblum and J. Rovnyak, *Hardy Classes and Operator Theory,* Oxford University Press, New York, 1985.

[42] M. Rotea and A. E. Frazho, Bounds on Solutions to $H^\infty$ Algebraic Riccati Equations and $H^2$ properties of $H^\infty$ Central Solution, *Systems and Control Letters,* **19** ,No 5 (1992), to appear.

[43] M. A. Rotea and P. P. Khargonekar, $H_2$-optimal control with an $H_\infty$-constraint: the state-feedback case, *Automatica,* **27**, No. 2, (1991), pp. 307-316.

[44] W. J. Rugh, *Nonlinear System Theory: The Volterra/Wiener Approach,* Johns Hopkins University Press, 1981.

[45] A. H. Sayed, T. Kailath, H. Lev-Ari and T. Constantinescu, Efficient Recursive Solutions of Rational Interpolation Problems, *Preprint.*

[46] B. Sz.-Nagy and C. Foias, Dilation des commutants d' opérateurs, *C. R. Acad. Sci. Paris, série A,* **266** (1968), pp. 493-495.

[47] B. Sz.-Nagy and C. Foias, *Harmonic Analysis of Operators on Hilbert Space,* North-Holland Publishing Co., Amsterdam, 1970.

[48] B. Sz.-Nagy and A. Korani, Relations d'in problème de Nevanlinna et Pick avec la théorie des opérateurs de l'espace Hilbertien, *Acta Sci. Math.,* **7** (1956), pp. 295-302.

[49] D. Williamson, *Digital Control and Implementation, Finite Wordlength Considerations,* Prentice Hall, 1991.

[50] K. Zhou and P. P. Khargonekar, An algebraic Riccati equation approach to $H_\infty$ optimization, *Systems and Control Letters,* **11** (1988), pp. 85-92.

School of Aeronautics and Astronautics, Purdue University
West Lafayette, IN 47907, U.S.A.

**MSC 1992 :** Primary, 47A20, 47A57, 93B36.

Operator Theory:
Advances and Applications, Vol. 64




# ON A COMPLETION PROBLEM FOR MATRICES

Israel Gohberg and Caixing Gu

In this paper the following completion problem for square matrices is solved and analyzed. Given a square matrix with unidentified last row, except the diagonal entry. Describe the invariant polynomials of all possible completions of this matrix.

## 1 Introduction

Let $H$ and $K$ be two Banach spaces and $L(H,K)$ be the space of linear bounded operators acting from $H$ into $K$. In this paper we consider the problem of describing the possible spectra and invariants of operators of the form

$$M_X = M_X(A,B,C) = \begin{bmatrix} A & B \\ X & C \end{bmatrix}, \quad X \in L(H,K), \tag{1}$$

where $A \in L(H,H)$, $B \in L(K,H)$ and $C \in L(K,K)$ are given operators and $X$ can be any operator in $L(H,K)$. This problem is motivated by [1] and [2] and also by other spectral completion problems (see [7], [9], [10], [14] and references therein).

We will use here the approach suggested in [1], [6] and [9] which consists firstly of a classification problem with the aim of simplifying the given data with the help of a class of triangular similarities of the following form

$$\begin{bmatrix} L_{11} & 0 \\ L_{21} & L_{22} \end{bmatrix} \begin{bmatrix} A & B \\ X & C \end{bmatrix} \begin{bmatrix} L_{11} & 0 \\ L_{21} & L_{22} \end{bmatrix}^{-1}$$

$$= \begin{bmatrix} L_{11}AL_{11}^{-1} - L_{11}BL_{22}^{-1}L_{21}L_{11}^{-1} & L_{11}BL_{22}^{-1} \\ (L_{21}A + L_{22}X)L_{11}^{-1} - (L_{21}B + L_{22}C)L_{22}^{-1}L_{21}L_{11}^{-1} & (L_{21}B + L_{22}C)L_{22}^{-1} \end{bmatrix};$$

The second step is to study the spectra and spectral problems for all possible completions, in particular the problem of computing the minimal spectral radius for all completions.

In this paper we will restrict ourselves to the case when $K$ is one-dimensional. The paper consists of four sections. In Section 2 we state the main theorems when $H$ is finite dimensional. In Section 3 we assume $(A, B)$ is a full range (controllable) pair and give simple and direct proofs of the results of Section 3 in this special case. Section 4 is basically devoted to the proofs of the results stated in Section 2. Finally if $H$ is infinite dimensional, by applying the results from Section 3, we compute in Section 5 the minimal spectral radius for all completions $M_X$ for the important case of a full range pair $(A, B)$.

The authors thank Ciprian Foias and Sorin Rubinstein for several valuable remarks and suggestions.

# 2 Main theorems in the finite dimensional case

Throughout Sections 2, 3 and 4 we keep the notations used in the introduction and assume $H$ is a finite dimensional vector space over the complex field and $dim\, K = 1$. Let $n$ denote the dimension of $H$. We begin with the definition of upper equivalence of two matrices.

**DEFINITION**. The matrices $M_X(A, B, C)$ and $M_Y(A', B', C')$ are called upper equivalent if there exists a $2 \times 2$ block lower triangular invertible matrix

$$L = \begin{bmatrix} L_{11} & 0 \\ L_{21} & L_{22} \end{bmatrix} \tag{2}$$

such that $LM_X(A, B, C)L^{-1}$ and $M_Y(A', B', C')$ have the same block upper triangular part. We call such $L$ a lower similarity of $M_X(A, B, C)$ and $M_Y(A', B', C')$. This definition is influenced by [1], [6] and [9].

Since the block upper triangular part of $LM_X(A, B, C)L^{-1}$ does not depend on $X$, this is an equivalence relation and divides the $2 \times 2$ block matrices into classes of equivalence. Given $(A, B, C)$, in this section we find a complete set of invariants and a simple *canonical form* for the matrices which belong to the class of upper equivalence of $M_X(A, B, C)$.

In order to do this, first recall that two matrix polynomials $G_1(\lambda)$, $G_2(\lambda)$ are called *equivalent* if there exist two matrix polynomials $E(\lambda)$ and $F(\lambda)$ with nonzero constant determinants, such that

$$E(\lambda)G_1(\lambda)F(\lambda) = G_2(\lambda). \tag{3}$$

By Smith's theorem, any matrix polynomial $G(\lambda)$ of size $l \times k$ (for our purpose we assume here that $l \leq k$) is equivalent to the canonical form (Smith form), i.e, there exist two matrix polynomials $E(\lambda)$ and $F(\lambda)$ with their determinant nonzero constants, such that

$$E(\lambda)G_1(\lambda)F(\lambda) = \begin{bmatrix} \delta_1(\lambda) & \cdots & 0 & \cdots & 0 \\ \vdots & \ddots & \vdots & \vdots & \vdots \\ 0 & \cdots & \delta_l(\lambda) & \cdots & 0 \end{bmatrix},$$

where $\delta_i(\lambda)$, $i = 1, \cdots, l$, are monic plynomials such that $\delta_i(\lambda)$ is divisible by $\delta_{i-1}(\lambda)$ for $i = 2, \cdots, l$. The polynomials $\delta_i(\lambda)$, $i = 1, \cdots, l$, are called the invariant polynomials of $G(\lambda)$ and are uniquely determined by $G(\lambda)$; indeed, up to a constant, $\delta_i(\lambda)$ is equal to the $d_i(\lambda)/d_{i-1}(\lambda)$, where $d_i(\lambda)$ denotes the least common divisor of all *ith* order minors and $d_0(\lambda) = 1$ (see [8], p.314); moreover, two matrix polynomials are equivalent if and only if they have the same set of invariant polynomials.

Next recall that two matrix pencils $\lambda[I\ 0] - [A\ B]$ and $\lambda[I\ 0] - [A'\ B']$ are called to be *strictly equivalent* if there exist invertible operators $S \in L(H, H)$ and $T \in L(H \oplus K, H \oplus K)$ such that

$$S(\lambda[I\ 0] - [A\ B])T = \lambda[I\ 0] - [A'\ B']. \tag{4}$$

Since $K$ is one dimensional, in fact $\lambda[I\ 0] - [A\ B]$ and $\lambda[I\ 0] - [A'\ B']$ are *strictly equivalent* if and only if they are equivalent (see [6]). Hence $\lambda[I\ 0] - [A\ B]$ and $\lambda[I\ 0] - [A'\ B']$ are *strictly equivalent* if and only if they have the same set of invariant polynomials .

**THEOREM 1.** *Let $(A, B, C)$ and $(A', B', C',)$ be given. Then the matrices $M_X(A, B, C)$ and $M_Y(A', B', C'')$ are upper equivalent if and only if $\lambda[I\ 0] - [A\ B]$ and $\lambda[I\ 0] - [A'\ B']$ have the same set of invariant polynomials and $trA + C = trA' + C'$.*

Now we can write down a simple representative for each class of upper equivalence.

**THEOREM 2.** *Let $(A, B, C)$ be given. Let $\alpha_1 | \alpha_2 | \cdots | \alpha_n$ be the invariant polynomials of $\lambda[I\ 0] - [A\ B]$. Let*

$$\alpha_1 \cdots \alpha_n = (\lambda - \lambda_1)^{s_1} \cdots (\lambda - \lambda_m)^{s_m}, \ s = s_1 + \cdots + s_m < n, \tag{5}$$

*and*

$$\beta = \frac{trA + C - (s_1\lambda_1 + \cdots + s_m\lambda_m)}{(n+1) - (s_1 + \cdots + s_m)}. \tag{6}$$

*Let $\alpha_i = 1, i = 1, \cdots, l$, and the degree of $\alpha_{l+1}$ is larger than 1. Then*

$$s = n - rank \begin{bmatrix} B & AB & \cdots & A^{n-1}B \end{bmatrix},$$

*and $M_X$ is upper equivalent to the following canonical form*

$$\begin{bmatrix} A_0 & B_0 \\ 0 & \beta \end{bmatrix}, \tag{7}$$

*with*

(i) $A_0 = diag(A_1, A_2)$, $B_0 = \begin{bmatrix} B_1 \\ 0 \end{bmatrix}$, *and* $B_1$, $A_1$ *and* $A_2$ *are matrices of sizes* $(n-s)\times 1$, $(n-s)\times(n-s)$ *and* $s \times s$, *respectively,*

$$(ii)\ A_1 = \begin{bmatrix} \beta & 1 & & 0 \\ & \ddots & \ddots & \\ & & \ddots & 1 \\ 0 & & & \beta \end{bmatrix}, \quad B_1 = \begin{bmatrix} 0 \\ \vdots \\ 0 \\ 1 \end{bmatrix},$$

*and*

(iii) $A_2 = diag(R_{l+1}, \cdots, R_n)$, *where* $R_i$ *is the companion matrix of the invariant factor* $\alpha_i$, $i = l+1, \cdots, n$.

**THEOREM 3.** *Let* $(A, B, C)$ *be given. Let* $\alpha_1 \,|\, \alpha_2 \,|\, \cdots \,|\, \alpha_n$ *be the invariant polynomials of* $\lambda[I\ 0] - [A\ B]$. *Let*

$$\alpha_1 \cdots \alpha_n = (\lambda - \lambda_1)^{s_1} \cdots (\lambda - \lambda_m)^{s_m}, \; s = s_1 + \cdots + s_m < n,$$

*and*

$$\beta = \frac{trA + C - (s_1\lambda_1 + \cdots + s_m\lambda_m)}{(n+1) - (s_1 + \cdots + s_m)}.$$

*Let* $\Delta$ *denote a set of* $n+1$ *(not necessarily different) complex numbers. Then there exists* $X$ *such that* $\sigma(M_X) = \Delta$ *if and only if*

$$\Delta = G(A, B) \cup \{\beta_1, \cdots, \beta_{n+1-s}\},$$

*where* $G(A, B) = \{\lambda_1, \cdots, \lambda_1, \cdots, \lambda_m, \cdots, \lambda_m\}$ *with* $\lambda_i$ $(1 \le i \le m)$ *repeating* $s_i$ *times, and* $\beta_1, \cdots, \beta_{n+1-s}$ *complex numbers such that* $\beta_1 + \cdots + \beta_{n+1-s} = (n+1-s)\beta$.

As an immediate consequence of the above theorem, we have the following formula for computing the minimal spectral radius for all completions.

**THEOREM 4.** *Let* $(A, B, C)$ *be given and let* $\alpha_1 \,|\, \alpha_2 \,|\, \cdots \,|\, \alpha_n$ *be the invariant polynomials of* $\lambda[I\ 0] - [A\ B]$. *Then*

$$r_{min} = \min_{X \in L(H,K)} r(M_X) = \max\{|\lambda_1|, \cdots, |\lambda_m|, \left|\frac{trA + C - (s_1\lambda_1 + \cdots + s_m\lambda_m)}{(n+1) - (s_1 + \cdots + s_m)}\right|\}. \quad (8)$$

*where*

$$\alpha_1 \cdots \alpha_n = (\lambda - \lambda_1)^{s_1} \cdots (\lambda - \lambda_m)^{s_m}.$$

We remark that $\lambda_i, i = 1, \cdots, m$ in the above theorem, are in $\sigma(A)$, the spectrum of $A$.

# 3 The full range case

First recall that a pair $(A, B)$ is said to be full range if the matrix $\begin{bmatrix} B & AB & \cdots & A^{n-1}B \end{bmatrix}$ is right invertible. Then by a classical result [12], a pair $(A, B)$ is full range if and only if all the invariant polynomials of $\lambda[I\ 0] - [A\ B]$ are equal to the constant 1. Therefore the following result follows from Theorem 1 and 2.

**THEOREM 5.** *Let $(A, B, C)$ and $(A', B', C')$ be given. Assume the pair $(A, B)$ is full range. Then $M_X(A, B, C)$ and $M_Y(A', B', C')$ are upper equivalent if and only if $(A', B')$ is full range and $trA + C = trA' + C'$. Moreover $M_X(A, B, C)$ is upper equivalent to the following $(n+1) \times (n+1)$ matrix*

$$\begin{bmatrix} \beta & 1 & & 0 \\ & \ddots & \ddots & \\ & & \ddots & 1 \\ 0 & & & \beta \end{bmatrix}, \tag{9}$$

*with $\beta = (trA + C)/(n+1)$.*

**PROOF.** Here we will present a direct proof of this theorem (in which we do not refer to results from system theory). For completeness we will reproduce some well known arguments from system theory. If $M_X(A, B, C)$ and $M_Y(A', B', C')$ are upper equivalent, it is obvious that $trA + C = trA' + C'$. To show that the pair $(A', B')$ is full range, we note that

$$\begin{bmatrix} L_{11} & 0 \\ L_{21} & L_{22} \end{bmatrix} \begin{bmatrix} A & B \\ X & C \end{bmatrix} \begin{bmatrix} L_{11} & 0 \\ L_{21} & L_{22} \end{bmatrix}^{-1}$$

$$= \begin{bmatrix} L_{11}AL_{11}^{-1} - L_{11}BL_{22}^{-1}L_{21}L_{11}^{-1} & L_{11}BL_{22}^{-1} \\ (L_{21}A + L_{22}X)L_{11}^{-1} & (L_{21}B + L_{22}C)L_{22}^{-1}L_{21}L_{11}^{-1} & (L_{21}B + L_{22}C)L_{22}^{-1} \end{bmatrix} = \begin{bmatrix} A' & B' \\ Y & C' \end{bmatrix}$$

implies that

$$A' = N^{-1}(A + BF)N, \quad B' = N^{-1}BM,$$

where $N = L_{11}^{-1}, F = -L_{22}^{-1}L_{21}, M = L_{22}$. Therefore the matrix

$$\begin{bmatrix} B' & A'B' & \cdots & (A')^{n-1}B' \end{bmatrix}$$

$$= N^{-1} \begin{bmatrix} B & AB + s_1B & \cdots & A^{n-1}B + s_1A^{n-2}B + \cdots + s_{n-1}B \end{bmatrix} diag(M, \cdots, M)$$

for some complex numbers $s_1, \cdots, s_{n-1}$, is right invertible; thus $(A', B')$ is full range.

Next we prove that if $(A, B)$ is full range, $M_X(A, B, C)$ is upper equivalent to the matrix (9). This also proves the converse part of the first statement in the theorem. To do this, we will construct a lower similarity $L$ in the form of (2). Let $\delta_0, \cdots, \delta_{n-1}$ be such that

$A^nB+\sum_{i=0}^{n-1}\delta_i A^iB=0$. We first define $T=[T_{ij}]_{i,j=1,2}$ as follows: $T_{22}=1$, $T_{21}=[\delta_0\cdots\delta_{n-1}]$ and $T_{11}=[q_n,\cdots q_1]$ with $q_1=B$ and $q_j=Aq_{j-1}+\delta_{n+1-j}B$ for $j=2,\cdots,n$. By a direct computation we have

$$\begin{bmatrix} A & B \\ 0 & C \end{bmatrix}\begin{bmatrix} q_n & \cdots & q_1 & 0 \\ \delta_0 & \cdots & \delta_{n-1} & 1 \end{bmatrix}=\begin{bmatrix} 0 & q_n & \cdots & q_1 \\ \delta_0 C & \cdots & \delta_{n-1}C & C \end{bmatrix}$$

$$=\begin{bmatrix} q_n & \cdots & q_1 & 0 \\ \delta_0 & \cdots & \delta_{n-1} & 1 \end{bmatrix}\begin{bmatrix} 0 & 1 & & 0 \\ & \ddots & \ddots & \\ & & \ddots & 1 \\ \delta_0 C & \cdots & -\delta_{n-2}+\delta_{n-1}C & -\delta_{n-1}+C \end{bmatrix}$$

Note that $-\delta_{n-1}=trA$. Therefore

$$T^{-1}\begin{bmatrix} A & B \\ X & C \end{bmatrix}T=\begin{bmatrix} 0 & 1 & & 0 \\ & \ddots & \ddots & \\ & & \ddots & 1 \\ y_1 & \cdots & y_n & trA+C \end{bmatrix}, \tag{10}$$

for some $y=[y_1\cdots y_n]$. Next we define the lower triangular $(n+1)\times(n+1)$ matrix $S=[S_{ij}]_{i,j=1.n+1}$ as follows:

$$S_{ij}=\binom{i-1}{j-1}\beta^{i-j},\ 1\le j\le i, i=1,\cdots,n+1,$$

$$S_{ij}=0,\ j>i,\ i=1,\cdots,n+1.$$

By direct computation we can check that $S$ is a lower similarity of the matrix on the right hand side of (10) and the matrix (9). Now let $L=TS$, then we obtain that $L^{-1}$ is a lower similarity of $M_X(A,B,C)$ and the matrix (9). ⊣

**COROLLARY 6.** *Let $(A,B,C)$ be give with the pair $(A,B)$ of full range and let $\{\beta_1,\cdots,\beta_{n+1}\}$ denote a set of $n+1$ (not necessarily different) complex numbers. Then there exists an operator $X\in L(H,K)$ such that $\sigma(M_X(A,B,C))=\{\beta_1,\cdots,\beta_{n+1}\}$ if and only if $\beta_1+\cdots+\beta_{n+1}=trA+C$.*

**PROOF.** By the proof of Theorem 5, we may assume that $M_X(A,B,C)$ has the following form

$$\begin{bmatrix} 0 & 1 & & 0 \\ & \ddots & \ddots & \\ & & \ddots & 1 \\ x_1 & \cdots & x_n & trA+C \end{bmatrix}.$$

It is well known that the characteristic polynomial of $M_X$ is

$$\lambda^{n+1} - (trA + C)\lambda^n - x_n\lambda^{n-1} + \cdots - x_1.$$

Hence $\{\beta_1, \cdots, \beta_{n+1}\}$ is the set of the eigenvalues of $M_X$ for some choice of $X$ if and only if $\beta_1 + \cdots + \beta_{n+1} = trA + C$. ⊣

Using Corollary 6 we can give a quick proof of the following result in [3].

**COROLLARY 7.** (Lemma A1.2 [3]) *Let $t_{12}, \cdots, t_{n-1n}$ be $n-1$ complex numbers different from zero. Let*

$$T = \begin{bmatrix} 0 & t_{12} & \cdots & t_{1n} \\ x_1 & 0 & \ddots & \vdots \\ \vdots & 0 & \ddots & t_{n-1n} \\ x_{n-1} & 0 & \cdots & 0 \end{bmatrix}$$

*Then for any monic polynomial $p(\lambda) = \lambda^n + a_{n-2}\lambda^{n-2} + \cdots + a_0$, there exist complex numbers $x_1, \cdots, x_{n-1}$ such that the characteristic polynomial of $T$ is equal to $p(\lambda)$.*

**PROOF.** Let $T'$ denote the transpose of $T$. Note that

$$QT'Q = \begin{bmatrix} A & B \\ X & 0 \end{bmatrix},$$

where

$$A = \begin{bmatrix} 0 & t_{n-1n} & \cdots & t_{2n} \\ 0 & 0 & \cdots & t_{2n-1} \\ \vdots & \vdots & \cdots & \vdots \\ 0 & 0 & \cdots & t_{23} \\ 0 & 0 & \cdots & 0 \end{bmatrix}, \quad B = \begin{bmatrix} t_{1n} \\ \vdots \\ t_{12} \end{bmatrix}, \quad X = \begin{bmatrix} x_{n-1} & \cdots & x_1 \end{bmatrix}$$

and $Q = [q_{ij}]_{n\times n}$ with $q_{in+1-i} = 1$ for $i = 1, \cdots, n$ and all other entries are equal to zero, and $QT'Q$ and $T$ have the same characteristic polynomial. Since $trA = 0$ and $C = 0$, by Corollary 6, it is enough to prove that the pair $(A, B)$ is full range. Straightforward computation shows that the matrix

$$Q\begin{bmatrix} B & AB & \cdots & A^{n-2}B \end{bmatrix} = \begin{bmatrix} t_{12} & & & \\ & t_{23}t_{12} & & \\ \vdots & & \ddots & \\ & & & t_{n-1n}\cdots t_{12} \end{bmatrix}$$

is lower triangular and invertible. The proof is complete. ⊣

The following result is an immediate consequence of Corollary 6.

**COROLLARY 8.** *Let $(A, B, C)$ be given, and assume that the pair $(A, B)$ is full range. Then*

$$\min_{X\in L(H,K)} r(M_X(A, B, C)) = \left|\frac{trA + C}{n+1}\right|. \quad (11)$$

# 4 The proof of the main theorems in the finite dimensional case

**PROOF OF THEOREM 1.** Assume there exist invertible operators $S$ and $T$ such that

$$S(\lambda[I\ 0]-[A\ B])T=\lambda[I\ 0]-[A'\ B'], \tag{12}$$

we write

$$T=\begin{bmatrix} T_{11} & T_{12} \\ T_{21} & T_{22} \end{bmatrix},$$

where the matrix $T_{11}$ is of size $n\times n$. Then

$$S(\lambda I-A)T_{11}-SBT_{21}=\lambda I-A'.$$

Hence $T_{11}=S^{-1}$ and $SAT_{11}+SBT_{21}=A'$. Equation (12) also implies that

$$S(\lambda I-A)T_{12}-SBT_{22}=-B',$$

thus $T_{12}=0$, $SBT_{22}=B'$.

It follows from the the condition $trA+C=trA'+C'$ and the above analysis that $T$ is $2\times 2$ block lower triangular and $T^{-1}M_X(A,B,C)T$ and $M_Y(A',B',C')$ have the same upper block triangular part. That is $M_X(A,B,C)$ is upper equivalent to $M_Y(A',B',C')$.

Conversely, if $M_X(A,B,C)$ is upper equivalent to $M_Y(A',B',C')$, let $L$ be a lower similarity of $M_X(A,B,C)$ and $M_Y(A',B',C')$; we write $L$ as in (2). Then it is easy to verify that (12) holds with $S=L_{11}^{-1}$ and $T=L$. This completes the proof.

**PROOF OF THEOREM 2.** By Theorem 1 we need only check that the invariant polynomials of $\lambda[I\ 0]-[A_0\ B_0]$ are $\alpha_1\,|\,\alpha_2\,|\cdots|\,\alpha_n$, and $trA_0+\beta=trA+C$. First we check that $trA_0+\beta=(n+1-s)\beta+\sum_{i=1}^m s_i\lambda_i=trA+C$ by the definition of $\beta$. Next we note that

$$\lambda[I\ 0]-[A_0\ B_0]\left(=\begin{bmatrix} \lambda I-A_1 & 0 & B_1 \\ 0 & \lambda I-A_2 & 0 \end{bmatrix}\right)$$

is equivalent to

$$\begin{bmatrix} \lambda I-A_1 & B_1 & 0 \\ 0 & 0 & \lambda I-A_2 \end{bmatrix}.$$

It is well known that the invariant polynomials of $\lambda[I\ 0]-[A_1\ B_1]$ are $\alpha_i=1$, $i=1,\cdots,n-s$ and the invariant polynomials of $\lambda I-A_2$ are $\rho_j=1$ $(j=1,2,...,l-n+s)$, and $\alpha_i$ $(i=l+1,\cdots n)$. Let $E_1(\lambda)$ and $F_1(\lambda)$ be such that $E_1(\lambda)(\lambda[I\ 0]-[A_1\ B_1])F_1(\lambda)$ is equal to the Smith canonical form of $\lambda[I\ 0]-[A_1\ B_1]$. Also $E_2(\lambda)$ and $F_2(\lambda)$ be such that $E_1(\lambda)(\lambda I-$

$A_2)F_1(\lambda)$ is equal to the Smith canonical form of $\lambda I - A_2$. Then $diag(E_1(\lambda), E_2(\lambda))$ and $diag(E_1(\lambda), E_2(\lambda))$ are two matrix polynomials with nonzero constant determinants and

$$\begin{bmatrix} E_1(\lambda) & 0 \\ 0 & E_2(\lambda) \end{bmatrix} \begin{bmatrix} [\lambda I - A_1 B_1] & 0 \\ 0 & \lambda I - A_2 \end{bmatrix} \begin{bmatrix} F_1(\lambda) & 0 \\ 0 & F_2(\lambda) \end{bmatrix}$$

$$= \begin{bmatrix} \alpha_1(\lambda) & \cdots & 0 & 0 \\ \vdots & \ddots & \vdots & \vdots \\ 0 & \cdots & \alpha_n(\lambda) & 0 \end{bmatrix}.$$

This completes the proof of Theorem 2.

**PROOF OF THEOREM 3.** We note that if $M_X(A, B, C)$ and $M_Y(A', B', C')$ are upper equivalent, then for a given $(n+1) \times (n+1)$ matrix $G$ there exists a $X$ such that $M_X(A, B, C)$ is similar to $G$ if and only if there exists a $Y$ such that $M_Y(A', B', C')$ is similar to $G$. Hence by Theorem 2 we can assume that $M_X$ has the following form,

$$M_X = \begin{bmatrix} A_1 & 0 & B_1 \\ 0 & A_2 & 0 \\ X_1 & X_2 & \beta \end{bmatrix}$$

where $\beta$ and $A_1, A_2$ are given by (6) and (7). It is clear that the characteristic polynomial of $M_X$ can be written as

$$\det(\lambda I - M_X) = \det(\lambda I - A_2) \det\left(\lambda I - \begin{bmatrix} A_1 & B_1 \\ X_1 & \beta \end{bmatrix}\right).$$

Now the final result follows from the definition of $A_2$ and Corollary 6, since the pair $(A_1, B_1)$ is of full range. The proof is complete.

Next let us describe all $X_0$ such that the minimum of $r(M_X)$ is reached at $X = X_0$. Let $(A, B, C)$ be given. Let $\alpha_1 \mid \alpha_2 \mid \cdots \mid \alpha_n$ be invariant polynomials of $\lambda[I\ 0] - [A\ B]$. Let

$$\alpha_1 \cdots \alpha_n = (\lambda - \lambda_1)^{s_1} \cdots (\lambda - \lambda_m)^{s_m}, \; s = s_1 + \cdots + s_m < n$$

and it is clear that for $r_{min} = \min_{X \in L(H,K)} r(M_X(A, B, C))$ the following equality holds

$$r_{min} = \max\{|\lambda_1|, \cdots, |\lambda_m|, \left|\frac{trA + C - (s_1\lambda_1 + \cdots + s_m\lambda_m)}{(n+1) - (s_1 + \cdots + s_m)}\right|\}.$$

Let $P[\lambda]$ denote the set of all monic polynomials of degree $k = (n+1) - s$ with the absolute values of its roots less than or equal to $r_{min}$ and the sum of its roots ( counting multiplicities) equal to $trA + C - (s_1\lambda_1 + \cdots + s_m\lambda_m)$. Then by Theorem 3, the number $r(M_X)$ reaches the minimum $r_{min}$ if and only if the characteristic polynomial of $M_X$ is equal to $\alpha_1 \cdots \alpha_n p(\lambda)$ for some $p(\lambda) \in P[\lambda]$. Next we express the above condition in terms of an equation in

$X = \begin{bmatrix} x_1 & \cdots & x_n \end{bmatrix}$. Let $p_i$ $(i = 1, \cdots, n+1)$, be the minor of order $n$ of $\lambda[I\ 0] - [A\ B]$ obtained by deleting the $i$-th column. Note that

$$det \begin{bmatrix} \lambda I - A & -B \\ -X & \lambda - C \end{bmatrix} = p_1(\lambda)(-1)^{n-1}x_1 + \cdots + p_n(\lambda)x_n + p_{n+1}(\lambda)(\lambda - C). \tag{13}$$

Hence the number $r(M_X)$ reaches the minimum $r_{min}$ if and only if $X$ satisfies the equation

$$p_1(\lambda)(-1)^{n-1}x_1 + \cdots + p_n(\lambda)x_n + p_{n+1}(\lambda)(\lambda - C) = \alpha_1 \cdots \alpha_n p(\lambda) \tag{14}$$

for some $p(\lambda) \in P[\lambda]$. To write the above equation (14) as a linear system, we note that $\alpha_1 \cdots \alpha_n$ is the greatest common divisor of $p_1(\lambda), \cdots, p_n(\lambda)$, $p_{n+1}(\lambda)$ (see [8], p.314). Hence we can write

$$p_i(\lambda) = \alpha_1 \cdots \alpha_n q_i(\lambda) = \alpha_1 \cdots \alpha_n \sum_{j=1}^{k-1} q_{ij}\lambda^j \quad i = 1, \cdots, n,$$

and

$$p_{n+1}(\lambda)(\lambda - C) = \alpha_1 \cdots \alpha_n \sum_{j=1}^{k} q_{0j}\lambda^j.$$

Also any $p(\lambda) \in P[\lambda]$ can be presented in the form

$$p(\lambda) = (\lambda - \beta_1) \cdots (\lambda - \beta_k) = \sum_{j=1}^{k} (-1)^j \sigma_j(\beta_1, \cdots, \beta_k)\lambda^{k-j},$$

where $\sigma_j(\beta_1, \cdots, \beta_k)$ is the $jth$ elementary symmetric polynomial in $\beta_1, \cdots, \beta_k$. Comparing coefficients of powers of $\lambda$ in (14), we see that (14) is equivalent to the following linear system,

$$\begin{bmatrix} q_{1k-1} & \cdots & q_{nk-1} \\ \vdots & \vdots & \vdots \\ q_{10} & \cdots & q_{n0} \end{bmatrix} \begin{bmatrix} (-1)^{n-1}x_1 \\ \vdots \\ x_n \end{bmatrix} = \begin{bmatrix} -\sigma_1(\beta_1, \cdots, \beta_k) \\ \vdots \\ (-1)^k \sigma_k(\beta_1, \cdots, \beta_k) \end{bmatrix} - \begin{bmatrix} q_{0k-1} \\ \vdots \\ q_{00} \end{bmatrix}. \tag{15}$$

Thus we have proved the following theorem.

**THEOREM 9.** *The set of all* $X = \begin{bmatrix} x_1 & \cdots & x_n \end{bmatrix}$ *such that* $r(M_X)$ *reaches the minimum* $r_{min}$ *consists of all* $X = \begin{bmatrix} x_1 & \cdots & x_n \end{bmatrix}$ *which satisfy (15) for some* $\beta_1, \cdots, \beta_k$ *such that* $|\beta_1|, \cdots, |\beta_k| \leq r_{min}$ *and* $\beta_1 + \cdots + \beta_k = trA + C - (s_1\lambda_1 + \cdots + s_m\lambda_m)$.

# 5 Infinite dimensional case

Let $A \in B(H)$, $B \in B(K, H)$ and $C \in L(K, K)$. In this section we deal with the case when $H$ is infinite dimensional and $dim\ K = 1$. In this case we are able only to compute the minimal spectral radius for all completions of $M_X(A,B,C)$ under the assumption that the pair $(A,B)$ is full range (see below for the definition).

Since $K$ is assumed to be of dimension one, for convenience we shall identify the operator $B$ with the vector $B \in H$. We say that a pair $(A,B)$ has full range if $B$ is a cyclic vector of $A$, i.e. the closed linear span of $\{A^iB, i = 0,1,2,\cdots\}$ is the whole space $H$. See also [5].

**LEMMA 10.** *Let $A_1, B_1, A_2$ and $B_2$ be matrices of sizes $n\times n$, $n\times 1$, $m\times m$ and $m\times 1$, respectively. Let*

$$A_3 = \begin{bmatrix} A_1 & 0 \\ 0 & A_2 \end{bmatrix}, \quad B_3 = \begin{bmatrix} B_1 \\ B_2 \end{bmatrix}.$$

*Then the pair $(A_3, B_3)$ is full range if and only if the following (i)-(ii) hold.*

*(i) $(A_1, B_1)$ and $(A_2, B_2)$ are full range.*

*(ii) The set $\sigma(A_1)\cap\sigma(A_2)$ is empty.*

*Here $\sigma(A_1)$, $\sigma(A_2)$ denote the spectrum of $A_1$ and $A_2$, respectively.*

**PROOF.** For sufficiency, we need to prove that $B_3, A_3B_3, \cdots, A_3^{n+m-1}B_3$ are linearly independent. Suppose this is not so. Let $p(\lambda)$ be a polynomial of a degree less than or equal to $n+m-1$ such that $p(A_3)B_3 = 0$. That is

$$p(A_1)B_1 = 0, \quad p(A_2)B_2 = 0.$$

It follows from the above that

$$p(A_1)A_1^iB_1 = 0,\ p(A_2)A_2^iB_2 = 0,\ i = 0,1,2,\cdots$$

Since $B_1$ and $B_2$ are cyclic vectors of $A_1$ and $A_2$, respectively, then $p(A_1) = 0$ and $p(A_2) = 0$. Let $p_1(\lambda)$ and $p_2(\lambda)$ be the minimal polynomials of $A_1$ and $A_2$, respectively ($p_1(\lambda)$ and $p_2(\lambda)$ are in fact the characteristic polynomials of $A_1$ and $A_2$, respectively, because of (i)). Then $p(\lambda)$ is divisible by $p_1(\lambda)p_2(\lambda)$, which is a contradiction since the degree of $p_1(\lambda)p_2(\lambda)$ $(= n+m)$ is bigger than the degree of $p(\lambda)$.

For necessity, let $(A_3, B_3)$ be full range, then it is easy to see that the minimal polynomial of $A_3$ is equal to the characteristic polynomial of $A_3$. Hence (ii) holds. Since $A_3^iB_3 = \begin{bmatrix} A_1^iB_1 \\ A_2^iB_2 \end{bmatrix}$ and the linear span of $B_3, A_3B_3, \cdots, A_3^{n+m-1}B_3$ is of dimension $n+m$, (i) holds. ⊣

**THEOREM 11.** *Let $A \in B(H)$, $B \in L(K,H)$. Assume $(A,B)$ is full range, and $C$ is a fixed complex number. Then*

$$r_{inf}(A,B,C) := \inf_{x \in B(H,K)} r(M_X(A,B,C)) = r_e(A) \tag{16}$$

*where $r_e(A)$ is the essential spectral radius of $A$.*

**PROOF.** Let $B(0, r_e(A))$ denote the closed disk of radius $r_e(A)$ with center at origin. We divide the proof into two cases.

**Case 1**: $A$ has an infinite number of eigenvalues outside $B(0, r_e(A))$.

Now for each $m$, Let

$$\sigma_m = \{\lambda : |\lambda| \geq r_e(A) + \frac{1}{m}\} = \{\lambda_1, \cdots, \lambda_k\}.$$

Using Riesz decomposition, we have

$$H = H_1 \oplus H_2, \quad A = \begin{bmatrix} A_1 & E \\ 0 & A_2 \end{bmatrix}$$

where $H_2$ is the range of Riesz projection $E(\sigma_m, A)$. Let $n_i$ be the dimension of the range of $E(\{\lambda_i\}, A)$, $i = 1, \cdots, k$. Then $n_m := dim H_2 = n_1 + \cdots + n_k$. Let $B_i$ be the orthogonal projection of $B$ onto $H_i$ $(i = 1,2)$. We claim that $(A_2, B_2)$ is a full range pair. Note that

$$A^i B = f_i + A_2^i B_2$$

for some $f_i \in H_1$. Therefore the closed linear span of $\{A^i B, i = 0, 1, 2, \cdots\}$ is contained in $H_1$ plus the linear span of $\{A_2^i B_2, i = 0, 1, 2, \cdots\}$. By the assumption that $(A,B)$ is full range, the pair $(A_2, B_2)$ is full range. Hence by Corollary 8 there exists $X_2 \in B(H_2, K)$ such that

$$r(M_X(A_2, B_2, C)) = |\frac{trA_2 + C}{n_m + 1}| = |\frac{n_1\lambda_1 + \cdots + n_k\lambda_k + C}{n_m + 1}|$$

we define $X \in B(H,K)$ to be

$$X(h_1 + h_2) = X_2 h_2, \ h_1 \in H_1, h_2 \in H_2$$

That is $M_X$ has the following form

$$M_X = \begin{bmatrix} A_1 & E & B_1 \\ 0 & A_2 & B_2 \\ 0 & X_2 & C \end{bmatrix}$$

Therefore

$$r_{inf}(A,B,C) \leq r(M_X) = \max\{r(A_1), \frac{n_1|\lambda_1| + \cdots + n_k|\lambda_k| + |C|}{n_m + 1}\} \tag{17}$$

Now let $m \to \infty$, we see that the right side of the above inequality approaches $r_e(A)$. The other direction in (16) is straightforward since for any $x$, the spectrum of $M_X$ contains the essential spectrum of $A$.

**Case 2**: $A$ has only a finite number of eigenvalues outside $B(0, r_e(A))$.

As in case 1, let

$$\sigma_f = \{\lambda : |\lambda| \geq r_e(A) + \frac{1}{m}\} = \{\lambda_1, \cdots, \lambda_k\}.$$

By using Riesz decomposition, we have

$$H = H_1 \oplus H_2, \quad A = \begin{bmatrix} A_1 & E \\ 0 & A_2 \end{bmatrix}$$

where $H_2$ is the range of the Riesz idempotent $E(\sigma_f, A)$. Let $n_2 = dim H_2 \geq k$. Now fix any $\epsilon > 0, \epsilon < dist(\sigma_f, \sigma_e(A))$, where $dist$ denotes the distance between two sets. Note that $(A_2, B_2)$ is full range, where $B_2$ is the projection of $B$ onto $H_2$.

For $A_1$, let $\lambda$ be in the essential spectrum of $A_1$ with $|\lambda| = r_e(A)$, we claim that there exists a sequence of orthogonal unit vectors $\{e_n\}_{n=1}^{\infty}$ such that

$$||(A^* - \overline{\lambda})e_n|| < \frac{\epsilon}{8 \times 2^n}, \quad (B, e_n) \neq 0, \quad n = 1, 2, \cdots \tag{18}$$

Since $\lambda$ is in the essential spectrum of $A_1$, there exists a sequence of orthogonal unit vectors $e_n$ such that $||(A^* - \overline{\lambda})e_n|| \to 0$ as $n$ tends to $\infty$. If the number of vectors $e_n$ such that $(B, e_n) \neq 0$ is infinite, then we pass to this subsequence of $\{e_n\}_{n=1}^{\infty}$. If there is only a finite number of vectors $e_n$ such that $(B, e_n) \neq 0$, then there exists a $N$ such that $(B, e_n) = 0$ for $n \geq N$. Consider a new sequence $e'_n$ defined by

$$e'_n = \sqrt{(1 - \frac{\epsilon}{8 \times 2^n})} e_{N+n} + \frac{\epsilon}{8 \times 2^n} B, \; n = 1, 2, \cdots.$$

In this case by passing to the subsequence of $\{e'_n\}_{n=1}^{\infty}$, still denoted by $\{e_n\}_{n=1}^{\infty}$, we have that $\{e_n\}_{n=1}^{\infty}$ satisfies (18).

Now let $n$ be such that

$$|\frac{\sum_{i=1}^{n}(\lambda + \frac{\epsilon}{8 \times 2^i}) + \sum_{i=1}^{n_2} \lambda_i}{n + n_2}| < r_e(A) + \frac{\epsilon}{8} \tag{19}$$

where $\{\lambda_1, \cdots, \lambda_{n_2}\}$ are the eigenvalues of $A_2$ with multiplicities. Let $\beta_i = \lambda + \frac{\epsilon}{8 \times 2^i}$, $i = 1, 2, \cdots, n$ and $H'_1$ be the linear span of $\{e_i, i = 1, 2, \cdots, n\}$. We define $A'_1$ on $H_1 = H_0 \oplus H'_1$ by

$$A_1'^* e_i = \overline{\beta}_i e_i \; i = 1, 2, \cdots, n, \quad A_1'^* e = A^* e, \text{ if } e \in H_0 \tag{20}$$

For $A_2$, let $A'_2$ be such that $||A_2 - A'_2|| < \frac{\epsilon}{8}$, $A_2$ has $n_2$ distinct eigenvalues $\lambda'_1, \cdots, \lambda'_{n_2}$ and $(A'_2, B_2)$ is full range.

Next let $H = H_0 \oplus H_1' \oplus H_2$, and

$$A' = \begin{bmatrix} A_1' & E \\ 0 & A_2' \end{bmatrix} = \begin{bmatrix} A_0 & E_1 & E_2 \\ 0 & A_1'' & E_3 \\ 0 & 0 & A_2' \end{bmatrix}, \quad B = \begin{bmatrix} B_0 \\ B_1' \\ B_2 \end{bmatrix}.$$

By (20) it is clear that $A_1''$ is a diagonal matrix with diagonal elements equal to $\beta_1, \cdots, \beta_n$ and $(A_1'', B_1')$ is full range. Also $||A' - A|| < \frac{\epsilon}{4}$.

Now let $m$ be an integer such that $||\frac{E_3}{m}|| < \frac{\epsilon}{4}$. Note that for an invertible operator $S$

$$\begin{bmatrix} S & 0 \\ 0 & I \end{bmatrix} \begin{bmatrix} A' & B \\ X & C \end{bmatrix} \begin{bmatrix} S^{-1} & 0 \\ 0 & I \end{bmatrix} = \begin{bmatrix} SA'S^{-1} & SB \\ XS^{-1} & C \end{bmatrix}.$$

Therefore $r_{inf}(A', B, C) = r_{inf}(SA'S^{-1}, SB, C)$.For

$$S = \begin{bmatrix} I & 0 & 0 \\ 0 & m^{-1}I & 0 \\ 0 & 0 & I \end{bmatrix},$$

we have

$$SA'S^{-1} = \begin{bmatrix} A_0 & mE_1 & E \\ 0 & A_1'' & m^{-1}E_3 \\ 0 & 0 & A_2 \end{bmatrix}, \quad SB = \begin{bmatrix} B_0 \\ m^{-1}B_1' \\ B_2 \end{bmatrix}.$$

Let

$$A'' = \begin{bmatrix} A_0 & mE_1 & E \\ 0 & A_1'' & 0 \\ 0 & 0 & A_2' \end{bmatrix},$$

By Lemma 8 we see that the pair $\left(\begin{bmatrix} A_1'' & 0 \\ 0 & A_2' \end{bmatrix}, \begin{bmatrix} m^{-1}B_1' \\ B_2 \end{bmatrix}\right)$ is full range and also $||A'' - SA'S^{-1}|| < \frac{\epsilon}{4}$. But for $(A'', SB, C)$, as in (11) we have

$$r_{inf}(A'', SB, C) \leq \max\{r(A_0), |\frac{\sum_{i=1}^{n}(\lambda + \frac{\epsilon}{8\times 2^i}) + \sum_{i=1}^{n_2} \lambda_i'}{n + n_2}|\}$$

$$< r_e(A) + \frac{\epsilon}{4}$$

Hence

$$r_{inf}(A, B, C) \leq r_{inf}(A', B, C) + \frac{\epsilon}{4} = r_{inf}(SA'S^{-1}, SB, C) + \frac{\epsilon}{4}$$

$$\leq r_{inf}(A'', SB, C) + \frac{\epsilon}{4} + \frac{\epsilon}{4} \leq r_e(A) + \epsilon$$

By letting $\epsilon \to 0$, we finish the proof. ⊣

## REFERENCES

[1] J. A. Ball, I. Gohberg, L. Rodman and T. Shalom, On the eigenvalues of matrices with given upper triangular part, *Integral Equations Operator Theory* **13** (1990) 488-497.

[2] H. Bercovici, C. Foias and A. Tannenbaum, A spectral commutant lifting theorem, *Trans. Amer. Math. Soc.* **325** (1991) 741-763.

[3] C. Apostol, L. A. Fialkow, D. A. Herrero and D. Voiculescu, Approximation of Hilbert space operators, vol 2, Research Notes in Math., Vol 102, 1984.

[4] C. Davis, W. M. Kahan and H. F. Weinberger, Norm-preserving dilations and their applications to optimal error bounds, *SIAM J. Num. Anal.* **19** (1982) 445-469.

[5] A. Feintuch, On single input controllability for infinite dimensional linear systems, *J. Math. Anal. Appl.* **62** (1978) 538-546.

[6] I. Gohberg, M. A. Kaashoek and F. van Schagen, Similarity of operator blocks and canonical form, I. General results, feedback equivalence and Kronecker indices, *Integral Equations Operator Theory* **3** (1980) 350-396.

[7] I. Gohberg, M. A. Kaashoek and F. van Schagen, Eigenvalues of completions of submatrices, *Lin. Multilin. Alg.* **25** (1989) 55-70.

[8] I. Gohberg, P. Lancaster and L. Rodman, *Matrix Polynomials.* Academic Press, New York, 1982.

[9] I. Gohberg and S. Rubinstein, A classification of upper equivalent matrices the generic case, *Integral Equations Operator Theory* **14** (1991) 533-544.

[10] L. Gurvits, L. Rodman and T. Shalom, Controllability and completion of partial upper triangular matrices over rings, *Lin. Alg. Appl.* **172** (1992) 135-149.

[11] S. Parrott, On the quotient norm and Sz.-Nagy-Foias lifting theorem, *J. Funct. Anal.* **30** (1978) 311-328.

[12] H. H. Rosenbrock, State space and multivariable theory, Thomas Nelson, London, 1970.

[13] I. Zaballa, Matrices with prescribed rows and invariant factors, *Lin. Alg. Appl.* **87** (1987) 113-146.

[14] I. Zaballa, Matrices with prescribed invariant factors, *Lin. Multilin. Alg.* **27** (1990) 325-343.

Israel Gohberg
School of Mathematical Sciences
Raymond and Beverly Sackler
Faculty of Exact Sciences
Tel-Aviv University, Ramat Aviv, Israel.

Caixing Gu
Department of Mathematics
Indiana University Bloomington
Bloomington, IN 47405, U.S.A.

1991 *Mathematics Subject Classification*: 47A10, 15A09.

Titles previously published in the series

**OPERATOR THEORY: ADVANCES AND APPLICATIONS**
BIRKHÄUSER VERLAG

3. **K. Clancey, I. Gohberg:** Factorization of Matrix Functions and Singular Integral Operators, 1981, (3-7643-1297-1)
4. **I. Gohberg** (Ed.): Toeplitz Centennial, 1982, (3-7643-1333-1)
5. **H.G. Kaper, C.G. Lekkerkerker, J. Hejtmanek:** Spectral Methods in Linear Transport Theory, 1982, (3-7643-1372-2)
6. **C. Apostol, R.G. Douglas, B. Sz-Nagy, D. Voiculescu, Gr. Arsene** (Eds.): Invariant Subspaces and Other Topics, 1982, (3-7643-1360-9)
7. **M.G. Krein:** Topics in Differential and Integral Equations and Operator Theory, 1983, (3-7643-1517-2)
8. **I. Gohberg, P. Lancaster, L. Rodman:** Matrices and Indefinite Scalar Products, 1983, (3-7643-1527-X)
9. **H. Baumgärtel, M. Wollenberg:** Mathematical Scattering Theory, 1983, (3-7643-1519-9)
10. **D. Xia:** Spectral Theory of Hyponormal Operators, 1983, (3-7643-1541-5)
11. **C. Apostol, C.M. Pearcy, B. Sz.-Nagy, D. Voiculescu, Gr. Arsene** (Eds.): Dilation Theory, Toeplitz Operators and Other Topics, 1983, (3-7643-1516-4)
12. **H. Dym, I. Gohberg** (Eds.): Topics in Operator Theory Systems and Networks, 1984, (3-7643-1550-4)
13. **G. Heinig, K. Rost:** Algebraic Methods for Toeplitz-like Matrices and Operators, 1984, (3-7643-1643-8)
14. **H. Helson, B. Sz.-Nagy, F.-H. Vasilescu, D.Voiculescu, Gr. Arsene** (Eds.): Spectral Theory of Linear Operators and Related Topics, 1984, (3-7643-1642-X)
15. **H. Baumgärtel:** Analytic Perturbation Theory for Matrices and Operators, 1984, (3-7643-1664-0)
16. **H. König:** Eigenvalue Distribution of Compact Operators, 1986, (3-7643-1755-8)
17. **R.G. Douglas, C.M. Pearcy, B. Sz.-Nagy, F.-H. Vasilescu, D. Voiculescu, Gr. Arsene** (Eds.): Advances in Invariant Subspaces and Other Results of Operator Theory, 1986, (3-7643-1763-9)
18. **I. Gohberg** (Ed.): I. Schur Methods in Operator Theory and Signal Processing, 1986, (3-7643-1776-0)
19. **H. Bart, I. Gohberg, M.A. Kaashoek** (Eds.): Operator Theory and Systems, 1986, (3-7643-1783-3)
20. **D. Amir:** Isometric characterization of Inner Product Spaces, 1986, (3-7643-1774-4)

21. **I. Gohberg, M.A. Kaashoek** (Eds.): Constructive Methods of Wiener-Hopf Factorization, 1986, (3-7643-1826-0)
22. **V.A. Marchenko:** Sturm-Liouville Operators and Applications, 1986, (3-7643-1794-9)
23. **W. Greenberg, C. van der Mee, V. Protopopescu:** Boundary Value Problems in Abstract Kinetic Theory, 1987, (3-7643-1765-5)
24. **H. Helson, B. Sz.-Nagy, F.-H. Vasilescu, D. Voiculescu, Gr. Arsene** (Eds.): Operators in Indefinite Metric Spaces, Scattering Theory and Other Topics, 1987, (3-7643-1843-0)
25. **G.S. Litvinchuk, I.M. Spitkovskii:** Factorization of Measurable Matrix Functions, 1987, (3-7643-1883-X)
26. **N.Y. Krupnik:** Banach Algebras with Symbol and Singular Integral Operators, 1987, (3-7643-1836-8)
27. **A. Bultheel:** Laurent Series and their Pade Approximation, 1987, (3-7643-1940-2)
28. **H. Helson, C.M. Pearcy, F.-H. Vasilescu, D. Voiculescu, Gr. Arsene** (Eds.): Special Classes of Linear Operators and Other Topics, 1988, (3-7643-1970-4)
29. **I. Gohberg** (Ed.): Topics in Operator Theory and Interpolation, 1988, (3-7634-1960-7)
30. **Yu.I. Lyubich:** Introduction to the Theory of Banach Representations of Groups, 1988, (3-7643-2207-1)
31. **E.M. Polishchuk:** Continual Means and Boundary Value Problems in Function Spaces, 1988, (3-7643-2217-9)
32. **I. Gohberg** (Ed.): Topics in Operator Theory. Constantin Apostol Memorial Issue, 1988, (3-7643-2232-2)
33. **I. Gohberg** (Ed.): Topics in Interplation Theory of Rational Matrix-Valued Functions, 1988, (3-7643-2233-0)
34. **I. Gohberg** (Ed.): Orthogonal Matrix-Valued Polynomials and Applications, 1988, (3-7643-2242-X)
35. **I. Gohberg, J.W. Helton, L. Rodman** (Eds.): Contributions to Operator Theory and its Applications, 1988, (3-7643-2221-7)
36. **G.R. Belitskii, Yu.I. Lyubich:** Matrix Norms and their Applications, 1988, (3-7643-2220-9)
37. **K. Schmüdgen:** Unbounded Operator Algebras and Representation Theory, 1990, (3-7643-2321-3)
38. **L. Rodman:** An Introduction to Operator Polynomials, 1989, (3-7643-2324-8)
39. **M. Martin, M. Putinar:** Lectures on Hyponormal Operators, 1989, (3-7643-2329-9)
40. **H. Dym, S. Goldberg, P. Lancaster, M.A. Kaashoek** (Eds.): The Gohberg Anniversary Collection, Volume I, 1989, (3-7643-2307-8)
41. **H. Dym, S. Goldberg, P. Lancaster, M.A. Kaashoek** (Eds.): The Gohberg Anniversary Collection, Volume II, 1989, (3-7643-2308-6)
42. **N.K. Nikolskii** (Ed.): Toeplitz Operators and Spectral Function Theory, 1989, (3-7643-2344-2)
43. **H. Helson, B. Sz.-Nagy, F.-H. Vasilescu, Gr. Arsene** (Eds.): Linear Operators in Function Spaces, 1990, (3-7643-2343-4)

44. **C. Foias, A. Frazho:** The Commutant Lifting Approach to Interpolation Problems, 1990, (3-7643-2461-9)
45. **J.A. Ball, I. Gohberg, L. Rodman:** Interpolation of Rational Matrix Functions, 1990, (3-7643-2476-7)
46. **P. Exner, H. Neidhardt** (Eds.): Order, Disorder and Chaos in Quantum Systems, 1990, (3-7643-2492-9)
47. **I. Gohberg** (Ed.): Extension and Interpolation of Linear Operators and Matrix Functions, 1990, (3-7643-2530-5)
48. **L. de Branges, I. Gohberg, J. Rovnyak** (Eds.): Topics in Operator Theory. Ernst D. Hellinger Memorial Volume, 1990, (3-7643-2532-1)
49. **I. Gohberg, S. Goldberg, M.A. Kaashoek:** Classes of Linear Operators, Volume I, 1990, (3-7643-2531-3)
50. **H. Bart, I. Gohberg, M.A. Kaashoek** (Eds.): Topics in Matrix and Operator Theory, 1991, (3-7643-2570-4)
51. **W. Greenberg, J. Polewczak** (Eds.): Modern Mathematical Methods in Transport Theory, 1991, (3-7643-2571-2)
52. **S. Prössdorf, B. Silbermann:** Numerical Analysis for Integral and Related Operator Equations, 1991, (3-7643-2620-4)
53. **I. Gohberg, N. Krupnik:** One-Dimensional Linear Singular Integral Equations, Volume I, Introduction, 1992, (3-7643-2584-4)
54. **I. Gohberg, N. Krupnik:** One-Dimensional Linear Singular Integral Equations, Volume II, General Theory and Applications, 1992, (3-7643-2796-0)
55. **R.R. Akhmerov, M.I. Kamenskii, A.S. Potapov, A.E. Rodkina, B.N. Sadovskii:** Measures of Noncompactness and Condensing Operators, 1992, (3-7643-2716-2)
56. **I. Gohberg** (Ed.): Time-Variant Systems and Interpolation, 1992, (3-7643-2738-3)
57. **M. Demuth, B. Gramsch, B.W. Schulze** (Eds.): Operator Calculus and Spectral Theory, 1992, (3-7643-2792-8)
58. **I. Gohberg** (Ed.): Continuous and Discrete Fourier Transforms, Extension Problems and Wiener-Hopf Equations, 1992, (ISBN 3-7643-2809-6)
59. **T. Ando, I. Gohberg** (Eds.): Operator Theory and Complex Analysis, 1992, (3-7643-2824-X)
60. **P.A. Kuchment:** Floquet Theory for Partial Differential Equations, 1993, (3-7643-2901-7)
61. **A. Gheondea, D. Timotin, F.-H. Vasilescu** (Eds.): Operator Extensions, Interpolation of Functions and Related Topics, 1993, (3-7643-2902-5)
62 **T. Furuta, I. Gohberg, T. Nakazi** (Eds.): Contributions to Operator Theory and its Applications. The Tsuyoshi Ando Anniversary Volume, 1993, (3-7643-2928-9)
63. **I. Gohberg, S. Goldberg, M.A. Kaashoek:** Classes of Linear Operators, Volume 2, 1993, (3-7643-2944-0)
64. **I. Gohberg** (Ed.): New Aspects in Interpolation and Completion Theories, 1993, (ISBN 3-7643-2948-3)